HEATING, VENTILATING AND AIR CONDITIONING LIBRARY VOLUME III

Radiant Heating

Water Heaters • Ventilation

Air Conditioning

Heat Pumps • Air Cleaners

by James E. Brumbaugh

Macmillan Publishing Company
New York

Collier Macmillan Publishers
London

Macmillan Publishing Company
866 Third Avenue, New York, NY 10022
Collier Macmillan Canada, Inc.

Library of Congress Cataloging in Publication Data

Brumbaugh, James E.
    Heating, ventilating and air conditioning library.

    Includes indexes.
    1. Heating.   2. Ventilation.   3. Air conditioning.
I. Title.

TH7222.B78   1984   697   83-7064

ISBN 0-672-23389-4 (v. 1)
ISBN 0-672-23390-8 (v. 2)
ISBN 0-672-23391-6 (v. 3)
ISBN 0-672-23388-6 (set)

Macmillan books are available at special discounts for bulk purchases
for sales promotions, premiums, fund-raising, or educational use.
For details, contact:

        Special Sales Director
        Macmillan Publishing Company
        866 Third Avenue
        New York, NY 10022

10 9 8 7 6 5

Printed in the United States of America

Foreword

The purpose of this series is to provide the layman with an introduction to the fundamentals of installing, servicing, and repairing the various types of equipment used in residential heating, ventilating, and air conditioning systems. Consequently, it was not written only for the engineer or skilled serviceman, but also for the average homeowner. A special effort was made to remain consistent with the terminology, definitions, and practices of the various professional and trade associations involved in the heating, ventilating, and air conditioning fields.

Volume 1 begins with a description of the principles of thermal dynamics and ventilation, and proceeds from there to a general description of the various heating systems used in residences and small commercial buildings. Volume 2 contains descriptions of the working principles of various kinds of equipment and other components used in these systems. Volume 3 includes detailed instructions for installing, servicing, and repairing these different types of equipment and components. Those sections in Volume 3 that deal with air conditioning follow a similar format.

The author wishes to acknowledge the cooperation of the many organizations and manufacturers for their assistance in supplying valuable data in the preparation of this book. Every effort was made to give credit and courtesy lines for materials and illustrations used in this edition.

JAMES E. BRUMBAUGH

Contents

chimneys—stoves, ranges, and heaters—installation instructions—operating instructions

CHAPTER 4

CHAPTER 5

CHAPTER 6

CHAPTER 7

curves—general ventilation—determining air intake—screen efficiency—
static pressure—local ventilation—exhaust-hood design recommendations—fan motors—troubleshooting fan motors—fan selection—
fan installation—fan installation checklist—air volume control—noise
control— fan applications—attic ventilating fans—kitchen exhaust
systems—bathroom exhaust systems—air circulators

CHAPTER 8

CHAPTER 9

CHAPTER 10

CHAPTER 11

refrigerants—liquid refrigerant control devices—automatic expansion valves—thermostatic expansion valves—float valves—capillary tubes—refrigerant piping—troubleshooting refrigerant piping—filters and dryers—pressure-limiting controls—reversing valves—water-regulating valves—automatic controls—system troubleshooting—general servicing and maintenance—pumping down—purging—evacuating the system—charging—silver brazing repairs

CHAPTER 12

Heating cycle—cooling cycle—heat sink—defrost cycle—types of heat pumps—heat pump installations—heat pump components—heat pump compressor—four-way reversing valve—heat pump coils—air filters—auxiliary electric heaters—condensation drain line—automatic controls—refrigerant lines—installation recommendations—operating instructions—heating instructions—cooling instructions—heat pump maintenance—troubleshooting heat pumps

CHAPTER 13

Humidifiers—spray humidifiers—pan humidifiers—stationary-pad humidifiers—rotary humidifiers—automatic controls—installation instructions—operating and maintenance suggestions—troubleshooting humidifiers—dehumidifiers—adsorption dehumidifiers—spray dehumidifiers—refrigeration dehumidifiers—automatic controls—installation suggestions—operating and maintenance suggestions—troubleshooting refrigeration dehumidifiers

CHAPTER 14

Electronic air cleaners—automatic controls—clogged filter indicator—performance lights—sail switch—in-place water-wash controls—cabinet model control panels—installation instructions—electrical wiring—maintenance instructions—replacing tungsten ionizing wires—troubleshooting electronic air cleaners—air washers—air filters—dry air filters—viscous air filters—filter installation and maintenance

CHAPTER 1

Radiant Heating

Radiant heating involves the transfer of heat energy by means of waves (thermal radiation). In radiant panel heating systems, one or more interior surfaces (ceiling, walls, floor) will radiate heat, thereby reducing the body heat caused by radiation. This working principle distinguishes radiant heating from *conventional* steam, hot-water, and warm-air heating systems, which reduce the body heat loss caused by convection. The principles of radiant heating are described in greater detail in the section below.

Residential radiant heating commonly takes the form of either panel heating systems or radiant baseboard systems. The radiant baseboard systems use metal baseboard units located along the outer walls of a room. The source of the heat is hot water circulated through the baseboard units by a boiler. Additional information about radiant baseboard systems is contained in Chapter of Volume 1 (Hot-Water Heating Systems) and Chapter 2 of the present volume (Radiators, Convectors, and Unit Heaters).

This chapter is primarily concerned with a description of radiant panel heating (or panel heating), which can be defined as a form of radiant heating in which large surfaces are used to radiate heat at relatively low temperatures.

PRINCIPLES OF RADIANT HEATING

Heat is lost from the human body through radiation, convection, and evaporation. Radiation heat loss represents the transfer of energy by means of electromagnetic waves. The convection loss is the heat carried away by the passage of air over the skin and clothing. The evaporation loss is the heat used up in converting moisture on the surface of the skin into vapor.

Heat transfer, whether by convection or radiation, follows the same physical laws in the radiant heating system as in any other; that is, heat flows from the warmer to the cooler exposure at a rate directly proportional to the existing temperature difference.

The natural tendency of warmed air to rise makes it apparent that this induced air current movement is greater at the cooler floor and exterior walls of the average heated enclosure than at its ceiling. It is through absorption by these air currents that the radiant panel releases the convection component of its heat transfer into the room air.

The average body heat loss is approximately 400 Btu per hour; total radiation and convection account for approximately 300 to 320 Btu of it. Because this is obviously the major portion, the problem of providing comfort is principally concerned with establishing the proper balance between radiation and convection losses.

It is important to understand that bodily comfort is obtained in radiant heating by maintaining a proper balance between radiation and convection. Thus, if the air becomes cooler and accordingly the amount of heat given off from the body by convection *increases*, then the body can still adjust itself to a sense of comfort if the heat given off from the body by radiation is *decreased*. The amount given off from the body by radiation can be decreased by raising the temperature of the surrounding surfaces, such as the walls, floor, and ceiling. For comfort, the body demands that if the amount of heat given off by convection increases, the heat given off by radiation must decrease, and vice versa.

The principles involved in radiant heating exist in such commonplace sources of heat as the open fireplace, outdoor campfires, electric spot heaters, and similar devices. In each of these examples, no attempt is made to heat the air or enclosing surfaces surrounding the individual. In fact, the temperature of the air and surrounding surfaces may be very low, but the radiant heat from the fireplace or campfire will still produce a sensation of comfort (or even discomfort from excess heat) to those persons within range. This situation can occur even though a conventional thermometer may indicate a temperature well below freezing. Radiant heat rays do not perceptibly heat the atmosphere through which they pass. They move from warm to colder surfaces where a portion of their heat is absorbed.

TYPES OF RADIANT HEATING SYSTEMS

There are several methods of applying radiant heating in buildings. Among the most popular are:

1. Circulating hot water through pipe coils located in the ceiling, floor, and/or walls.
2. Circulating warm air through shallow ducts or passages located in the walls, under the floor, or in the ceiling.
3. Using electrically heated metal plates, panels, or cables.
4. Using electrically heated wall tapestrips or portable screens.

In most radiant heating installations, an entire ceiling, floor, or wall is used to radiate heat to the interior spaces. Prefabricated radiant heating panels can be purchased and installed in a structure, but it is a much more common practice to build the heating panels during the initial construction stages. In operation, the warmed panel surfaces transfer heat to the surrounding air by convection, and to all solid surfaces by radiation alone. The radiant heat ray passes through the air-filled space without giving up heat to it.

This chapter is primarily concerned with a description of the installation and operation of *hot-water* radiant panel heating systems.

Radiant Panel Heating

Heating a room with panels is very similar to lighting a room. Heat radiation is exactly like light radiation except that it has a longer wavelength. If a room is lighted by means of a large number of small units distributed uniformly over the ceiling, the room is lighted more nearly uniformly than if it was lighted by means of a single unit of equal capacity. Similarly, if the entire ceiling is the heating panel, the room is heated more uniformly than if only a fractional part of the ceiling was used as the heating panel.

The surface area of a heating panel should be made as large as possible. For example, if the ceiling is being used as the heating panel, it is best to use the entire ceiling. The same is true of floor panels. Generally, only the upper portions of walls are used as heating panels because the lower portions are often blocked by furniture.

Heating panels may be located in ceilings, walls, or floors. Each location has different advantages and disadvantages as a heating surface.

Ceiling Panels

The advantage of a ceiling panel is that its heat emissions are not affected by drapes or furniture. As a result, the entire ceiling area can be used as a heating panel. Ceiling panels are recommended for rooms or space with 7-ft. ceilings or higher. A ceiling panel should never be installed in a room with a low ceiling (under 7 ft.) because it may produce an undesirable heating effect on the head.

In multiple-story construction, the use of ceiling panels appears to be more desirable from both the standpoint of physical comfort and overall economy. The designed utilization of the upward heat transmission from ceiling panels to the floor of the area immediately above will generally produce moderately tempered floors. Supplementing this with automatically controlled ceiling panels will result in a very efficient radiant heating system. Except directly below roofs or other unheated areas, this design eliminates the need for the intermediate floor insulation sometimes used to restrict the heat transfer from a ceiling panel exclusively to the area immediately below. It must be remembered, however, that when intermediate floor insulations are omitted, the space above a heated ceiling will not be entirely independent with respect to temperature control, but will necessarily be influenced by the conditions in the space below. Typical ceiling installations are shown in Fig. 1-1, 1-2, 1-3, and 1-4.

Apartment buildings and many office and commercial structures should find the ceiling panel method of radiant heating most desirable. In offices and stores, the highly variable and changeable furnishings, fixtures, and equipment favor the construction of ceiling panels, to say nothing of the advantage of being able to make as many partition alterations as desired without affecting the efficiency of the heating system.

Floor Panels

Floor panels are usually easier to install than either ceiling or wall panels. Using floor panels is the most effective method of eliminating cold floors in slab construction. Another advantage of heating with floor panels is that much of the radiated heat is

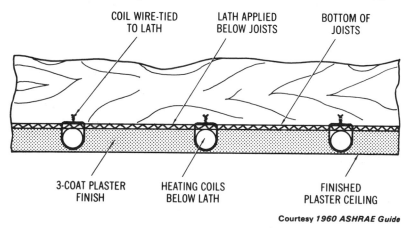

COIL WIRE-TIED
TO LATH

LATH APPLIED
BELOW JOISTS

BOTTOM OF
JOISTS

3-COAT PLASTER
FINISH

HEATING COILS
BELOW LATH

FINISHED
PLASTER CEILING

Courtesy *1960 ASHRAE Guide*

Fig. 1-1. Coils in plaster below lath.

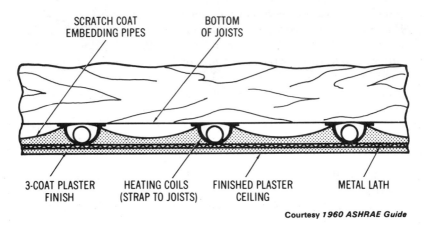

SCRATCH COAT EMBEDDING PIPES

BOTTOM OF JOISTS

3-COAT PLASTER FINISH

HEATING COILS (STRAP TO JOISTS)

FINISHED PLASTER CEILING

METAL LATH

Courtesy 1960 ASHRAE Guide

Fig. 1-2. Coils in plaster above lath.

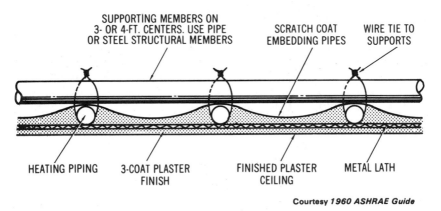

SUPPORTING MEMBERS ON 3- OR 4-FT. CENTERS. USE PIPE OR STEEL STRUCTURAL MEMBERS

SCRATCH COAT EMBEDDING PIPES

WIRE TIE TO SUPPORTS

HEATING PIPING

3-COAT PLASTER FINISH

FINISHED PLASTER CEILING

METAL LATH

Courtesy 1960 ASHRAE Guide

Fig. 1-3. Coils in plaster above lath in suspended ceiling.

delivered to the lower portions of the walls. The principal disadvantage of using floor panels is that furniture and other objects block portions of the heat emission.

Floor panels are recommended for living or working areas constructed directly on the ground, particularly one-story structures. Partial ceiling or wall treatment may be used as a supplement wherever large glass or door exposures are encountered. Typical floor installations are shown in Figs 1-5 and 1-6.

16

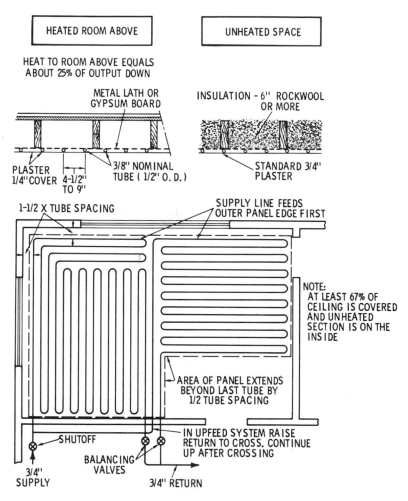

| HEATED ROOM ABOVE | UNHEATED SPACE |

HEAT TO ROOM ABOVE EQUALS
ABOUT 25% OF OUTPUT DOWN

METAL LATH OR
GYPSUM BOARD

INSULATION - 6" ROCKWOOL
OR MORE

PLASTER
1/4" COVER 4-1/2" TO 9"

3/8" NOMINAL
TUBE (1/2" O. D.)

STANDARD 3/4"
PLASTER

1-1/2 X TUBE SPACING

SUPPLY LINE FEEDS
OUTER PANEL EDGE FIRST

NOTE:
AT LEAST 67% OF
CEILING IS COVERED
AND UNHEATED
SECTION IS ON THE
INSIDE

AREA OF PANEL EXTENDS
BEYOND LAST TUBE BY
1/2 TUBE SPACING

SHUTOFF

IN UPFEED SYSTEM RAISE
RETURN TO CROSS. CONTINUE
UP AFTER CROSSING

BALANCING
VALVES

3/4"
SUPPLY

3/4" RETURN

Fig. 1-4 Diagram for a typical ceiling panel.

Wall Panels

Walls are not often used for heating panels because large sections of the wall area are often broken by windows and doors. Furthermore, the heat radiation from heating coils placed in the lower sections of a wall will probably be blocked by furniture. As a result, wall panels are generally used to supplement ceiling or floor panels in a radiant panel heating system.

Wall panels are commonly used as supplementary heating in bathrooms and in rooms in which there are a number of large picture windows. In the latter case, the heating coils are installed in the walls opposite the windows. Wall panels will probably not be necessary if the room has good southern exposure. Typical wall installations are shown in Figs. 1-7 and 1-8.

DESIGNING A RADIANT HEATING SYSTEM

It is *very* important that the design of a radiant panel heating system be correct at the outset. The fact that the coils or cables are permanently embedded in concrete, plaster, or some other material makes corrections or adjustments very difficult and expensive.

Many manufacturers that produce radiant panel heating equipment have devised simplified and dependable methods for designing this type of heating system. In most cases, the manu-

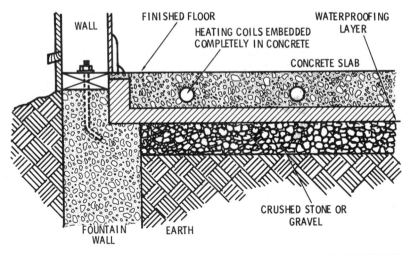

Courtesy *1960 ASHRAE Guide*

Fig. 1-5. Coils in floor slab on grade.

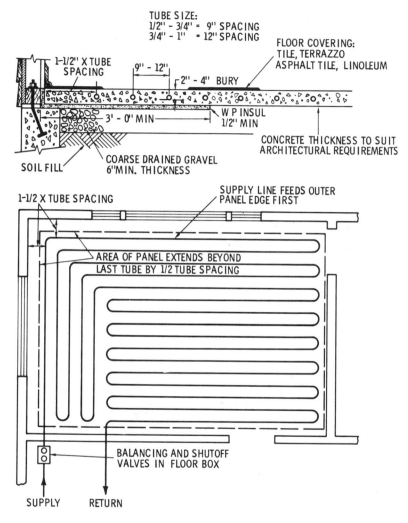

TUBE SIZE:
1/2" - 3/4" = 9" SPACING
3/4" - 1" = 12" SPACING

FLOOR COVERING:
TILE, TERRAZZO
ASPHALT TILE, LINOLEUM

1-1/2" X TUBE SPACING

9" - 12"

2" - 4" BURY

3' - 0" MIN

W P INSUL
1/2" MIN

CONCRETE THICKNESS TO SUIT
ARCHITECTURAL REQUIREMENTS

COARSE DRAINED GRAVEL
6"MIN. THICKNESS

SOIL FILL

SUPPLY LINE FEEDS OUTER
PANEL EDGE FIRST

1-1/2 X TUBE SPACING

AREA OF PANEL EXTENDS BEYOND
LAST TUBE BY 1/2 TUBE SPACING

BALANCING AND SHUTOFF
VALVES IN FLOOR BOX

SUPPLY   RETURN

Fig. 1-6. Diagram for a typical radiant floor.

facturer will send you a catalog and any other available materials to assist you in calculating the requirements of your particular installation. Addresses of manufacturers can be obtained from your local library.

19

PANEL TEMPERATURES

When planning a radiant panel heating system, you must take into consideration the *maximum allowable temperatures* of the heating panels. These temperatures will vary according to the location of the panel.

The maximum allowable temperature of a ceiling panel depends upon its height. A maximum allowable temperature of 100°F is recommended for panels installed in 7-ft. ceilings. This is probably the most common ceiling height used in modern home construction. For each foot increase in height, the maximum allowable temperature should be increased by 10°F. Thus, a temperature of 110°F is recommended for an 8-ft. ceiling, 120°F for a 9-ft. ceiling, and 130°F for a 10-ft ceiling.

A temperature of 85°F is recommended for floor panels. This lower maximum allowable temperature results from the more

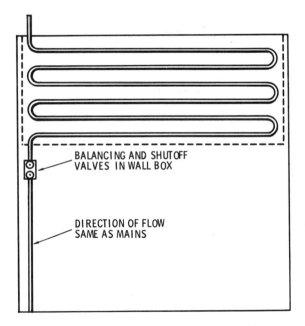

Fig. 1-7. Typical wall installation. Panel is installed as high on the wall as possible.

effective insulating characteristics of the materials used to cover the floor coils. Lower temperatures are recommended for floor panels to protect furniture, carpets, and floor finishes from heat damage. Another important reason for using lower temperatures in floor panels is that the feet are more sensitive to heat than the

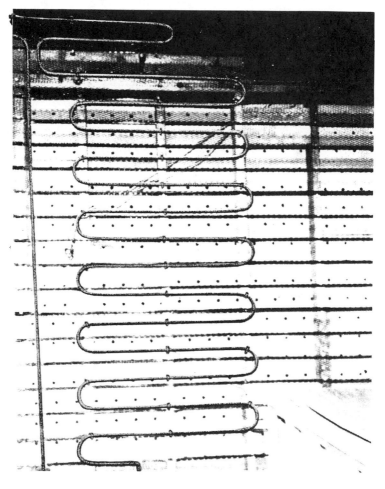

Fig. 1-8. Bathroom wall panels extended to floor.

upper portions of the body. As a result, lower temperatures in the floor panels produce the same results as slightly higher ones do in ceiling and wall panels.

The maximum allowable temperature for wall panels should be about 100° to 110°F. The temperature will depend upon how the wall panel is used. A lower temperature is recommended if the panel is used to supplement the heat from a ceiling or floor radiant heating panel. A higher maximum allowable temperature is necessary if the wall panel is the only source of heat.

The maximum allowable temperatures given in the preceding paragraphs are only recommended temperatures. The specific maximum allowable temperatures used in an installation will depend upon a number of variables, such as the type of material covering the coils, ducts, or heating cables; the location of the panel (ceiling, floor, or wall); and the amount of heat loss.

It must be remembered that the maximum allowable temperature is the temperature of the heat *radiated* from the panel. The temperature of the water *in* the coils will be slightly higher. The temperature of the water is reduced to some degree by the insulating effect of the material covering the coils, by the depth at which the coils are buried, and by the temperature drop in the coils caused by normal heat loss. For these reasons, it is customary to speak in terms of the *average* temperature of the water circulating in the coils. In many installations, the average water temperature will vary between 120°F and 130°F.

An important factor to consider when designing a hot-water radiant panel heating system is the temperature drop (i.e., the temperature difference) that occurs in the coils of the heating panel. As the hot water passes through the coils of a panel, it loses a certain amount of heat; a factor that results in the outlet water having a lower temperature than the inlet water. This temperature difference ranges between 20°F and 30°F.

The degree of temperature drop in hot-water radiant heating panels must be limited, or the efficiency of the heating system will suffer. On the other hand, maintaining the maximum allowable temperature of the panel is also essential to the efficiency of the heating system. The temperature drop and the maximum allowable temperature must be adjusted so that each will operate within the desired temperature range.

RADIANT PANEL
INSULATION REQUIREMENTS

The design of a hot-water or a warm-air radiant panel heating system is such that the *interior* surfaces of the outside walls of a structure are generally warmer than when a convection heating system is used. As a result, extra insulation is not required. As a matter of fact, a tightly constructed, well-insulated structure will be uncomfortably warm with this type of heating system because the lack of sufficient air infiltration allows the heat to accumulate until it becomes excessive.

The insulation requirements for electric radiant panel heating are essentially the same as those required for other radiant panel heating when the system consists of electric cables imbedded in the walls of the structure. Ceiling and floor panels may require additional insulation. Local codes and regulations should be consulted for additional information.

SYSTEM COMPONENTS

A hot-water radiant panel heating system generally consists of the following basic components:

1. Hot-water boiler.
2. Circulating pump.
3. Expansion tank.
4. Automatic controls.
5. Valves and piping.
6. Air vents.

These system components are described in the sections that follow. Figs. 1-9 and 1-10 illustrate some typical arrangements of these components.

BOILER AND CIRCULATING PUMP

The boilers used in hot-water panel heating are generally the conventional gas-fired or oil-fired units found in other hot-water

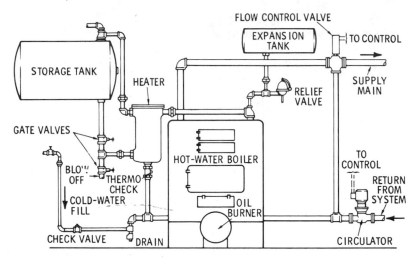

Fig. 1-9. Piping diagram for a typical small radiant heating system.

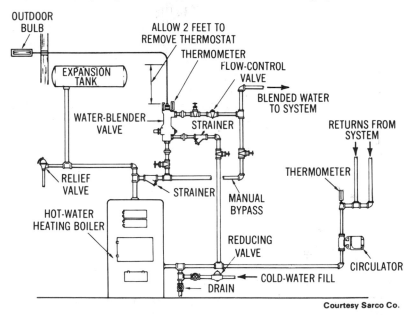

Courtesy Sarco Co.

Fig. 1-10. Piping diagram of a typical radiant heating installation in which an automatic water blender valve is controlled by an outside thermostat.

heating systems. These are compact boilers designed for installation in closets or in similar locations where space is limited.

The size of the circulating pump selected for a radiant panel heating system will depend upon the pressure drop in the system and the rate at which water must circulate. The circulation rate of the water is determined by the heating load and the *design* temperature drop of the system, and is expressed in gallons per minute (gpm).

This can be calculated by using the following formula:

$$\text{gpm} \times \frac{\text{Total Heating Load}}{\text{T} \times 60 \times 8}$$

The *total heating load* is calculated for the structure and is expressed in Btu per hour. A value of 20°F is generally used for the design temperature drop (T) in most hot-water radiant panel heating systems. The other two values in the formula are the minutes per hour (60) and the weight (in pounds) of a gallon of water (8).

By way of example, the rate of water circulation for a structure with a total heating load of 30,000 Btu per hour may be calculated as follows:

$$\text{gpm} = \frac{30,000 \text{ Btu/hr}}{20 \times 60 \times 8}$$

$$= \frac{30,000}{9,600} = 3.13$$

AUTOMATIC CONTROLS

While any thermostatic method of control will function with a radiant panel heating system, the most desirable method is that based on continuously circulating hot water. The temperature of the water should be automatically adjusted to meet outdoor conditions, but the circulation itself controlled by interior limiting thermostats rather than the simple off-on method of circulating hot water at a fixed temperature. Standard equipment of both types is readily available (Figs. 1-11, 1-12, and 1-13).

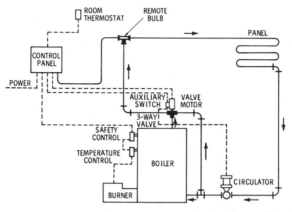

Fig. 1-11. Automatic heat control system for light radiant heat panels.

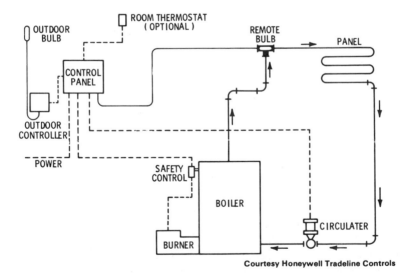

Fig. 1-12. Automatic heat control system for heavy radiant heat panels.

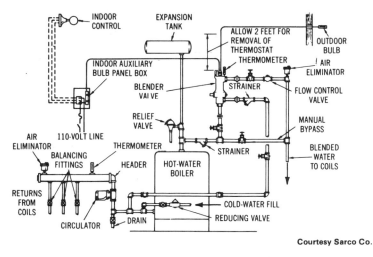

Courtesy Sarco Co.

Fig. 1-13. Heating system using indoor-outdoor controls.

NONCONTINUOUS WATER CIRCULATION

The flow of water in some radiant heating systems is controlled by the pump. When the room thermostat calls for heat, the booster pump starts and rapidly circulates heated water through the radiant panels until the heat requirement is satisfied. The pump is then shut off by the thermostat. In some systems, a flow-control valve is forced open by the flow of water through the pipes as long as the pump is running, permitting free circulation of heated water through the system. When the pump stops, the control valve closes, preventing circulation by gravity, which might cause overheating.

The principal disadvantage of a system with this off-on control is that it results in temperature lag and causes the panels to intermittently heat and cool.

CONTINUOUS WATER CIRCULATION

The continuous circulation of water through radiant heating panels is made possible by means of an outdoor-indoor control.

In this arrangement, hot water from the boiler is admitted to the system in modulated quantities when the temperature of the circulating water drops below the heat requirement of the panels. This modulated bleeding of water into the panel is accomplished through a bypass valve. When no additional heat is required, the valve is closed. When more heat is required, the valve is gradually opened by the combined action of the outdoor temperature bulb and a temperature bulb in the supply main. This system gives control by the method of varying the temperature of the water.

INSTALLATION REQUIREMENTS

The successful operation of any hot-water heating system requires the incorporation of design provisions that assure an even and balanced flow of water through the pipes or coils of the installation.

The procedure for designing a panel heating system was outlined by the *ASHRAE 1960 GUIDE* as follows:

1. Determine the total rate of heat loss per room in the structure.
2. Determine the available area for panels in each room.
3. Determine the output required by each panel to replace the heat loss.
4. Determine the required surface temperature for each panel.
5. Determine the required heat input to the panel (should equal heat output).
6. Determine the most efficient and economical means of supplying heat to the panel.
7. Install adequate insulation on the reverse side and edges of the panel to prevent undesirable heat loss.
8. Install the panels opposite room areas where the greater heat loss occurs.

Never use ceiling panels in rooms or spaces with low ceilings. *Always* keep floor temperatures at or below recommended limits.

PANEL TESTING PROCEDURES

Radiant heating coils should be tested for leaks after they have been secured in position but before they are covered with plaster, concrete, or some other material. Both a compressed-air test and a hydraulic pressure test are used for this purpose.

The compressed-air test requires a compressor, a pressure gauge, and a shutoff valve. The idea is to inject air under pressure into the radiant heating system and watch for a pressure drop on the gauge. A continually dropping pressure is an indication of a leak somewhere in the system.

The pressure gauge is attached to one of the radiant heating coils, and the shutoff valve is placed on the *inlet* side of the gauge in a valve-open position. The air compressor is then connected and compressed air is introduced into the system under approximately 100 psi. After the introduction of the air, the shutoff valve is closed and the compressor is disconnected. The system is now a closed one. If there are no leaks, the air pressure reading on the gauge will remain at approximately 100 psi. A steady drop in the air-pressure reading means a leak exists somewhere in the system. A leak can be located by listening for the sound of escaping air. Another method is to use a solution of soap and water and watch for air bubbles.

The hydraulic pressure test requires that the coils be filled with water and the pressure in the coils be increased to approximately 275 to 300 psi. Care must be taken that *all* air is removed from the coils before the system is closed. The system is then closed, and the gauge is watched for any change in pressure. A leak in the system will be indicated by a steady drop in pressure on the gauge. The source of the leak can be located by watching for the escaping water.

After a leak has been located, the coil should be repaired or replaced and a new test run on the system.

AIR VENTING REQUIREMENTS

A common defect encountered in hot-water system design is improper venting. The flow of water should be automatically kept free of air binding throughout the system. Air in the pipes or pipe coils almost always results in a reduction of heat.

A practical method of venting is shown in Fig. 1-14. The key to this method is the use of automatic air traps. Each air trap should be located in an area readily accessible for repair. The air trap

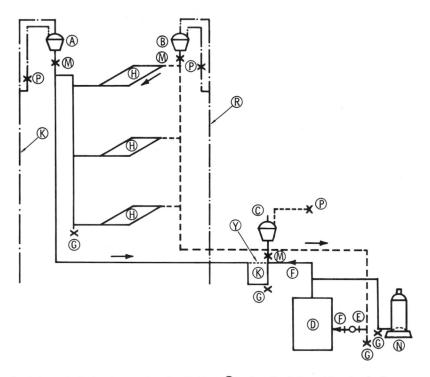

Symbols:——Indicates downward grade of tubing. Ⓐ automatic air trap at top of main flow riser; Ⓑ automatic air trap at top of main return riser: Ⓒ automatic air trap at top of special loop Ⓚ required by possible obstruction and when small size vent by-pass is also not permissible at Ⓨ; Ⓓheater; Ⓔ pump; Ⓕ check valve; Ⓖ drain valve; Ⓗ heating panel coil; Ⓚ loop in main flow (See Ⓒ); Ⓜ trap shut-off valve (for repair); Ⓝ expansion tank; Ⓟ manual test cock (air trap); Ⓡ open and automatic vent tube (1/2 in. copper).

Note--By reversing direction of grade at Ⓗ air trap Ⓑ can be eliminated. Same riser vent layout should be used for up-feed systems. Test cocks Ⓟ should be located accessible for occasional use. Open ends of vent tubes Ⓡ (normally dry) can discharge visibly into nearest drain or sink.

Fig. 1-14. An automatic vent radiant heating system.

test cock should be placed where it can be easily operated. Both the air trap and the air trap test cock must be located where they are not subject to freeze-up as both are noncirculating except during venting operation (automatic or manual).

COILS AND COIL PATTERNS

Hot-water heating panels are available as prefabricated units, or they can be constructed at the site. The two basic coil patterns used in radiant panel heating are: (1) the continuous heating coil, and (2) the grid heating coil (Figs. 1-15 and 1-16). The latter is

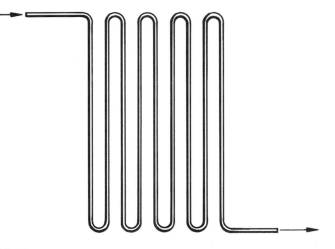

Fig. 1-15. Continuous heating coil pattern.

recommended when a low pressure drop is required, or where it is necessary to have a constant temperature over the entire panel surface. Sometimes the two heating coil patterns are combined (Fig. 1-17). This is usually the case in large heating installations where there is more than one coil.

In a well-designed hot-water panel heating system, the linear travel from the heating unit and pump should be the same for each of the panels (Fig. 1-18). This will result in the flow through each panel being in natural balance.

31

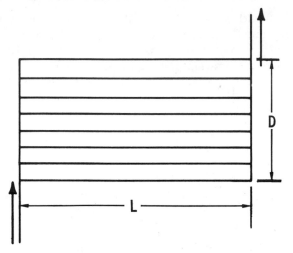

Fig. 1-16. A grid-type heating coil used where a low-pressure drop is desired or a constant panel surface temperature is required.

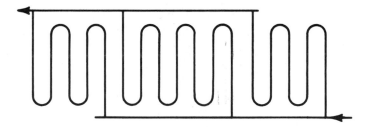

Fig. 1-17. A combination grid and continuous coil pattern.

INSTALLING COILS

The coils used in hot-water radiant heating installations are most commonly made of copper tubing (both the hard and soft varieties), steel, or wrought-iron pipe. This tubing or pipe is available in a variety of lengths and diameters, depending upon the material.

Bending devices or jigs are commonly employed to bend the tubing or pipe into the coil pattern. Only soft copper tubing can

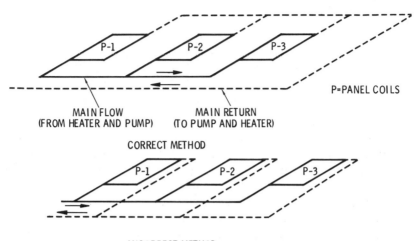

P=PANEL COILS

MAIN FLOW
(FROM HEATER AND PUMP)

MAIN RETURN
(TO PUMP AND HEATER)

CORRECT METHOD

INCORRECT METHOD

Fig. 1-18. Correct and incorrect method of laying out a forced hot-water
distribution system. The travel from pump and heater should be
the same through P1, P2, and P3 as shown in the correct method.

be easily bent by hand. It is recommended that a tube bender of
this type be made for each of the different center-to-center spacings needed for the various panel coils in the installation.

Soft copper tubing is commonly available in coil lengths of 40
ft., 60 ft., and 100 ft. When the tubing is uncoiled, it should be
straightened in the trough of a straightener jig. For convenience
of handling, the straightener should not be more than 10 ft. long.

The lighter metals, such as hard or soft copper tubing, should
be used in ceiling panels. Try to avoid using wrought-iron pipe
(or even steel) for this purpose. The heavier weight of these
materials can cause problems. Fig. 1-19 shows an example of
prefabricated panel made of ¼-in. copper tubing. The tubing is
constructed with metal lath for use with plaster construction.

Whenever possible, continuous lengths of tubing should be
used with as few fitting connections as possible. Coils of 60 ft. or
100 ft. are best for this purpose, and are generally preferred for
floor panels.

The spacing between tubing or pipe should be uniform and
restricted to 12 in. or less (Fig. 1-6). Solder joints are used to make
connections.

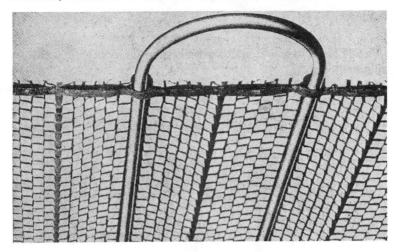

Fig. 1-19. Prefabricated radiant panel made of ¼-in. copper tubing in metal lath for use with plaster construction.

TYPES OF VALVES

The principal valves used in a hot-water radiant heating system are:

1. Pressure-relief valves.
2. Pressure-reducing valves.
3. Flow-control valves.
4. Water-blending valves.
5. Water-tempering valves.
6. Balancing valves.
7. Zoning valves.

Pressure-relief valves and pressure-reducing valves are standard equipment in all hot-water heating systems. A *pressure-relief valve* is used to prevent excessive and dangerous pressure from building up in the system. It is generally located on or near the boiler. A *pressure-reducing valve* is designed to reduce the pressure of the water entering the system, and to maintain the pressure at a specific minimum setting (usually about 12 lb.). This valve is generally placed between gate valves in the cold-water supply line.

A *flow-control valve* is a check valve used to prevent thermal circulation of the water through the heating system when the thermostat has not called for circulation. Some typical locations of flow-control valves are shown in Figs. 1-9, 1-10, and 1-13. Flow-control valves are not used when the radiant panel is below the level of the boiler.

A *water-blending valve* (or *water blender*) is a thermostatic three-way valve used to recirculate a variable portion of the return water and at the same time add a sufficient quantity of hot boiler water to maintain the required water temperature in the coils.

The *Sarcotherm* water blender shown in Figs. 1-20 and 1-21 is frequently used in hot-water radiant panel heating installations to provide system water temperature modulation in accordance with outside temperatures. This type of control is necessary in installations where continuous water circulation is desired. A limit control consisting of an indoor thermostat sensitive to radiant heat prevents overheating without stopping the circulation of the water.

Water temperature adjustments are made on the water blender shown in Fig. 1-20 by loosening the locknut (J) and turning the adjustment knob (K) until the desired temperature is obtained. In operation, temperature buildup causes the thermostatic element (A) to push the piston (B) downward. This causes hot water to enter the valve at 1 and return water from the system through inlet 2. The hot and cooler return water pass through the mixing valve (S), then over the inside thermostatic element (A), and to the system through outlet 3.

Some water blenders are equipped with an indoor temperature compensation device that prevents the kind of overheating caused by time lag between indoor and outdoor temperature changes (Fig. 1-21). Typical installations in which water blenders are used are shown in Figs. 1-10 and 1-13.

If the boiler in a radiant panel heating system is also used to heat the domestic water supply, a *tempering valve* should be installed to reduce the temperature of the water. Without a tempering valve, the domestic hot-water supply will be excessively hot.

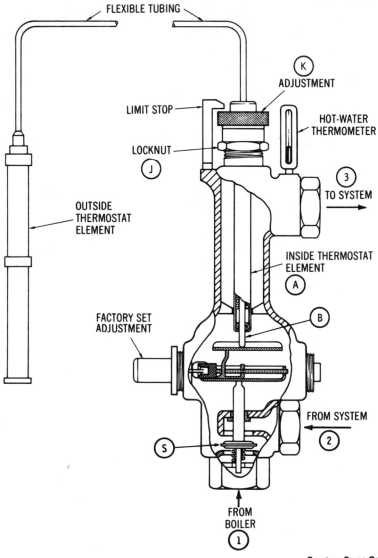

FLEXIBLE TUBING

K
ADJUSTMENT

LIMIT STOP

HOT-WATER
THERMOMETER

LOCKNUT

J

3
TO SYSTEM

OUTSIDE
THERMOSTAT
ELEMENT

INSIDE THERMOSTAT
ELEMENT

A

FACTORY SET
ADJUSTMENT

B

FROM SYSTEM

2

S

FROM
BOILER

1

Courtesy Sarco Co.

Fig. 1-20. Operating principles of an automatic water blender valve controlled by an outside thermostatic element.

A *balancing valve* should be installed in either the supply or return line of a panel in order to balance the heat output of each panel in the same zone. Figs. 1-4, 1-6, 1-7, and 1-13 illustrate typical locations of balancing valves in radiant panel heating systems.

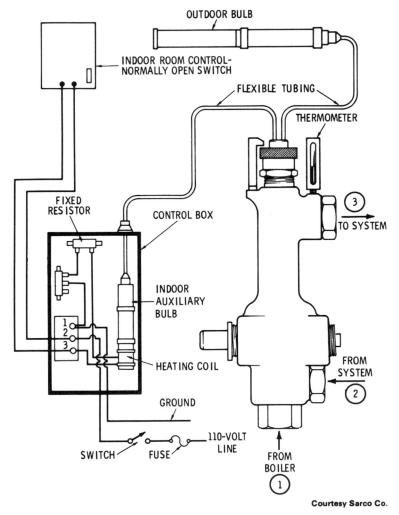

Fig. 1-21. An automatic water blender valve with indoor compensation feature.

A radiant panel heating system will often use a number of motorized *zoning valves* in order to maintain a uniform temperature throughout the various rooms and spaces in a structure. A typical installation is shown in Fig. 1-22.

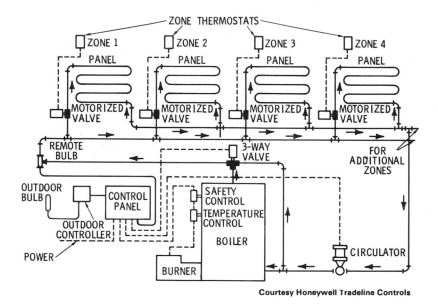

Courtesy Honeywell Tradeline Controls

Fig. 1-22. A typical control system for a multiple-zone radiant heating system.

Read Chapter 9 of Volume 2 (Valves and Valve Installation) for a detailed description of the operating principles of the various valves used in heating and cooling installations.

WARM-AIR PANELS

Air can also be used as a heat conveying medium in radiant panel heating. The warm air is circulated through ducts or spaces located in the ceiling, walls, or floor. Both forced warm-air furnaces and heat pumps have been used to heat the air in this type of installation.

Warm-air ceiling panels are often used to heat office buildings and structures of similar size. In some installations, the ceiling panel is constructed by suspending a false ceiling a few inches below the actual ceiling of the structure. The space between the two ceilings is then sealed, and warm air is delivered to the enclosed space through a standard warm-air duct (Fig. 1-23).

Two examples of warm-air floor panel installations are illustrated in Figs. 1-24 and 1-25. In both cases, the air passages are enclosed in concrete.

Warm-air panels should be constructed in accordance with recommendations given in Manual 7-A of the National Warm Air Heating and Air Conditioning Association.

ELECTRIC RADIANT HEATING PANELS

Electric resistance cables can be embedded in ceilings, walls, or floors during construction in much the same manner as hot-water piping. Another method is to attach prefabricated electric heating panels to interior room surfaces. The latter method is ideal for older structures. Two examples of electric panel construction are illustrated in Figs. 1-26 and 1-27.

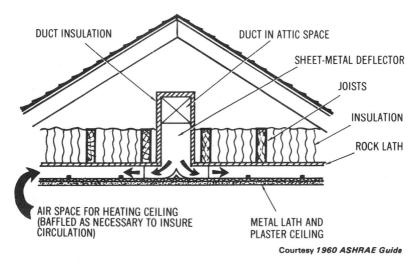

Courtesy *1960 ASHRAE Guide*

Fig. 1-23. Warm-air plaster ceiling construction.

39

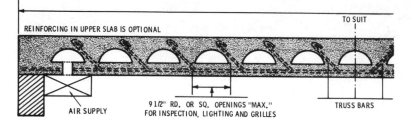

REINFORCING IN UPPER SLAB IS OPTIONAL

TO SUIT

AIR SUPPLY

9 1/2" RD. OR SQ. OPENINGS "MAX."
FOR INSPECTION, LIGHTING AND GRILLES

TRUSS BARS

Fig. 1-24. Hollow concrete floor

The use of embedded resistors or cables is generally limited to ceilings and floors in the United States. Most local codes prohibit electric wall panels of this type of construction because of the possible danger to the cables by nails driven into the wall to hang pictures or plants.

ADVANTAGES OF RADIANT PANEL HEATING

One of the principal advantages of radiant panel heating is that it produces a very uniform heat. This type of heating is also characterized by reduced air temperatures and the absence of significant air currents or drafts. All these factors contribute to a comfortable and healthy environment.

A properly designed radiant panel heating system is efficient and economical to operate. In hot-water radiant panel heating systems, the low operating temperatures of the water result in lower maintenance costs. This is also true of both warm-air and electric radiant panel heating.

DISADVANTAGES OF RADIANT PANEL HEATING

The principal disadvantage of a radiant panel heating system can be summarized as follows:

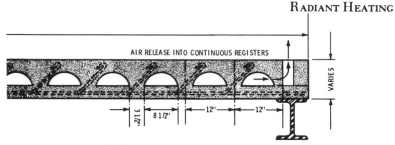

Courtesy Airfloor Co. of California, Inc.

serving as a common heat plenum.

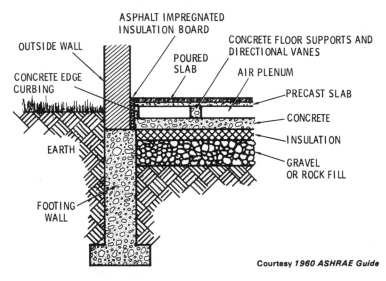

Courtesy *1960 ASHRAE Guide*

Fig. 1-25. Warm-air floor panel construction.

1. High installation cost.
2. Lack of ventilation.
3. Limitation of panel size.

Another very important disadvantage of most radiant panel heating systems is the difficulty of making design corrections or adjustments after the final construction stage.

A disadvantage of *hot-water* radiant panel heating is the need to provide for separate cooling and dehumidification. Cold water from a water chiller can be circulated through the coils or

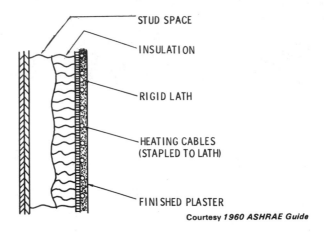

STUD SPACE

INSULATION

RIGID LATH

HEATING CABLES
(STAPLED TO LATH)

FINISHED PLASTER

Courtesy *1960 ASHRAE Guide*

Fig. 1-26. Electric heating cables in plaster.

piping, but this type of installation is more expensive than a split system and is generally not as efficient.

RADIANT PANEL COOLING

Hot-water radiant panel heating systems are capable of providing both heating and cooling independently of air movement. For the heating cycle, hot water is circulated through the pipe coils. For the cooling cycle, cold water (*above* the dew point) is circulated, and the heating cycle is reversed. By keeping the water temperature above 65°F, harmful moisture condensation is avoided.

Radiant panel cooling results only in the removal of sensible heat, and there is sometimes an uncomfortable feeling of dampness. As a result, a separate means of dehumidification is often necessary. Often this can be quite expensive because it may require the installation of a separate dehumidification unit and air conduits or ducts to the various rooms and spaces in the structure.

The central unit of a radiant panel cooling system is often a separate water chiller operating in parallel with a hot-water source heating boiler. A three-way valve is often used in the piping joining the two units.

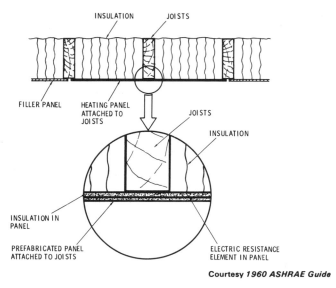

INSULATION JOISTS

FILLER PANEL HEATING PANEL
ATTACHED TO
JOISTS

JOISTS

INSULATION

INSULATION IN
PANEL

PREFABRICATED PANEL
ATTACHED TO JOISTS

ELECTRIC RESISTANCE
ELEMENT IN PANEL

Courtesy *1960 ASHRAE Guide*

Fig. 1-27. Prefabricated electric panels.

A much less expensive method of cooling a structure equipped
with a hot-water heating system (whether it be the baseboard or
radiant panel) is by adding forced air cooling. The forced air
cooling system will consist of a cooling unit, condenser, and the
necessary supply and return air ducts. This arrangement is
referred to as a *split system.*

CHAPTER 2

Radiators, Convectors, and Unit Heaters

The two basic methods by which heat-emitting units transfer heat to their surroundings are: (1) radiation and (2) convection. *Radiation* is the transmission of thermal energy by means of electromagnetic rays. In other words, an object is warmed by heat waves radiating from a hot surface. *Convection* is the transfer of heat by natural or forced movement (circulation) of the air across a hot surface. In actual practice, heat-emitting units will transfer heat partially by radiation (up to 30 percent) and partially by convection (70 to 90 percent).

The output of heat-emitting units is expressed in terms of Btu per hour (Btuh), in square feet of equivalent direct radiation (EDR), or in 1000 Btu per hour (MBh). The required radiation of an installation is determined on the basis of the Btu per hour capacity of each heat-emitting unit. See "Determining Required Radiation" in this chapter.

The selection of a heat-emitting unit will depend upon the type of heating system, the cost, the required capacity, and the application. For example, electric unit heaters should only be used where the cost of electricity is especially low. These heaters are generally associated with high operating costs. On the other hand, they are relatively inexpensive, and their installation cost is low because no separate piping or boiler is required. Each type of heat-emitting unit will have similar advantages and disadvantages that you must carefully consider before choosing the type most suited for the installation.

The principal heat-emitting units used in heating systems are:

1. Cast-iron radiators.
2. Convectors.
3. Baseboard units.
4. Finned-tube units.
5. Unit heaters.

CAST-IRON RADIATORS

A *cast-iron radiator* is a heat-emitting unit that transmits a portion of its heat by radiation and the remainder by convection. An *exposed* radiator (or freestanding radiator) transmits approximately half of its heat by radiation, the exact amount depending on the size and number of the sections. The balance of the emission is by conduction to the air in contact with the heating surface, and the resulting circulation of the air warms by convection.

Cast-iron radiators have been manufactured in column and tubular types (Figs. 2-1 and 2-2). Column and large-tube radiators (with 2½-in. spacing per section) have been discontinued. The small-tube radiator with spacings of 1¾-in. per section is now the prevailing type. Ratings for various cast-iron radiators are given in Tables 2-1, 2-2, and 2-3 courtesy of the American Society of Heating, Refrigerating and Air-Conditioning Engineers.

Wall and window radiators are cast-iron units designed for specific applications. Wall radiators are hung on the wall and are especially useful in installations where the floor must remain

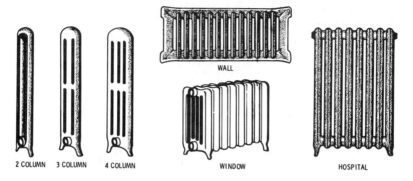

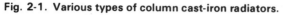

Fig. 2-1. Various types of column cast-iron radiators.

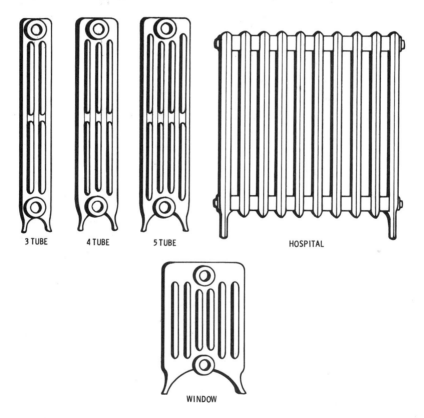

Fig. 2-2. Various types of tubular cast-iron radiators.

Table 2-1. Column-Type Cast-Iron Radiators

Height In.	Generally Accepted Rating per Section[a]					
	One Column		Two Column		Three Column	
	Sq. Ft.	Btuh	Sq. Ft.	Btuh	Sq. Ft.	Btuh
15			1 ½	360		
18					2 ¼	540
20	1 ½	360	2	480		
22			2 ¼	540	3	720
23	1 ⅔	400	2 ⅓	560		
26	2	480	2 ⅔	640	3 ¾	900
32	2 ½	600	3 ⅓	800	4 ½	1080
38	3	720	4	960	5	1200
45			5	1200	6	1440

Height In.	Four Column		Five Column		Six Column	
	Sq. Ft.	Btuh	Sq. Ft.	Btuh	Sq. Ft.	Btuh
13					3	720
16					3 ¾	900
18	3	720	4 ⅔	1120	4 ½	1080
20					5	1200
22	4	960				
26	5	1200	7	1680		
32	6 ½	1560				
38	8	1920	10	2400		
45	10	2400				

*These ratings are based on steam at 215°F and air at 70°F. They apply only to installed radiators exposed in a normal manner; not to radiators installed behind enclosures, grilles, or under shelves.

Courtesy *1960 ASHRAE Guide*

clear for cleaning or other purposes. Window radiators are located beneath a window on an exterior wall. The heat waves radiating from the surface of the unit provide a very effective barrier against drafts.

Attempts to improve the appearances of cast-iron radiators by painting them, covering them, or recessing them in walls also succeed in reducing their heating efficiency. An unpainted, uncovered, freestanding radiator is always more efficient.

Radiator Valves

Various valves are required for the efficient operation of radiators. The choice of valve will depend upon the requirements of the particular system.

Table 2-2. Large-Tube Cast-Iron Radiators

Sectional, cast-iron, tubular-type radiators of the large-tube pattern, that is, having tubes approximately $1\frac{3}{8}$ in. in diameter, $2\frac{1}{2}$ in. on centers.

Number of Tubes per Section	Catalog Rating per Section*		Height	Width	Section Center Spacing†	Leg Height‡ to Tapping
	Sq. Ft.	Btuh	In.	In.	In.	In.
3	$1\frac{3}{4}$	420	20	$4\frac{5}{8}$	$2\frac{1}{2}$	$4\frac{1}{2}$
	2	480	23		$2\frac{1}{2}$	$4\frac{1}{2}$
	$2\frac{1}{3}$	560	26		$2\frac{1}{2}$	$4\frac{1}{2}$
	3	720	32		$2\frac{1}{2}$	$4\frac{1}{2}$
	$3\frac{1}{2}$	840	38		$2\frac{1}{2}$	$4\frac{1}{2}$
4	$2\frac{1}{4}$	540	20	$6\frac{1}{4}$–$6\frac{13}{16}$		
	$2\frac{1}{2}$	600	23		$2\frac{1}{2}$	$4\frac{1}{2}$
	$2\frac{3}{4}$	660	26		$2\frac{1}{2}$	$4\frac{1}{2}$
	$3\frac{1}{2}$	840	32		$2\frac{1}{2}$	$4\frac{1}{2}$
	$4\frac{1}{4}$	1020	38		$2\frac{1}{2}$	$4\frac{1}{2}$
5	$2\frac{2}{3}$	640	20	9–$8\frac{9}{16}$	$2\frac{1}{2}$ §	$4\frac{1}{2}$
	3	720	23		$2\frac{1}{2}$ §	$4\frac{1}{2}$
	$3\frac{1}{2}$	840	26		$2\frac{1}{2}$ §	$4\frac{1}{2}$
	$4\frac{1}{3}$	1040	32		$2\frac{1}{2}$ §	$4\frac{1}{2}$
	5	1200	38		$2\frac{1}{2}$ §	$4\frac{1}{2}$
6	3	720	20	9–$10\frac{3}{8}$	$2\frac{1}{2}$	$4\frac{1}{2}$
	$3\frac{1}{2}$	840	23		$2\frac{1}{2}$	$4\frac{1}{2}$
	4	960	26		$2\frac{1}{2}$	$4\frac{1}{2}$
	5	1200	32		$2\frac{1}{2}$	$4\frac{1}{2}$
	6	1440	38		$2\frac{1}{2}$	$4\frac{1}{2}$
7	$2\frac{1}{2}$	600	14	$11\frac{3}{8}$–$12\frac{13}{16}$	$2\frac{1}{2}$	3
	3	720	17		$2\frac{1}{2}$	3
	$3\frac{2}{3}$	880	20		$2\frac{1}{2}$	3 or $4\frac{1}{2}$

*These ratings are based on steam at 215°F and air at 70°F. They apply only to installed radiators exposed in a normal manner; not to radiators installed behind enclosures, grilles, or under shelves.

†Maximum assembly 60 sections. Length equals number of sections times $2\frac{1}{2}$ in.

‡Where greater than standard leg heights are required, this dimension shall be 6 in., except for 7-tube sections, in heights from 13 to 20 in., inclusive, for which this dimension shall be $4\frac{1}{2}$ in. Radiators may be furnished without legs.

§For 5-tube hospital-type radiation, this dimension is 3 in.

Courtesy *1960 ASHRAE GUIDE*

The four principal function provided by valves operating in conjunction with radiators are:

1. Admission and throttling of the steam or hot-water supply.
2. Expulsion of the air liberated on condensation.

Table 2-3. Small-Tube Cast-Iron Radiators

Number of Tubes per Section	Section Dimensions						
	Catalog Rating per Section*		A Height‡	B Width		C Spacing†	D Leg Height‡
				Min	Max		
	Sq. Ft.	Btuh	In.	In.	In.	In.	In.
3§	1.6	384	25	3¼	3½	1¾	2½
4§	1.6	384	19	4⁷⁄₁₆	4¹³⁄₁₆	1¾	2½
	1.8	432	22	4⁷⁄₁₆	4¹³⁄₁₆	1¾	2½
	2.0	480	25	4⁷⁄₁₆	4¹³⁄₁₆	1¾	2½
5§	2.1	504	22	5⅝	6⁵⁄₁₆	1¾	2½
	2.4	576	25	5⅝	6⁵⁄₁₆	1¾	2½
6§	2.3	552	19	6¹³⁄₁₆	8	1¾	2½
	3.0	720	25	6¹³⁄₁₆	8	1¾	2½
	3.7	888	32	6¹³⁄₁₆	8	1¾	2½

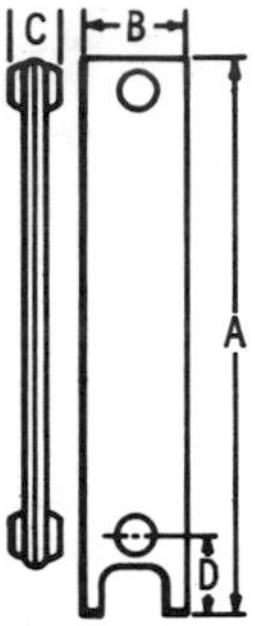

*These ratings are based on steam at 215°F and air at 70°F. They apply only to installed radiators exposed in a normal manner; not to radiators installed behind enclosures, grilles, or under shelves.

†Length equals number of sections times 1¾ in.

‡Overall height and leg height, as produced by some manufacturers, are one inch greater than shown in Columns A and D. Radiators may be furnished without legs. Where greater than standard leg heights are required, this dimension shall be 4½ in.

§Or equal.

Courtesy *1960 ASHRAE Guide*

3. Expulsion of air from spaces being filled by steam or hot water.

4. Expulsion of the condensation.

Radiator valves (packed or packless), manual or automatic air valves, and thermostatic expulsion valves (traps) are used to perform the above functions.

The packed-type radiator valve is an ordinary low-pressure steam valve having a stuffing box and a fibrous packing to prevent leakage around the stem (Fig. 2-3). The objection to this type of valve is the frequent need for adjustment and renewal of the packing to keep the joint tight. These valves also require many turns of the stem to fully open.

The packless radiator valve is one that has no packing of any kind. Sealing is obtained by means of a diaphragm (Fig. 2-4) or a bellows (Fig. 2-5). On each valve there is no connection between

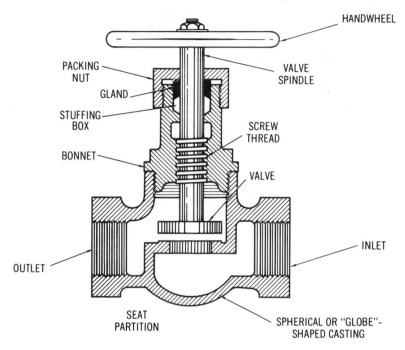

Fig. 2-3. Typical packed radiator valve.

the actuating element (stem and screw) and the valve being sealed hermetically; hence there can be no leakage. With the diaphragm arrangement a spring is used to open the valve, whereas with bellows construction there is no spring—a shoulder on the end of the stem works in a bearing on the valve inside the bellows.

Some so-called packless radiator valves actually employ spring discs to secure a tight joint. Although called "packless," the spring discs form a metallic equivalent of packing.

Both manual and automatic air valves are used to remove air from radiators. The manual valves are not well adapted for this function because they usually receive only irregular attention. Air is constantly forming in the radiator and should be removed as it forms. After the air valve remains closed for some time, the radiator gradually fills with air (or becomes air bound), as shown in Fig. 2-6, with the air at the bottom and the steam at the top. On

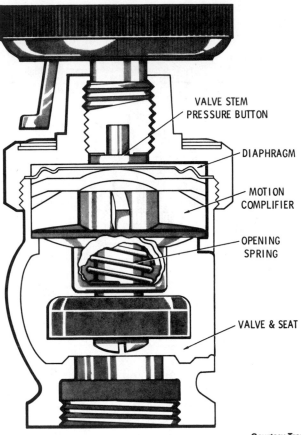

VALVE STEM
PRESSURE BUTTON

DIAPHRAGM

MOTION
COMPLIFIER

OPENING
SPRING

VALVE & SEAT

Courtesy Trane Co.

Fig. 2-4. Diaphragm-type packless radiator valve.

Fig. 2-5. Bellows-type packless
radiator valve.

Courtesy Sarco Co.

opening the valve (Fig. 2-7), the air is pushed out by the incoming steam. The radiator is gradually filled with steam until it begins to come out of the air valve (Fig. 2-8). At this point, the air valve should be closed.

An *automatic air valve* is one form of the thermostatic valve (Fig. 2-9). Automatic operation is made possible by a bimetallic

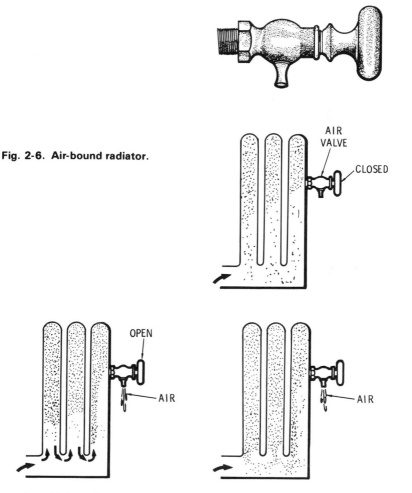

Fig. 2-6. Air-bound radiator.

Fig. 2-7. Air is pushed out by incoming steam.

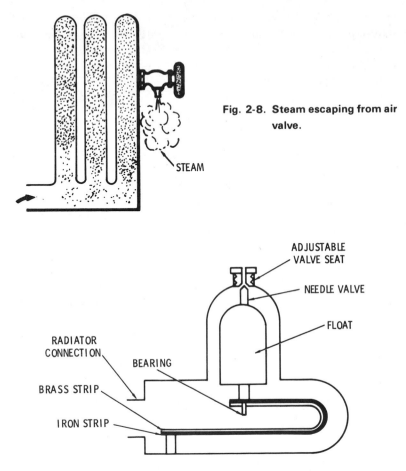

Fig. 2-8. Steam escaping from air valve.

Fig. 2-9. Working components of an automatic air valve.

element contained in the valve. The principles generally employed to secure automatic action are:

1. Expansion and contraction of metals.
2. Expansion and contraction of liquids.
3. Buoyancy of flotation.
4. Air expansion.

When the relatively cold air passes through the valve, the metal strips lie in contracted position (with legs close together and the valve open) allowing air to escape (Fig. 2-10). The steam then enters the valve, and its higher temperature causes the metal strips in the bimetallic element to expand. The brass strip expands more than the iron strip, and causes the end containing the valve spindle to rise and close (Fig. 2-11). When the strips are fully expanded, the valve is closed, shutting off the escape of steam (Fig. 2-12). In case the radiator becomes flooded with water, the latter entering will cause the float to push up the valve and prevent the escape of water (Fig. 2-13).

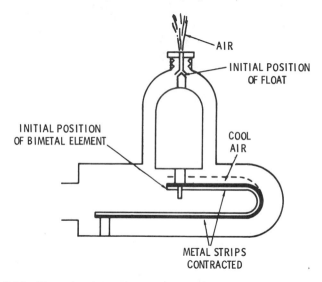

Fig. 2-10. Bimetal strips contracted and valve open.

Because an automatic air valve is used only for expelling air from a radiator, it should be distinguished from a thermostatic expulsion valve. A *thermostatic expulsion valve* opens to air and condensation, and closes to steam. The low temperature of the air and condensation causes the bimetallic element to contract and open the valve, whereas the relatively high temperature of the steam causes the element to expand and close the valve.

Although a thermostatic expulsion valve is sometimes referred to as a "trap," this term is more correctly used to indicate a larger unit not connected to a radiator and having the capacity to drain

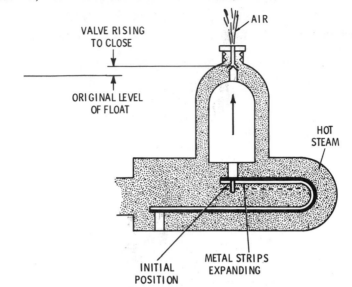

Fig. 2-11. Bimetal strip expanding.

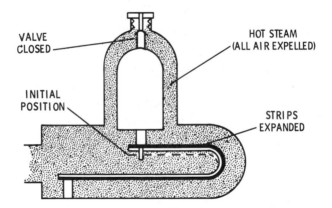

Fig. 2-12. Bimetal strips fully expanded and valve closed.

condensation from large mains. As distinguished from the thermostatic valve, a trap handles only condensation and not air.

A bellows charged with a liquid is used on some of the thermostatic valves as an actuating element instead of the bimetallic device. The operating principle of a Trane bellows-type thermostatic valve is illustrated in Figs. 2-14, 2-15, 2-16, and 2-17.

Additional information about valves and valve operating principles is contained in Chapter 9 of Volume 2 (Valves and Valve Installation).

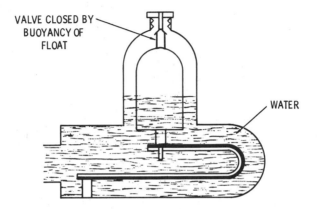

VALVE CLOSED BY
BUOYANCY OF
FLOAT

WATER

Fig. 2-13. Valve closed by buoyancy of water.

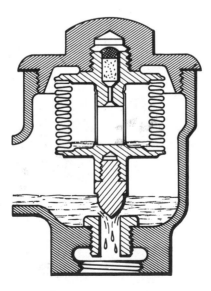

Fig. 2-14. Condensation is being discharged from heating unit.

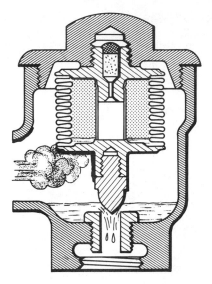

Fig. 2-15. Steam enters trap. Pressure within the bellows increases and causes it to expand.

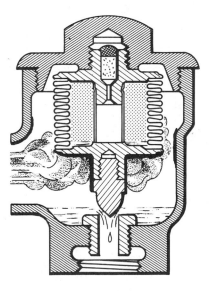

Fig. 2-16. All condensation is drained from unit.

Radiator Piping Connections

Some typical radiator piping connections are shown in Figs. 2-18, 2-19, 2-20, 2-21, and 2-22. The important thing to remember when connecting a radiator is to allow for movement of the risers and runouts. This movement is caused by the expansion and contraction resulting from temperature changes in the piping.

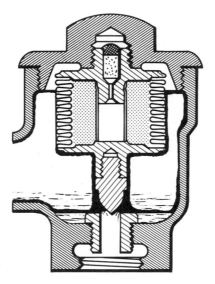

Fig. 2-17. Steam completely surrounds the bellows.

Radiator Efficiency

Radiator efficiency is important to the operating characteristics of the heating system. The following recommendations are offered as a guide for obtaining higher radiator operating efficiency:

1. A radiator *must* be level for efficient operation. Check it with a carpenter's level. Use wedges or shims to restore it to a level position.
2. Make sure the radiators have adequate air openings in the enclosure or cover. The openings must cover at least 40 percent of the total surface of the unit.

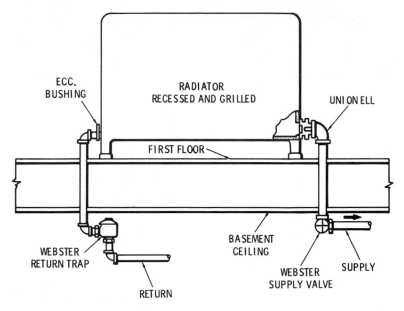

Fig. 2-18. Radiator supply and return connections for first-floor installation.

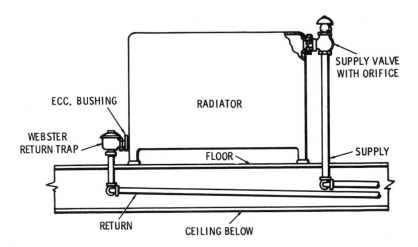

Fig. 2-19. Radiator supply and return connections for upper-floor installation.

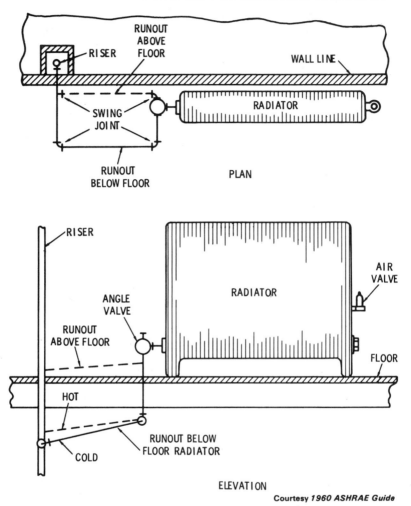

Fig. 2-20. One-pipe radiator connections.

3. Unpainted radiators give off more heat than painted ones. If the radiator is painted, strip the paint from the front, top, and sides. The radiator will produce 10 to 15 percent more heat at a lower cost.

4. Check the radiator air valve. If it is clogged, the amount of

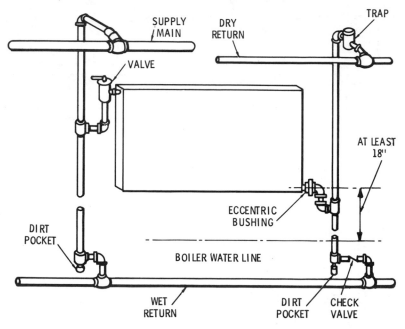

Fig. 2-21. Two-pipe connections to radiator installed on wall.

heat given off by the radiator will be reduced. Instructions for cleaning air valves are given in "Troubleshooting Radiators" (see below).

5. Radiators must be properly vented. This is particularly true of radiators located at the end of long supply mains. Instructions for venting radiators are given in "Vents and Venting" (see below).

6. Never block a radiator with furniture or drapes. Nothing should block or impede the flow of heat from the radiator.

7. Placing sheet metal or aluminum foil against the wall behind the radiator will reflect heat into the room.

DETERMINING REQUIRED RADIATION

Repeated tests have shown that the amount of heat given off by ordinary cast-iron radiators per degree difference in temperature between the steam (or water) in the radiator and the sur-rounding air to be about 1.6 Btu per square foot of heating sur-face per hour.

A cast-iron radiator will give off heat at the rate of 240 Btu per hour when supplied with steam at 2½ lb. pressure (220°F) with a surrounding air temperature of 70°F. It is determined as follows:

$$(220 - 70) \times 1.6 = 240 \text{ Btu}$$

One sq. ft. of steam radiation equals 1.6 sq. ft. of hot-water radiation or 1.4 sq. in. of warm-air pipe area. Tables 2-1, 2-2, and 2-3 list the heating surfaces for various column and tubular cast-iron radiators.

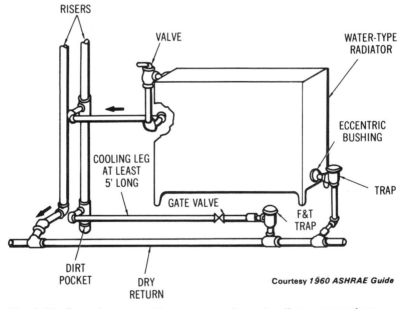

Fig. 2-22. Two-pipe top and bottom opposite-end radiator connections.

There is no IBR Code covering recessed radiation. As a result, manufacturers of recessed heat-emitting units must rate and cer-tify their own product. The Weil-McLain Company is typical of these manufacturers. Its certified ratings are determined from a series of tests conducted in accordance with the IBR Testing and

Rating Code for Baseboard Type Radiation whenever the provisions of the code should be applied. The Weil-McLain ratings include a 15 percent addition for heating effect and barometric pressure correction factor allowed by the IBR code. Other manufacturers use similar rating methods.

VENTS AND VENTING

Each volume of water contains a small percentage of air at atmospheric pressure mechanically mixed with it. This air is liberated during vaporization and causes some problems for the circulation of steam in the system. As steam starts to fill a heating system, it can enter the radiators, convectors, or baseboard units only as fast as the air escapes. For this reason, some means must be provided to vent this air from the system.

Adjustable air valves are often found in systems fired by automatic oil or gas burners. This type of air valve permits the adjustment of radiators varying in size and/or distance from the furnace or boiler so that radiators heat at an equal rate.

Nonadjustable air valves are not recommended because the larger radiators will still contain air after the smaller ones have been completely vented. The same problem occurs with the last radiator on a long main. Often the air valve on this radiator does not have enough time to rid the system of air before the *on* period is completed and the thermostat shuts off the burner. One method of handling this problem is by installing a large-size quick vent at the end of the long main (Fig. 2-23).

Double or triple venting is an extreme method of solving the problem of a persistently cold radiator (Fig. 2-24). There is usually only one opening for an air valve on a radiator. A second opening can be added by using a ⅛-in. pipe tap and a drill of the proper size.

If a multiple valve arrangement for a radiator fails to produce the desired results, the only other possibility is to lengthen the burner *on* period. This can be accomplished on oil burners by altering the differential.

Air must also be vented from hot-water heating systems. Trapped air will cause these systems to operate unsatisfactorily,

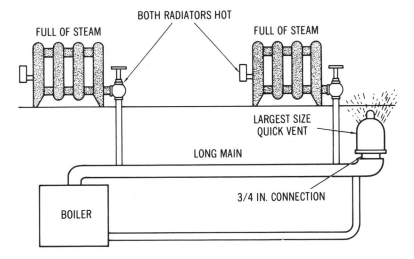

Fig. 2-23. Large-size quick-vent installed at end of long main.

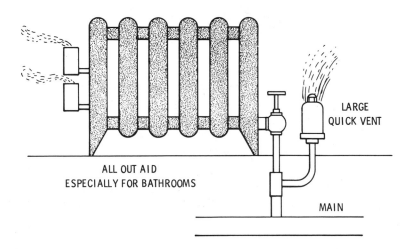

Fig. 2-24. Triple venting a radiator.

and a means should be provided to eliminate it. Manually operated air valves located at the highest levels in the heating system and automatic air valves placed at critical points will usually vent most of the trapped air.

TROUBLESHOOTING RADIATORS

If a radiator in a hot-water or steam heating system is not producing enough heat (or not producing heat at all), it may not be the fault of the radiator. Check the room thermostat and the automatic fuel-burning equipment (gas burner, oil burner, or coal stoker) to determine if they are malfunctioning. Methods for doing this are detailed in the appropriate chapters of this book. If you are satisfied that they are operating properly, the problem is probably in the radiator.

Hot water or steam enters a radiator at an inlet in the bottom and must rise against the pressure of the air contained in the radiator. A radiator is equipped with an automatic or manual air valve at the top to allow the air to escape and consequently permit the water or steam to rise.

In radiators equipped with automatic air valves, rising water or steam usually has enough force to push the air in the radiator out through the valve. The valve is automatically closed by a thermostatic control when it comes in contact with the hot water or steam. If a radiator equipped with an automatic air valve is not producing enough heat, the valve may be clogged. This can be checked by closing the shutoff valve at the bottom of the radiator and unscrewing the air valve. If air begins to rush out, open the radiator shutoff valve to see if it will heat up. An increase of heat is an indication that the air valve is clogged. Close the radiator shutoff valve again, remove the air valve, and clean it by boiling it in a solution of water and baking soda for 20 or 30 minutes. The radiator should now operate properly.

A radiator equipped with a manual air valve should be bled of air at the beginning of each heating season. It should also be bled if it fails to heat up properly. This is a very simple operation. Open the manual air valve until water or steam begins to run out. The water or steam running out indicates that all the air has been eliminated from the radiator.

Sometimes radiators are painted to improve their appearance. When a metallic paint (such as aluminum or silver) is used, the heating efficiency is reduced by 15 to 20 percent. If you must paint a radiator, use a nonmetallic paint for all surfaces *facing* the room. Dirty surfaces will also reduce the heating efficiency of a radiator. A good cleaning will eliminate the problem.

CONVECTORS

A *convector* is a heat-emitting unit that heats primarily by convection. In other words, most of the heat is produced by the movement of air around and across a heated metal surface. The air movement across this surface can be gravity induced or forced. As a result, convectors are classified as either *gravity air convectors* or *forced-air convectors*.

Small, upright gravity and forced-air convectors are commonly found in older heating installations (Fig. 2-25). The design of this type of unit was probably influenced by cast-iron radiators. A much more efficient convector is the finned-tube baseboard unit (see "Finned-Tube Baseboard Units").

Fig. 2-25. Typical convector.

Courtesy American-Standard

CONVECTOR PIPING CONNECTORS

The piping connections for a typical gravity convector are shown in Fig. 2-26. Supply connections to the convector heating element are made at the top, bottom, or end of the inlet header. Return connections are made at the bottom or end of the header at the opposite end of the unit. Fig. 2-27 illustrates two recommended piping connections for convectors used in a hot-water

67

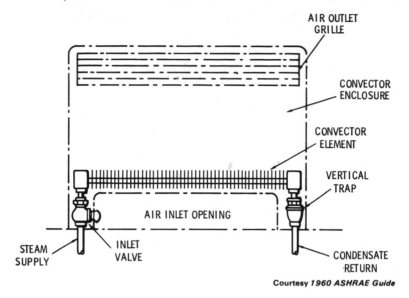

Courtesy *1960 ASHRAE Guide*

Fig. 2-26. Typical convector connections.

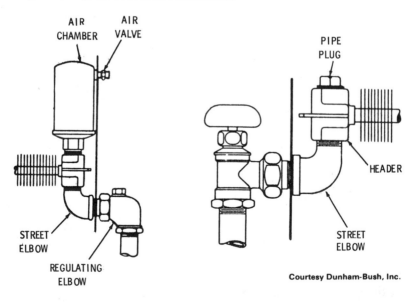

Courtesy Dunham-Bush, Inc.

Fig. 2-27. Convector piping connections in a hot-water heating system.

heating system. Typical convector piping connections for units used in steam heating systems are shown in Figs. 2-28 and 2-29.

Gravity convector piping connections are very similar to those used with radiators, except that the lines must be sized for a greater condensation rate. The usual method for determining convector capacities is to convert them to the equivalent square feet of direction radiation (EDR):

$$EDR = \frac{\text{Convector Rating (Btuh)}}{240 \text{ Btu}}$$

The 240 Btu figure represents the amount of heat in Btu given off by ordinary cast-iron radiators per square foot of heating surface per hour under average conditions.

The connections used for a *forced* convector are similar to those used with unit heaters (see "Unit Heater Piping Connections").

BASEBOARD UNITS

Baseboard units are designed to be installed along the bottom of walls where they replace sections of the conventional baseboard (Fig. 2-30). Locating them beneath windows or along exterior walls is a particularly effective method of eliminating cold drafts.

Baseboard units are frequently used in steam and hot-water heating systems (Figs. 2-31 and 2-32). In a series-loop system, supply and branch piping can be eliminated by using the baseboard unit to replace sections of the piping. In other words, there is no need for supply and return branches between the baseboard unit and the mains.

Baseboard units are also available with electric heating elements controlled by a centrally located wall-mounted thermostat or a built-in thermostat. Each unit is actually a separate heater (see "Electric Baseboard Heaters"), but they can be joined and wired together to form a baseboard heating system.

The Institute of Boiler and Radiator Manufacturers (IBR) has

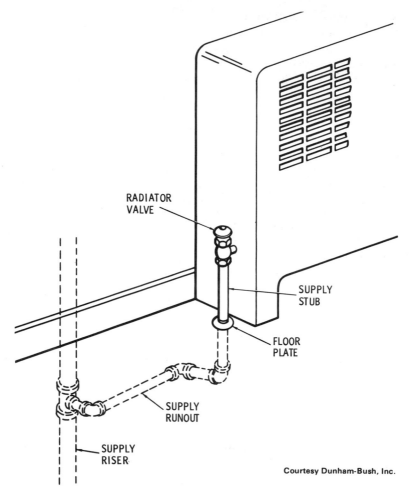

RADIATOR
VALVE

SUPPLY
STUB

FLOOR
PLATE

SUPPLY
RUNOUT

SUPPLY
RISER

Courtesy Dunham-Bush, Inc.

Fig. 2-28. Typical convector connections in a steam heating system.

established testing and rating methods of baseboard heat-emitting units. The output for baseboard units is rated in Btu per hour per linear foot.

Steam and Hot-Water Baseboard Units

Two types of baseboard units are used in steam and hot-water (hydronic) heating systems: those with separate fins attached to the tubing and those with the fins cast as an integral part of the unit. The integral-fin units are made of cast iron. Those with separate fins are made of nonferrous metals such as copper, aluminum, or alloys.

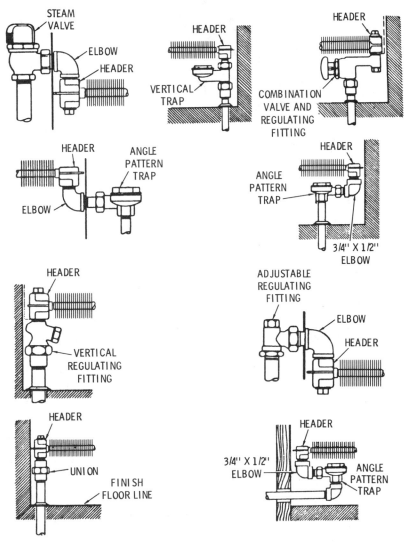

Fig. 2-29. Convector piping connections in a steam heating system.

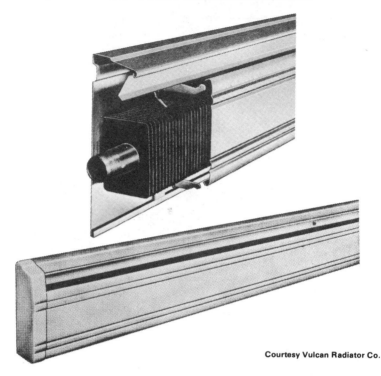

Courtesy Vulcan Radiator Co.

Fig. 2-30. Typical baseboard heating installation.

Separate Fin-and-Tube Baseboard Units.—An example of a baseboard unit with the fins attached to the tubing is shown in Fig. 2-30. The size and length of the tube, as well as the number, size, and spacing of the metal fins, will vary from one manufacturer to the next. Fin shapes, sizes, and spacings can be ordered to specification from the manufacturer. Basically the fins are either square or rectangular (Fig. 2-33). Some will have special design features, such as bent ends (Fig. 2-34).

The assembly of the heating element is covered by a sheet-metal enclosure. Openings are cut into the face of the cove to increase air circulation. When steam or hot water is passed through the tube, the heat is transmitted to the fins by conduction through the metals. The heat transmitted to the fins is transferred to the air by convection.

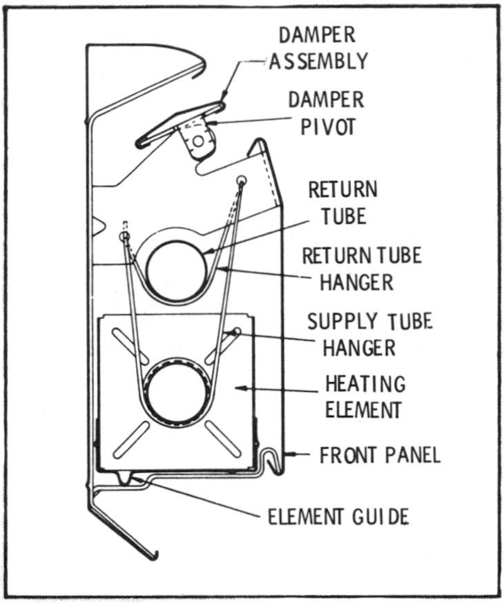

DAMPER
ASSEMBLY

DAMPER
PIVOT

RETURN
TUBE

RETURN TUBE
HANGER

SUPPLY TUBE
HANGER

HEATING
ELEMENT

FRONT PANEL

ELEMENT GUIDE

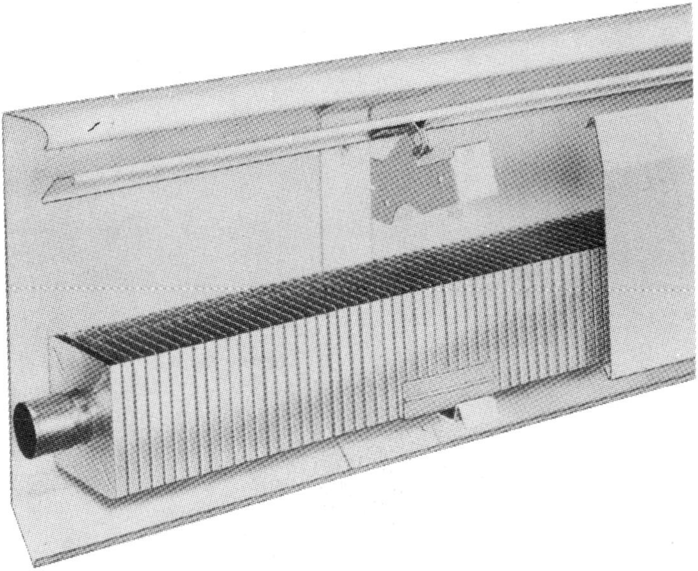

Courtesy Weil-McLain Co.

Fig. 2-31. Design features of a hot-water heating baseboard unit.

73

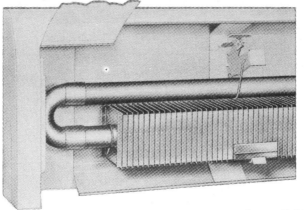

Courtesy Weil-McLain Co.

Fig. 2-32. Return tube installation for hot-water heating system.

Courtesy Vulcan Radiator Co.

Fig. 2-33. Typical fin shapes.

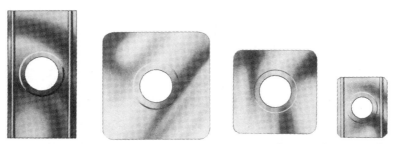

Courtesy Weil-McLain Co.

Fig. 2-34. Heating-element fins with flared ends.

The heating-element tube is available in a variety of sizes, including ¾, 1, 1¼, and 1½ in. (Fig. 2-35). On short pipe runs, the ¾-in. tube is recommended in order to ensure water velocities in the turbulent flow range. On long runs and with loop systems, it may be desirable to use 1- or 1¼-in. tube sizes.

Use of the small tube sizes results in lower cost for connecting piping in a run, for valves, expansion joints, balancing cocks, and fittings. There is also a lower heating-element cost involved with the small tube sizes.

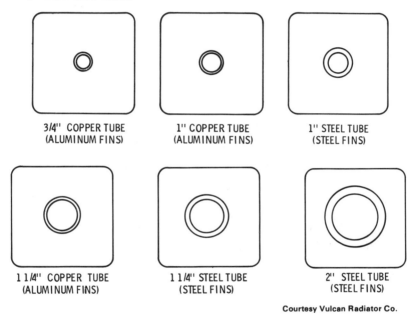

Courtesy Vulcan Radiator Co.

Fig. 2-35. Heating-element tube sizes.

The fins and tubing can be made of the same or different metals (Figs. 2-36 and 2-37). The following combinations are among those possible:

1. Copper fins on copper tubing.
2. Aluminum fins on copper tubing.
3. Aluminum fins on aluminum tubing.
4. Aluminum fins on steel tubing.

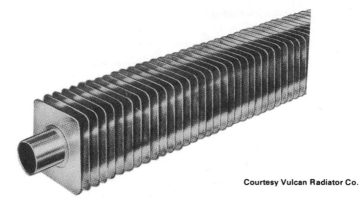

Courtesy Vulcan Radiator Co.

Fig. 2-36. Copper-aluminum heating elements.

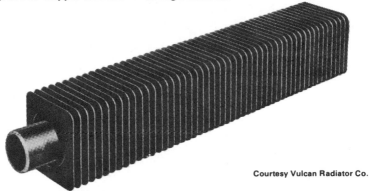

Courtesy Vulcan Radiator Co.

Fig. 2-37. Steel heating elements.

5. Stainless-steel fins on stainless-steel tubing.
6. Cupro-nickel fins on cupro-nickel tubing.

These different metals exhibit different heating characteristics and all are not suitable for the same application. For example, steel heating elements are recommended for high-temperature water systems. Copper-aluminum elements, on the other hand, work well with water temperatures up to 300°F. Copper-aluminum heating elements also produce a high heat output, but the copper tube has a very high rate of expansion (higher than steel). All these factors have to be taken into consideration when selecting a suitable heating element for the installation.

Integral Fin-and-Tube Baseboard Units

A cast-iron baseboard heater with the fins cast as an integral part of the unit is shown in Figs. 2-38 and 2-39. These baseboard heating units can be used in series-loop, one-pipe, and two-pipe (reverse-return) forced hot-water heating systems, and in two-pipe steam or vapor heating systems. They are not recommended for use in one-pipe steam heating systems.

Courtesy Burnham Corp.

Fig. 2-38. Cast-iron baseboard heater with fins cast as an integral part of the unit.

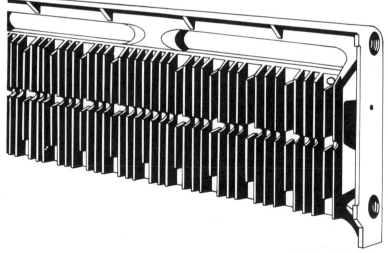

Courtesy Burnham Corp.

Fig. 2-39. Integral fin-and-tube construction details.

There are several advantages to using cast-iron baseboard heating units. Because the fins are a part of the casting, they will not bend, dent, or come apart. Cast iron is corrosion resistant and will not expand or contract with temperature changes. This last feature eliminates the expansion and contraction noises that frequently occur with nonferrous, separate fin-and-tube units, which allows them to fit closer to the wall.

INSTALLING BASEBOARD UNITS

A steel, finned-tube baseboard heating element will expand as much as ⅛ in. per 10 ft. with a 70°F to 200°F temperature rise. A copper element will expand as much as ⅟₁₆ in. per 10 ft. under the same conditions. This potential expansion of the heating element must be provided for when installing the system or problems will arise. The following provisions are recommended:

1. Allow a clearance of at least ¼ in. around all piping which passes through floors or wall partitions.
2. Wrap pipe passing through a floor or wall partition with a felt or foam sleeve to act as a cushion (Fig. 2-40).

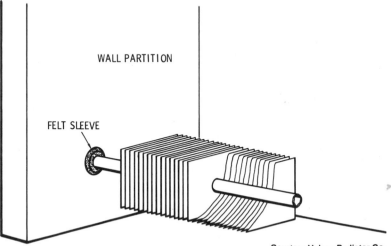

WALL PARTITION

FELT SLEEVE

Courtesy Vulcan Radiator Co.

Fig. 2-40. Clearance provided for expansion through wall partition.

3. Try to limit straight runs of pipe to a maximum of 30 ft. Wherever longer runs are necessary, install a bellows-type expansion joint near the center and anchor the ends.

4. Where a baseboard system extends around a corner, provide extra clearance at the ends (expansion will generally occur away from the corner) (Fig. 2-41).

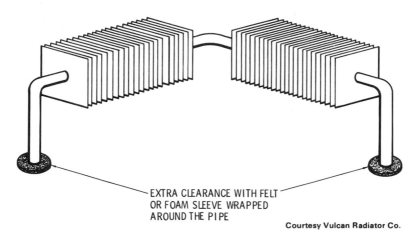

EXTRA CLEARANCE WITH FELT
OR FOAM SLEEVE WRAPPED
AROUND THE PIPE

Courtesy Vulcan Radiator Co.

Fig. 2-41. Extra clearance should be provided for an extension around a corner.

5. When a baseboard system is installed around three adjacent walls (forming a U), always use an expansion joint in the center leg of the U (Fig. 2-42).

6. When making piping connections, be sure to keep all radiator elements in proper vertical position so that fin edges will not touch other metal parts.

7. Be sure to support all mains and other piping runs adequately so that their weight will not cause bowing of the heating elements.

8. Install a felt sleeve around the pipe where it rests against a rigid hanging strap (Fig. 2-43).

Before installing the baseboard units, check the walls carefully for straightness. The baseboard units must be absolutely straight or their operating efficiency will be reduced. Therefore, it would

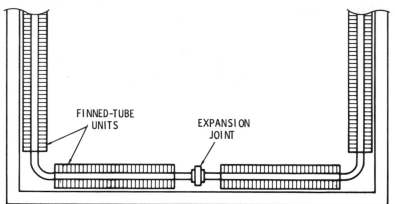

Fig. 2-42. Location of expansion joint.

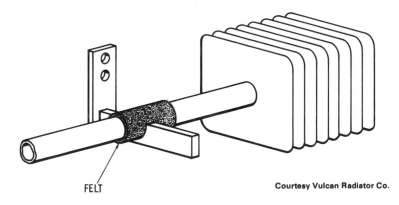

FELT

Courtesy Vulcan Radiator Co.

Fig. 2-43. Felt support for heating element.

be a mistake to use a wall to align the units if the wall were not straight. Shims may have to be used on wavy walls to keep the baseboard system straight.

Many baseboard units can be recessed the depth of the plaster. The back of each unit is nailed to the studs before plastering and the top of the baseboard unit hood serves as a plaster stop. The top accessories must be installed before plastering.

The procedure for installing a steam or hot-water baseboard heating system may be summarized as follows:

1. Place the backing of the unit against the wall surface or studs (in a recessed installation) and mark the location of the studs.
2. Punch holes in the backing and screw it to the studs (Fig. 2-44).
3. Install cover support brackets approximately 3 ft. apart (Fig. 2-45).
4. Place cradle hangers on rivet head and set the heating element on the cradles, making sure the fins are vertical and that all hangers are free to swing (Fig. 2-46).

Courtesy Vulcan Radiator Co.

Fig. 2-44. Unit back installed flush against wall surface or studs.

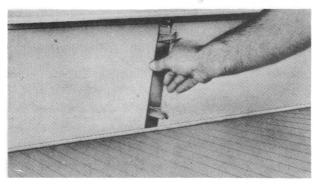

Courtesy Vulcan Radiator Co.

Fig. 2-45. Installing support brackets.

81

Courtesy Vulcan Radiator Co.

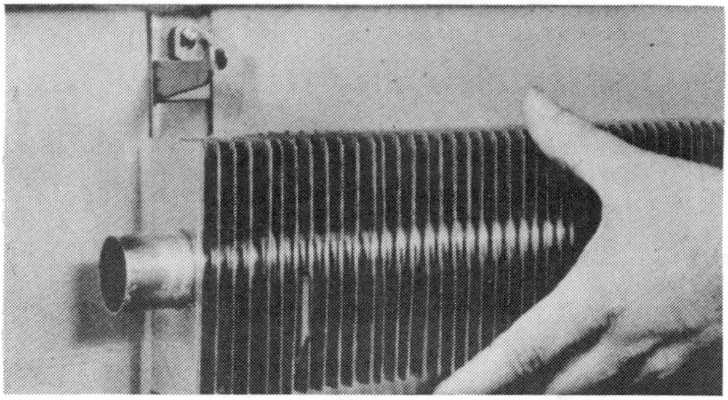

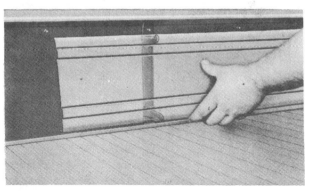

Fig. 2-46. Setting heating elements in place.

5. Complete piping and test for leaks before snapping on the fronts.
6. Snap the front cover onto the support brackets by hooking the top lip over the upper arms and snapping the bottom lip over the bottom arms (Fig. 2-47).

Courtesy Vulcan Radiator Co.

Fig. 2-47. Snapping on the front cover.

7. Add joining pieces and end enclosures where appropriate (Figs. 2-48 and 2-49).

Hot-water heating systems can be divided by temperature into the following three types:

1. High-temperature water systems 350°F to 450°F

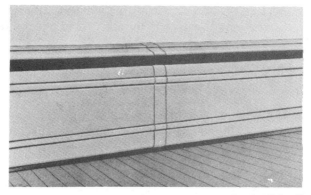

Courtesy Vulcan Radiator Co.

Fig. 2-48. Installing joint pieces.

Courtesy Vulcan Radiator Co.

Fig. 2-49. Installing corner and end enclosures.

| 2. Medium-temperature water systems | 250°F to 350°F |
| 3. Low-temperature | Below 250°F |

Although low-temperature hot-water heating systems are the predominant type used in the United States, there is a growing tendency to apply high-temperature water not only to underground distribution piping (as in district heating) but also to direct radiation.

When fin-tube baseboard units are used in a high-temperature hot-water heating system, care must be taken to keep the enclosure surface temperature to a minimum. This can be accomplished as follows:

1. Use the highest enclosure height practical.
2. Use heating elements only one row high.
3. Use a maximum fin spacing of 33 per foot.
4. Spread the heating element out along the entire wall length.
5. Limit the water temperature to that necessary to offset building heat loss.

Because of the higher heat level, the expansion of heating elements becomes more of a factor with high-temperature hot-water heating systems. As a result, more expansion joints should be used along the length of the heating elements than is the case with low-temperature water.

ELECTRIC BASEBOARD HEATERS

Electric baseboard heaters are designed for use in residential, commercial, industrial, and institutional heating systems. This chapter will be primarily concerned with a description of residential heaters. The electric heaters designed for use in commercial, industrial, or institutional installations differ from the residential type by being larger in capacity; otherwise, they are essentially identical in design (Fig. 2-50).

An electric baseboard heater contains one or more heating elements placed horizontally. Each electric heating element is a unit assembly consisting of a resistor, insulated supports, and terminals for connecting the resistor to the electric power supply.

By definition, a resistor is a material used to produce heat by passing an electrical current through it. Solids, liquids, and gases may be used as resistors, but solid resistors are the type most frequently used. Resistors may be made up of wire or metal ribbon, supported by refractory insulation, or embedded in refractory insulating material surrounded by a protective sheath of metal. A typical heating element used in Vulcan electric baseboard heaters is shown in Fig. 2-51. It consists of nichrome heater coils embedded in a ceramic core.

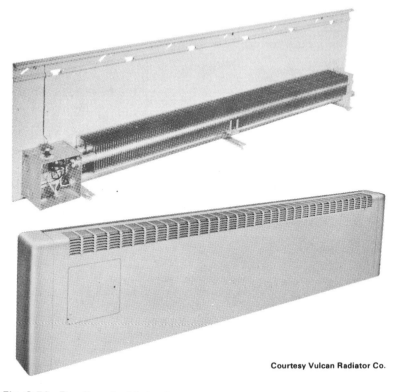

Courtesy Vulcan Radiator Co.

Fig. 2-50. Duraline electric heater.

Courtesy Vulcan Radiator Co.

Fig. 2-51. Heater core.

Each resistor is generally rated from 80 to 250 watts (270 to 850 Btu per hour) per linear foot of baseboard unit. The manufacturer will give the ratings for each unit (rather than per linear foot), but will also include its overall length in the data.

Either wall-mounted or built-in thermostats can be used to

85

control the heating elements in electric baseboard heaters. These thermostats (either low-voltage or line voltage types) are designed for single- or multiple-unit control with single- or two-stage heating elements. *Normally*, multiple-unit control is not feasible with two-stage units or with a built-in thermostat. It is recommended that single-stage heating units be used when it is necessary or desirable to control more than one unit from one thermostat.

Close control of temperatures can be obtained by installing a time-delay relay in the electric baseboard unit. This type of relay is particularly useful when a time delay is needed on switch make or break. Additional information about time-delay relays is included in Chapter 6 of Volume 2 (Other Automatic Controls).

A temperature-limit switch (thermal cutout) should be installed in each unit to prevent excessive temperature buildup. The maximum surface temperature for each baseboard enclosure should be limited to 190°F. Exceeding this limit may result in damage to the heater.

A typical temperature-limit switch used in an electric baseboard installation is the linear capillary type that assures constant protection over the entire length of the heater. An automatic resetting feature allows the heater to resume operation once conditions return to normal. The temperature-limit switch is usually located between the heater and the thermostat or disconnect switch in the wiring (Figs. 2-52 and 2-53). The only exception occurs when a built-in simultaneous switching thermostat is used (Fig. 2-54).

In a baseboard heating system, each temperature-limit switch must be wired to break the electrical circuit *only* to the heating element or elements of the unit on which it is installed.

Installing Electric Baseboard Heaters

Read the manufacturer's instructions carefully before attempting to install an electric baseboard heater. Particular attention should be paid to wiring instructions and required clearances. Check the local codes and ordinances before beginning any work. All installation must be made in accordance with local and national electrical codes, with the former taking precedence.

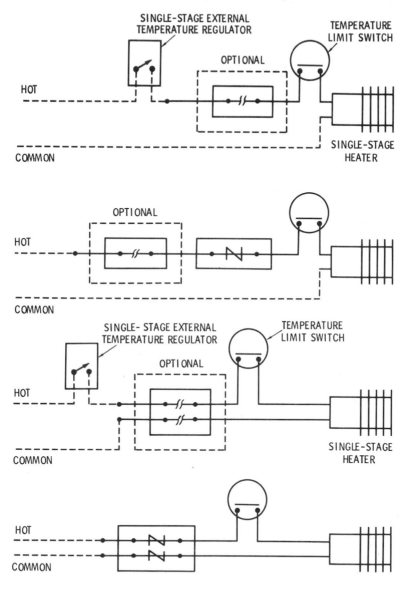

Courtesy Vulcan Radiator Co.

Fig. 2-52. Single-stage wiring diagrams.

87

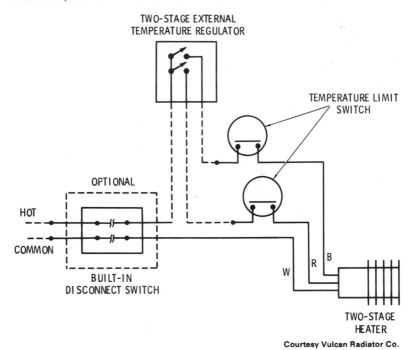

Courtesy Vulcan Radiator Co.

Fig. 2-53. Two-stage wiring diagram.

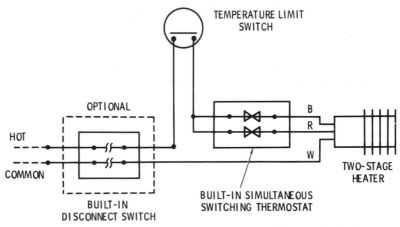

Courtesy Vulcan Radiator Co.

Fig. 2-54. Two-stage wiring diagram with built-in simultaneous switching thermostat.

A *minimum* clearance is generally required from the bottom of an electric heater to any obstructing surface, such as a floor, floor covering, ledge, or sill. The amount of clearance will depend upon the particular unit and manufacturer.

Check the heater operating voltage on the model before installing it. The watt output of an electric heating unit depends upon the supply (line) voltage. When the supply voltage equals the rated voltage of the unit, the output will equal the rated watts. *Never* connect a heater to a supply voltage greater than 5 percent above the marked operating voltage on the model label. A heater connected to a supply voltage that is *less* than the marked voltage on the model label will result in a watt output less than the model label rating. This will cause a reduced heating effect, which should be taken into consideration when determining heating requirements.

A change in watt output with respect to a variation in supply voltage may be conveniently calculated with the following formula:

$$2 \times (\text{supply voltage} - \text{rated voltage}) \times \text{amps}$$
$$\text{at rated voltage} = \text{change in watt output.}$$

For example, the change in watt output for a 120-volt/500-watt/4.2-amp heating unit connected to a 115-volt line can be calculated as follows:

$$2 \times (115\,\text{volt} - 120\,\text{volt}) \times 4.2\,\text{amp} =$$
$$2 \times (-5\,\text{volt}) \times 4.2\,\text{amp} =$$
$$-10\,\text{volt} \times 4.2\,\text{amp} =$$
$$-42\,\text{watt change in output.}$$

Thus, the output of this unit at 115 volts will be 458 watts (500 watts − 42 watts). The use of these calculations should be restricted to a + 10-volt difference in the supply voltage at 120 volts and a + 20-volt difference, or 208 volts, 240 volts, and 277 volts.

Some manufacturers of electric baseboard heating equipment produce heaters designed for installation as individual units or as components of a larger baseboard heating system. The procedure for installing individual units is as follows:

1. Locate the unit on the wall so that the *minimum* clearance is maintained between the bottom edge of the heater and the *finished* flooring (Fig. 2-55).
2. Position the unit so that the knock-outs in the junction box are aligned with the electrical rough-in location.
3. Locate the wall studs or mullions, and mark the location on the backing of the unit.

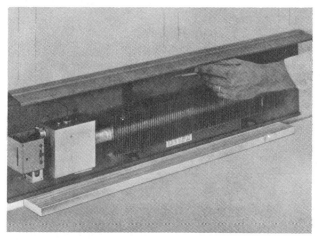

Courtesy Vulcan Radiator Co.

Fig. 2-55. Locating and marking the heater.

4. Drill or punch mounting holes in the backing a suitable distance below the hood and above the bottom edge.
5. Connect armored or nonmetallic sheathed cable through the knock-out in the backing or bottom of the junction box (Fig. 2-56). When entering through the backing, make up electrical connector, slide excess cable back into the wall space, and nail or screw the heater backing to the wall.
6. Make electrical connection of branch circuit wiring to the heater. Maintain continuity of grounding (see below).
7. Snap the heater cover over the brackets and at the same time align the thermostat operating shaft (built-in thermostats with the hole in the cover (Fig. 2-57).
8. Slide the end enclosures over each end of the unit (Fig. 2-58), and tighten the cover screws (Fig. 2-59).

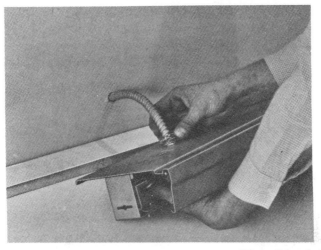

Courtesy Vulcan Radiator Co.

Fig. 2-56. Making the electrical connections.

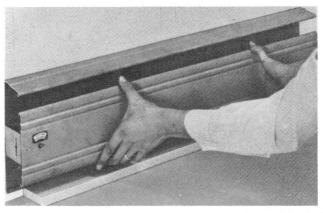

Courtesy Vulcan Radiator Co.

Fig. 2-57. Snapping cover over brackets.

For each electric heater unit, continuity of grounding must be maintained through circuit wiring devices. All branch circuit wiring and connections at the heater must be installed in accordance with local codes and regulations and the appropriate sections of the National Electrical Code.

Fig. 2-58. Typical end enclosure.

Courtesy Vulcan Radiator Co.

Courtesy Vulcan Radiator Co.

Fig. 2-59. Tightening cover screws.

Individual electric heaters can be mounted and wired to form a continuous baseboard heating system. The units used in a baseboard system should be ordered with a right-end mounted junction box to facilitate the wiring of adjacent unit. The instructions for installing individual electric heaters should be followed for each unit in the baseboard system (see Steps 1–7 above).

As shown in Fig. 2-60, straight line installation along a wall is simply a matter of butting the two units and connecting them with a joining piece (Fig. 2-61). Inside- and outside-corner installation methods are shown in Fig. 2-62 and 2-63.

Courtesy Vulcan Radiator Co.

Fig. 2-60. Straight line installation.

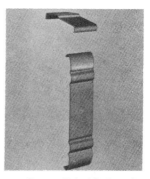

Fig. 2-61. Joining piece.

Courtesy Vulcan Radiator Co.

FLOOR RECESSED RADIATION

Floor recessed radiation is designed for installation below large glass areas that extend to the floor and do not permit the use of baseboard radiation. The heat-emitting unit illustrated in Fig. 2-64 is an example of recessed radiation used in a hydronic heating installation.

A typical hot-water recessed radiation heater consists of an enclosure, a finned element, two element glides, two rubber grommets, and a floor grille with or without dampers (Fig. 2-65).

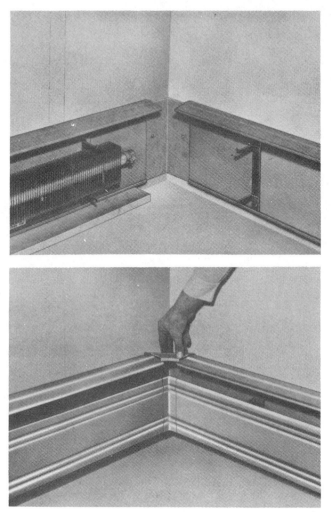

Fig. 2-62. Inside-corner installation.

The operating principle of this unit is relatively simple. The air entering the unit at the cool-air inlet is separated from the rising heated air by a baffle (Fig. 2-66). The baffle is designed to separate the cool inlet air from the rising heated air and to accelerate

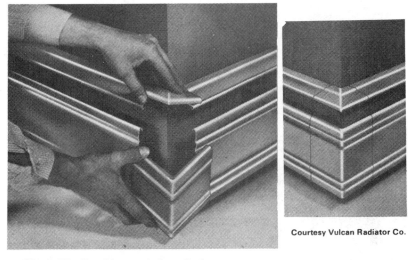

Courtesy Vulcan Radiator Co.

Fig. 2-63. Outside-corner installation.

Courtesy Weil-McLain Co.

Fig. 2-64. Typical floor recessed unit.

95

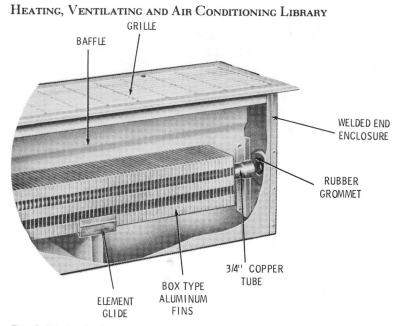

Fig. 2-65. Typical components of a recessed radiation unit.

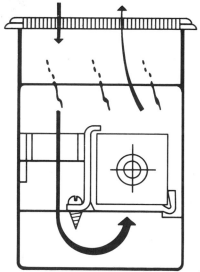

Courtesy Weil-McLain Co.

Fig. 2-66. Operating principle.

the flow of air through the unit for maximum heat output. The baffle is held in position by a guide on each end of the enclosure and a bracket in the center.

Figs. 2-67 and 2-68 show openings used to accommodate recessed floor units in a wood floor and in masonry construction. In a wood floor installation where the unit is to be installed parallel to the joists, an opening with the same dimensions as shown should be prepared by installing headers between joists.

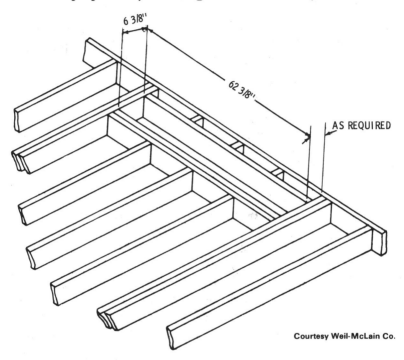

6 3/8"

62 3/8"

AS REQUIRED

Courtesy Weil-McLain Co.

Fig. 2-67. Wood floor opening.

UNIT HEATERS

A *unit heater* is an independent, self-contained device designed to supply heat to a given space. It is frequently referred to as a *space heater*.

The typical unit heater consists of a heating element, a fan or

97

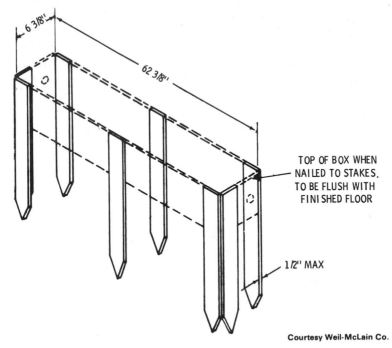

6 3/8"

62 3/8"

TOP OF BOX WHEN
NAILED TO STAKES.
TO BE FLUSH WITH
FINISHED FLOOR

1/2" MAX

Courtesy Weil-McLain Co.

Fig. 2-68. Stake arrangement for masonry floor opening.

propeller and motor, and a metal casing or enclosure. Unit heaters supply heat by forced convection using steam, hot water, gas, oil, or electricity as the heat source. The air is drawn into the unit heater and forced over the heating surfaces by a propeller or a centrifugal fan before it is expelled on a horizontal or vertical air current (Fig. 2-69).

Horizontal heaters have a specially designed broad-blade fan and are used for horizontal discharge space and spot heating assignments. When equipped with louver fin diffusers, the air stream can be directed in an unlimited number of diffusion patterns. A typical horizontal pipe suspended unit heater is shown in Fig. 2-70.

Forced-air unit heaters are small blower units that exhibit several times the capacity of a gravity circulation unit of the same size.

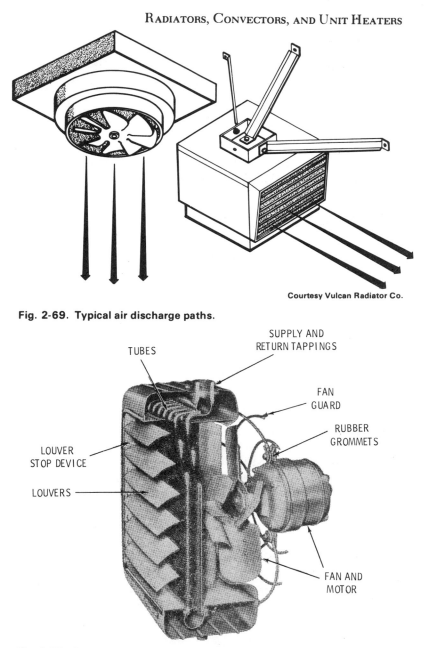

Courtesy Vulcan Radiator Co.

Fig. 2-69. Typical air discharge paths.

Fig. 2-70. Direct pipe suspended horizontal unit heater.

99

Unit Heater Piping Connections

The following general notes on piping propeller unit heaters are presented based on competent engineering installation practice:

1. Suspend unit heaters securely with provisions for easy removal.
2. Make certain units hang level vertically and horizontally.
3. Provide for expansion in supply lines. Note swing joints in suggested piping arrangements.
4. Provide unions adjacent to unit heaters in both supply and return laterals. Also provide shutoff valves in all supply laterals.
5. Use 45°-angle runoffs from all supply and return mains.
6. Make certain you have provided minimum clearances on all sides. Clearances will be specified in the heater manufacturer's installation literature.
7. When required, dirt pockets should be formed with pipe of the same size as the return tapping of the unit heater.
8. Pipe in the branching line should be the same size as the tapping in the trap.

Figs. 2-71 through 2-76 illustrate typical piping details for propeller-type horizontal unit heaters used in steam and hot-water heating systems.

The vapor steam system shown in Fig. 2-71 uses an overhead steam main and a gravity dry return. Strainer and float-type steam traps are used between the unit heater and the return main. A float-type air vent is installed below the float trap.

As shown in Fig. 2-72, the overhead steam main of a low-pressure or vacuum system can be vented and dripped independently. The steam supply to the unit heater is taken off the top of the main. The float vent is eliminated on systems using a vacuum pump.

Overhead supply and return mains are used (with bottom connections to the mains) in forced hot-water heating systems. The supply main is connected to the bottom of the unit heater. Either an automatic air vent or manual vent (pet cock) can be used at the high point on the return main (Fig. 2-73).

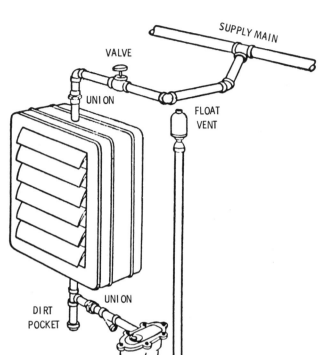

Fig. 2-71. Piping connections for a vapor steam heating system.

In the two-pipe steam system shown in Fig. 2-74, the steam supply of the unit heater is taken from the top of the main. The return from the unit heater is vented before dropping to the wet return. The distance to the water line of the boiler must be sufficient to allow for the pressure drop in the piping.

Both an overhead supply main and an overhead return are used in the high-pressure steam system shown in Fig. 2-75. The top of the bucket trap must be located below the return outlet of the coil for complete drainage of the condensation.

101

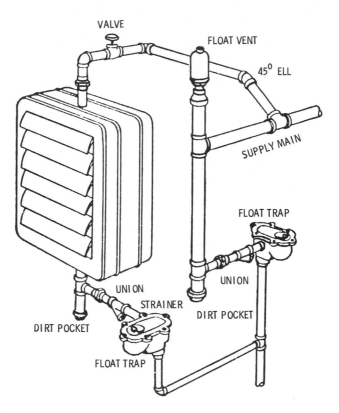

Fig. 2-72. Piping connections for a low-pressure or vacuum steam heating system.

Unit heater piping connections for a Trane high-pressure steam system are illustrated in Fig. 2-76. In this system, an overhead supply main is used with a lower return main. Where steam pressure fluctuates over a wide range, a swing-check valve should be placed in the return lateral between the strainer and the bucket trap to prevent reverse flow of the condensation or steam flashing when the pressure drops suddenly. The top of the bucket trap must be located below the return outlet of the coil for complete drainage of the condensation.

UNIT HEATER CONTROLS

Two control systems are used with unit heaters. The simpler type provides on-off operation of the fan. The other system provides modulating control of the heat source and continuous operation of the fan. It is considered the more effective control system.

Modulating control is obtained with either pneumatic or electric equipment. It allows a constant discharge of warm air and

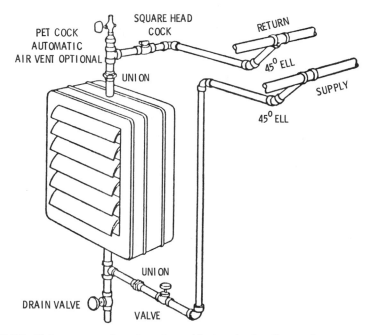

Fig. 2-73. Piping connections for a forced hot-water heating system.

eliminates intermittent blasts of hot air. The continuous circulation prevents air stratification, which occurs when the fan is off. A proportional room thermostat governs a valve that controls the heat source or a bypass around the heating limit. Fan operation is governed by an auxiliary switch or a limit thermostat. Either device is designed to stop the fan when the heat is shut off.

In an on-off control system, the fan motor is controlled by a room thermostat. A backup safety device in the form of a limit thermostat will stop the fan when heat is no longer being supplied to the unit heater.

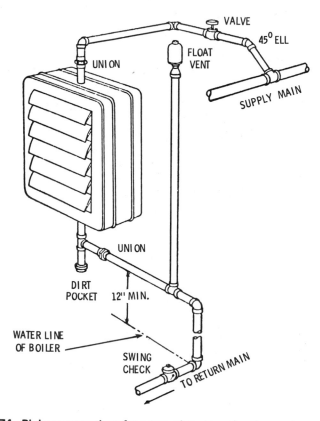

Fig. 2-74. Piping connections for a two-pipe steam heating system.

GAS-FIRED UNIT HEATERS

A gas-fired unit heater is designed to operate on either natural or propane gas (Fig 2-77). These units should be installed in a location where there will be sufficient ventilation for combustion and proper venting under normal conditions of use.

Local codes and regulations should be followed closely when installing a gas-fired unit heater. Additional useful information can be found in the current edition of *Installation of Gas Piping and Gas Appliances in Buildings* (ANSI-Z21.30).

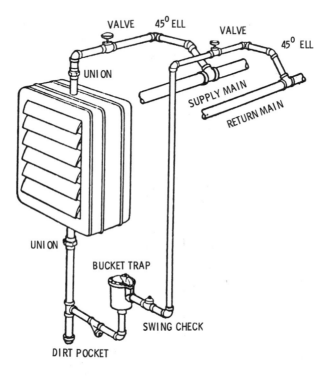

Fig. 2-75. Piping connections for a standard high-pressure steam heating system.

The automatic controls used to operate a gas-fired unit heater closely resemble those found on other types of gas heating equipment, such as furnaces, boilers, and water heaters. For example, the unit heater illustrated in Fig. 2-78 is equipped with a 24-volt gas valve with a built-in pilot relay, pilot gas adjustment, shutoff device for use with a thermocouple-type pilot for complete gas shutoff operation, and a gas-pressure regulator.

A separate gas-pressure regulator and pilot valve are generally supplied with control systems when the gas valve does not have a built-in gas-pressure regulator. The usual fan and limit controls are used on most gas-fired unit heaters. A wiring diagram for a typical control system is shown in Fig. 2-79.

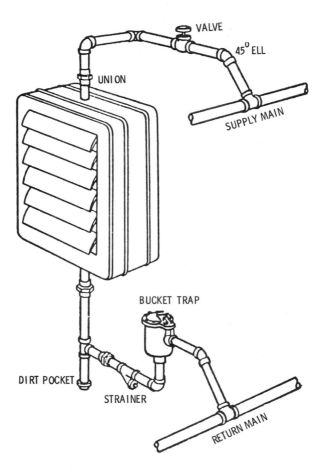

Fig. 2-76. Piping connections for a Trane high-pressure heating system.

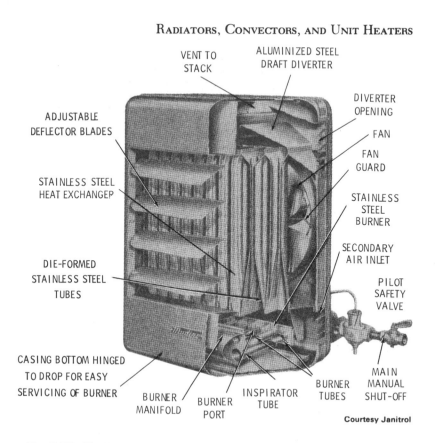

VENT TO
STACK

ALUMINIZED STEEL
DRAFT DIVERTER

DIVERTER
OPENING

FAN

FAN
GUARD

STAINLESS
STEEL
BURNER

SECONDARY
AIR INLET

PILOT
SAFETY
VALVE

ADJUSTABLE
DEFLECTOR BLADES

STAINLESS STEEL
HEAT EXCHANGEP

DIE-FORMED
STAINLESS STEEL
TUBES

CASING BOTTOM HINGED
TO DROP FOR EASY
SERVICING OF BURNER

BURNER
MANIFOLD

BURNER
PORT

INSPIRATOR
TUBE

BURNER
TUBES

MAIN
MANUAL
SHUT-OFF

Courtesy Janitrol

Fig. 2-77. Modine gas-fired unit heater showing principal components.

Fig. 2-78. Gas-fired unit heater.

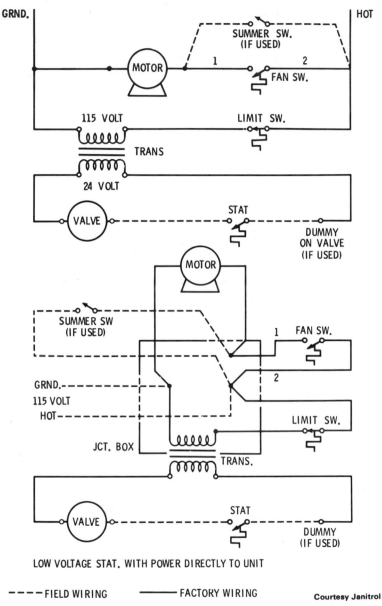

LOW VOLTAGE STAT. WITH POWER DIRECTLY TO UNIT

- - - - FIELD WIRING ———— FACTORY WIRING

Courtesy Janitrol

Fig. 2-79. Wiring diagram for a gas-fired unit heater.

OIL-FIRED UNIT HEATERS

Direct oil-fired, suspended unit heaters operate on the same principle as the larger oil-fired heating equipment (furnaces, boilers, water heaters, etc.). An oil burner located on the outside of the unit supplied heat to the combustion chamber, and a fan blows the heat into the room or space to be heated (Fig. 2-80).

Fig. 2-80. Oil-fired unit heater.

Courtesy National Oil Fuel Institute

An oil-fired unit heater should be located where it will heat efficiently and where it will receive sufficient air for combustion. Proper venting is also important. The waste products of the combustion process must be carried to the outdoors.

CHAPTER 3

Fireplaces, Stoves, and Chimneys

Until recently, fireplaces, stoves, ranges, wood heaters, and similar heating apparatus were the only source of heat for cooking, domestic hot water, and personal comfort. These apparatus generally burned solid fuels, such as wood, coke, or the different types of coal, but many later models could also be modified to burn fuel oil, gas (both natural and manufactured), and kerosene.

Central heating and the availability of clean, inexpensive, non-solid fuels (natural gas, oil, and electricity) have relegated these heating apparatus to a minor, decorative role except in remote rural areas of the country. At least this has been the situation until the late 1960's. The growing interest among people in leaving the cities and returning to the land, and the rising cost and potential scarcity of the more popular heating fuels, has brought renewed interest in the solid-fuel-burning fireplace, stove, range, and heater.

FIREPLACES

A fireplace is essentially a three-sided enclosure in which a fire is maintained. The heat from the fire enters the room from the open side of the enclosure. The traditional masonry fireplace is a recessed opening in the wall directly connected to a chimney. Modern prefabricated, freestanding fireplaces resemble stoves, but their operating principle is still that of the traditional fireplace; that is to say, heat is radiated from the opening out into the room.

Although a fireplace is not as complicated an apparatus as the furnace or boiler of a modern central heating system, its location, dimensions, and construction details still require careful planning if maximum heating efficiency and trouble-free operation are desired. These aspects of fireplace design and construction are examined in the sections that follow.

Fireplace Location

The location of the fireplace is determined by the location of the chimney. Unfortunately, chimney location—particularly in newer homes—is too often dictated by how the chimney will look, rather than how it will work. As a result, the chimney is often made too low or located where it may be obstructed by a section of the house or building. The top of the chimney should be at least 2 ft. higher than the roof.

Fireplace Dimensions

The principal components of a masonry fireplace are described in the next section (see "Fireplace Construction Details"). Those components important to the operation of a fireplace are:

1. The opening.
2. The throat.
3. The smoke chamber (and shelf).
4. The damper.
5. The flue.

These components are interdependent and must be properly dimensioned with respect to one another or the fireplace will not operate properly. Table 3-1 lists typical dimensions for finished masonry fireplaces, and these dimensions are identified in Figs. 3-1, 3-2, and 3-3.

The dotted lines and letters *(F-F, E-E,* and *T-T)* in Fig. 3-1 are used to indicate the dimensions of the throat, smoke chamber, and flue. The throat (or damper opening is identified by the line *F-F* in Fig. 3-1. The throat should be approximately 6 to 8 in. (or more) above the bottom of the lintel. The area of the throat should be not less than that of the flue, and its length should be equal to the width of the fireplace opening.

Starting approximately 5 in. above the throat (i.e. at line *E-E),* the inner wall surfaces should gradually slope inward approximately 30° to the beginning of the chimney flue (line *T-T* in Fig. 3-1). The smoke chamber is the area between the throat and the

Table 3-1. Recommended Dimensions for a Finished Masonry Fireplace

(Letters at heads of columns refer to Figs. 3-1, 3-2, and 3-3.)

Opening		Depth, d	Mini-mum back (hori-zontal) c	Vertical back wall, a	Inclined back wall, b	Outside dimensions of standard rectangular flue lining	Inside diameter of standard round flue lining
Width, w	Height, h						
Inches	Inches	Inches	Inches	Inches	Inches	Inches	Inches
24	24	16–18	14	14	16	8½ by 8½	10
28	24	16–18	14	14	16	8½ by 8½	10
24	28	16–18	14	14	20	8½ by 8½	10
30	28	16–18	16	14	20	8½ by 13	10
36	28	16–18	22	14	20	8½ by 13	12
42	28	16–18	28	14	20	8½ by 18	12
36	32	18–20	20	14	24	8½ by 18	12
42	32	18–20	26	14	24	13 by 13	12
48	32	18–20	32	14	24	13 by 13	15
42	36	18–20	26	14	28	13 by 13	15
48	36	18–20	32	14	28	13 by 18	15
54	36	18–20	38	14	28	13 by 18	15
60	36	18–20	44	14	28	13 by 18	15
42	40	20–22	24	17	29	13 by 13	15
48	40	20–22	30	17	29	13 by 18	15
54	40	20–22	36	17	29	13 by 18	15
60	40	20–22	42	17	29	18 by 18	18
66	40	20–22	48	17	29	18 by 18	18
72	40	22–28	51	17	29	18 by 18	18

beginning of the flue (i.e., between lines *E-E* and *T-T* in Fig. 3-1). The smoke shelf is the area extending from the throat to the line of the flue wall (Fig. 3-2). This dimension will vary depending upon the depth of the firebox. The damper opening (throat) should *never* be less than the flue area.

Fireplace Construction Details

Construction details of a typical masonry fireplace are shown in Fig. 3-4. Because this type of fireplace is recessed in a wall, it is easier and less expensive to build it while the structure is under construction.

The principal components of a masonry fireplace are:

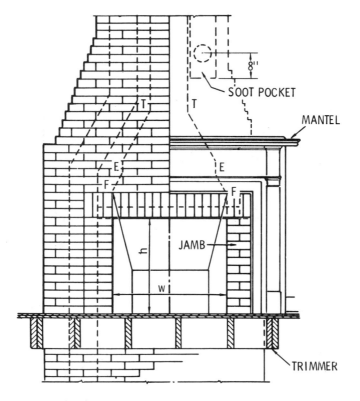

Fig. 3-1. Fireplace elevation.

1. Firebox.
2. Lintel.
3. Mantel.
4. Hearth.
5. Ashpit.
6. Ash dump.
7. Clean-out door.
8. Smoke chamber.

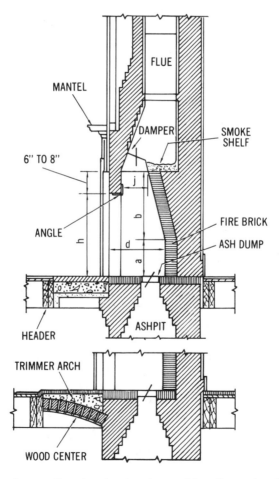

Fig. 3-2. Fireplace section showing two types of hearth construction.

115

9. Smoke shelf.
10. Throat.
11. Damper.
12. Flue and flue lining.

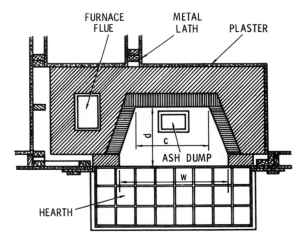

Fig. 3-3. Fireplace plan.

Firebox, Lintel, and Mantel

That portion of the fireplace in which the fire is maintained is sometimes called the *firebox*. This is essentially a three-sided enclosure with the open side facing the room. The area (in square inches) of the fireplace opening is directly related to the area of the chimney flue. The rule-of-thumb is to make the fireplace opening approximately twelve times the area of the flue.

As shown in Fig. 3-5, the *lintel* spans the top of the fireplace opening. This is a length of stone or metal used to support the weight of the fireplace superstructure. The *mantel* is a horizontal member extending across the top of the fireplace where it generally serves a decorative function. Sometimes the terms "mantel" and "lintel" are used synonymously. When this is the case, the mantel is used in the form of a wood beam, stone, or arch and functions as a lintel to support the masonry above the fireplace opening.

116

Fireplace Hearth

The *hearth* is the surface or pavement in the fireplace on which the fire is built. The hearth is sometimes made flush with the floor and will extend out in front of the fireplace opening where it

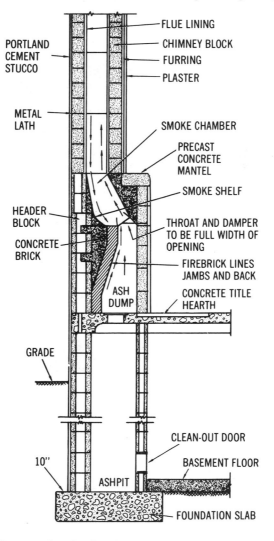

Fig. 3-4. Construction details of a masonry fireplace.

provides protection to the floor or floor covering against flying embers or ashes (Fig. 3-3). The recommended length of the hearth is the width of the fireplace opening *plus* 16 in.

Ash Dump, Ashpit, and Clean-out Door

An *ashpit* is a chamber located beneath the fireplace where the ashes from the fire are collected. An *ash dump* should be installed in the hearth toward the back of the fireplace. This is a metal plate that is pivoted so that the ashes can be easily dropped into the ashpit (Fig. 3-6). Ashes are removed from the ashpit through a *clean-out door* (Fig. 3-7).

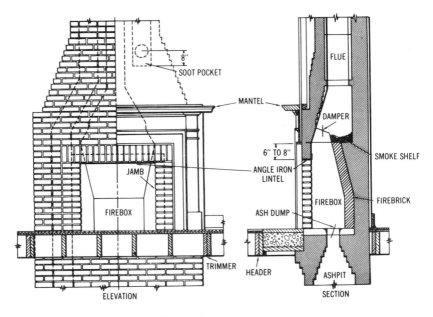

Fig. 3-5. Firebox, lintel, and mantel.

Courtesy Portland Stove Foundry Co.

Fig. 3-6. Centrally pivoted ash dump.

Smoke Chamber

The *smoke chamber* of the fireplace is a passage leading from the throat to the chimney flue (Fig. 3-4). It should be constructed so that it slopes at a 45° angle from the smoke shaft to the flue opening. The smoke chamber should also be constructed so that a smoke shelf extends straight back from the throat to the back of the flue line. The purpose of the smoke shelf is to prevent downdrafts from flowing into the fireplace. The damper should be mounted on the smoke shelf so that it covers the throat when it is closed. When open, the damper functions as a baffle against downdrafts.

Fireplace Dampers

One of the most frequent complaints about fireplaces is smoke backing up into the room. This condition, particularly if it persists, can usually be traced to a problem with the damper.

On very rare occasions, a fireplace and chimney will be built without a damper. When this is the case, the construction should be modified to incorporate a suitable damper; otherwise it will be built without a damper. When this is the case, the construction should be modified to incorporate a suitable damper; otherwise it will be impossible to build and maintain a fire in the fireplace.

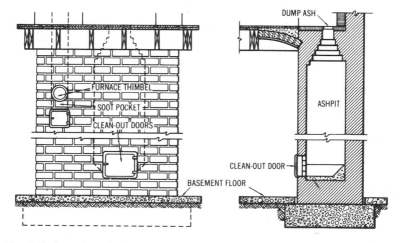

Fig. 3-7. Location of ash dump, ashpit, and clean-out door.

The damper should be installed at the front of the fireplace and should be wide enough to extend all the way across the opening. The two most common locations for a fireplace damper are shown in Figs. 3-8 and 3-9. If the damper is placed higher up

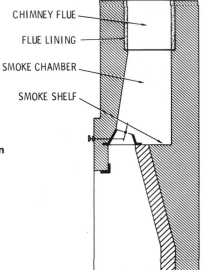

CHIMNEY FLUE

FLUE LINING

SMOKE CHAMBER

SMOKE SHELF

Fig. 3-8. Damper mounted on edge of smoke shelf.

Courtesy Portland Stove Foundry Co.

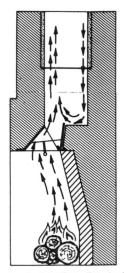

Fig. 3-9. Damper installed just above the fireplace opening.

Courtesy Portland Stove Foundry Co.

in the chimney, it is controlled with a long operating handle extending to just under the edge of the fireplace opening.

A rotary-control damper designed for use in the throat of a fireplace is shown in Fig. 3-10. This particular damper has a tilting lid that rotates on its lower edge rather than on centrally located pivots like other rotary dampers (Fig. 3-11). The lid is operated by a brass handle attached to a rod that extends through the face brick or tiling over the fireplace opening. Dampers equipped with ratchet-control levers are shown in Fig. 3-12.

MODIFIED FIREPLACES

A *modified fireplace* consists of a heavy metal manufactured unit designed to be set in place and concealed by brickwork or

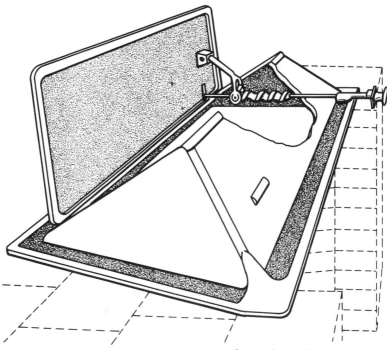

Courtesy Portland Stove Foundry Co.

Fig. 3-10. Rotary damper with lid that rotates on its lower edge.

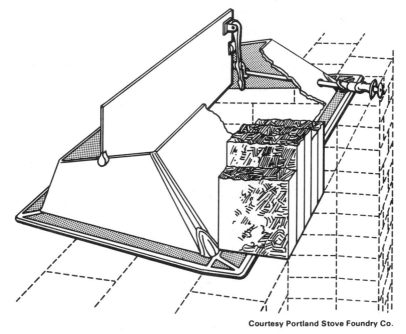

Courtesy Portland Stove Foundry Co.

Fig. 3-11. Rotary damper with lid that tilts on centrally located pivots.

other masonry construction (Figs. 3-13, 3-14, and 3-15).

Modified fireplaces are usually more efficient than the all-masonry type because provisions are made for circulating the heated air. As shown in Fig. 3-16, cool air enters the inlets at the bottom and is heated when it comes into contact with the hot surface of the metal shell. The heated air then rises by natural circulation, and is discharged through the outlets at the top. These warm-air outlets may be located in the wall of the room containing the fireplace, in the wall of an adjacent room, or in a room on the second floor. Sometimes a fan is installed in the duct system to improve the circulation.

FREESTANDING FIREPLACES

A *freestanding fireplace* is a prefabricated metal unit sold at many hardware stores, building supply outlets, and lumberyards.

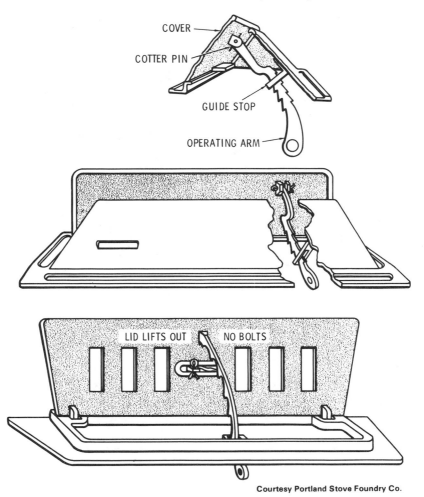

Courtesy Portland Stove Foundry Co.

Fig. 3-12. Damper equipped with ratchet-control levers.

Some typical examples of freestanding fireplaces are shown in Figs. 3-17 and 3-18. These fireplaces are exposed on all sides and their operation is essentially identical to stoves. They are available in a wide variety of styles and colors, and they are easy to install and operate. Because they are lightweight, no special supportive foundation is necessary, although it is a good idea to

123

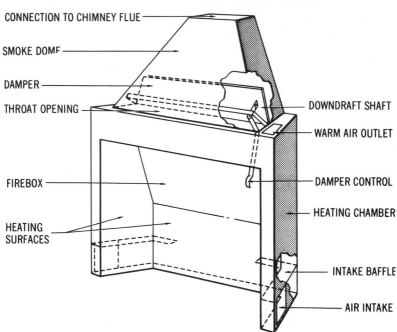

CONNECTION TO CHIMNEY FLUE

SMOKE DOME

DAMPER

THROAT OPENING

DOWNDRAFT SHAFT

WARM AIR OUTLET

FIREBOX

DAMPER CONTROL

HEATING CHAMBER

HEATING SURFACES

INTAKE BAFFLE

AIR INTAKE

Fig. 3-13. Prefabricated firebox used in the construction of a modified fireplace.

install a base under the unit to prevent stray sparks and embers from falling on the floor or floor covering. Most local codes and regulations require that a protective base be used, and the manufacturer's installation literature probably also recommends its use.

The protective base serves the same purpose as a hearth extension on a masonry or modified fireplace. The base should be made from some sort of noncombustible material and should be large enough in area to extend at least 12 in. beyond the unit in all directions (Fig. 3-19).

CHIMNEY DRAFT

A properly designed and constructed chimney is essential to a fireplace because it provides the draft necessary to remove the

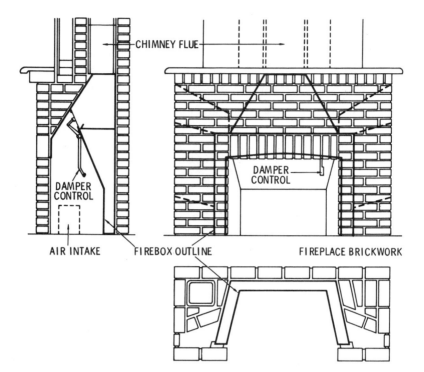

Fig. 3-14. Construction details of a modified fireplace.

smoke and flue gases. The motive power that produces this natural draft is the slight difference in weight between the column of rising hot flue bases *inside* the chimney and the column of colder and heavier air *outside* the chimney.

The intensity of the draft will depend upon the height of the chimney and the difference in temperature between the columns of air on the inside and the outside. It is measured in inches of a water column.

The theoretical draft of a chimney in inches of water at sea level can be determined with the following formula:

$$D = 7.00 \, H \left(\frac{1}{461 + T} - \frac{1}{461 + T_1} \right)$$

125

Where:

D = Theoretical draft.
H = Distance from top of chimney to grates.
T = Temperature of air outside the chimney.
T^1 = Temperature of gases inside the chimney.

For altitudes *higher* than sea level, the calculations obtained in the formula should be corrected by the factors listed in Table 3-2. Thus, if the structure is located at an altitude of approximately 1000 ft., the results obtained in the formula should be multiplied

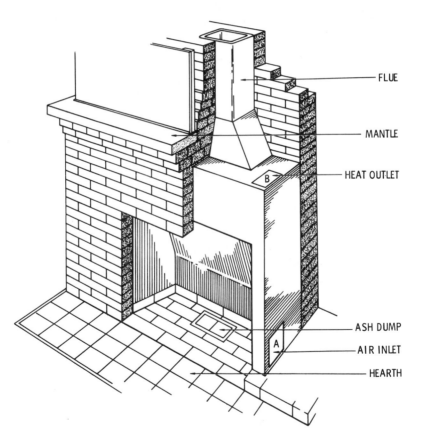

FLUE

MANTLE

HEAT OUTLET

ASH DUMP

AIR INLET

HEARTH

Fig. 3-15. Typical modified fireplace.

by the correction factor 0.966 to obtain the correct draft for that altitude.

Table 3-2. Correction Factors for Altitudes Above Sea Level

Altitudes (in feet)	Correction Factors
1000	0.966
2000	0.932
3000	0.900
5000	0.840
10,000	0.694

CHIMNEY CONSTRUCTION DETAILS

Because of its height and weight, a masonry chimney should be supported by a suitable foundation. This foundation should extend *at least* 10 in. below the bottom of the chimney, and a similar distance beyond its outer edge on all sides (Fig. 3-4). The foundation can be made more stable by making the base larger in area than the top. This can be accomplished by stepping down the footing at an angle.

CHIMNEY CAP

The *chimney cap* (or *capping)* is both a decorative and functional method of finishing off the top of the chimney (Fig. 3-20). Built of brick, concrete, or concrete slab, the chimney capping tends to counteract eddy currents coming from a high roof. The slab-type capping shown in Fig. 3-21 is an effective means of preventing downdrafts. Chimney cappings are usually designed to harmonize with the architecture of the structure.

CHIMNEY FLUES AND FLUE LININGS

The *chimney flue* is the passage through which the smoke, gases, and other products of the combustion process travel to the

127

outdoors. The flue lining should be smooth, clean, and nonporous so that it will be easy to clean and will not absorb moisture. The flue liner should have rounded corners; these are much more desirable than sharp corners because smoke and heat rise in a spiral motion.

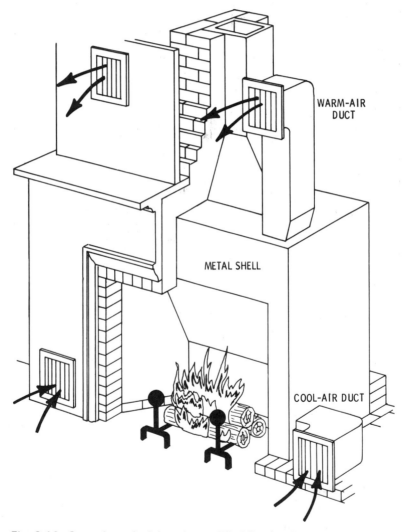

Fig. 3-16. Operating principles of a modified fireplace.

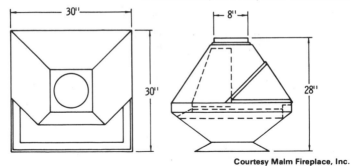

Courtesy Malm Fireplace, Inc.

Fig. 3-17. Wall-model freestanding fireplace.

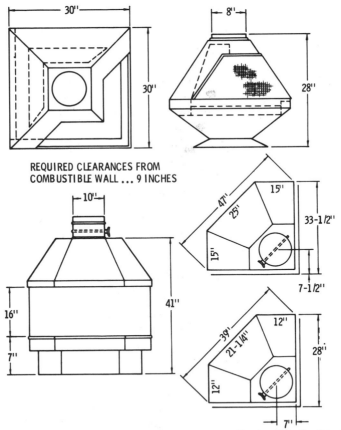

REQUIRED CLEARANCES FROM
COMBUSTIBLE WALL ... 9 INCHES

Courtesy Malm Fireplace, Inc.

Fig. 3-18. Corner-model freestanding fireplace.

129

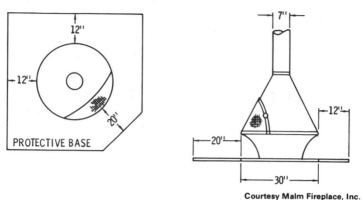

Courtesy Malm Fireplace, Inc.

Fig. 3-19. Dimensions of a typical protective base showing clearances of fireplace.

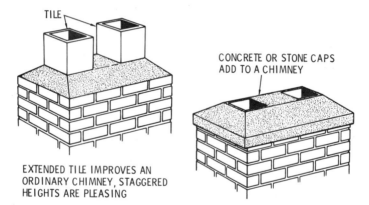

Fig. 3-20. Typical chimney caps.

Although most flues are lined with a fireproof material, chimneys with walls 8 in. or more in thickness do not need a flue lining. Fired-clay linings are used for flues in chimneys having walls less than 8 in. thick. Chimneys without a flue lining should be carefully pargeted and troweled so that the surface is as smooth as possible.

Flue construction is very important. If the flue is too small, it will restrict the passage and slow the rise of the flue gases. A flue that is too large will cause the fireplace to smoke when the fire is

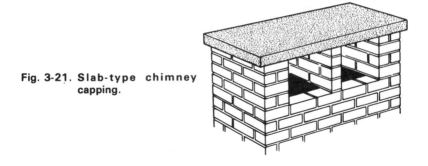

Fig. 3-21. Slab-type chimney capping.

first started. Because the flue is too large, the fire takes longer to heat the air and consequently the flue gases are initially slow to overcome the heavy cold air found in the chimney.

If the same chimney is used to serve fireplaces on two or more floors, *each* fireplace should have a separate flue (Fig. 3-22). This also holds true when a furnace (or boiler) and one or more fireplaces use the same chimney. Each heat source should be connected to a separate flue in the chimney.

Do not allow two (or more) flues to end on the same level at the top of the chimney. Doing so can result in smoke or gases from one flue being drawn down into the other one. One of the flues should be at least 6 in. higher than the other.

SMOKE PIPE

The *smoke pipe* is used to connect the heating equipment to the chimney flue. *Never* allow the smoke pipe to extend beyond the flue lining in the chimney or it will obstruct the flow of smoke and other gases (Figs. 3-23 and 3-24).

CLEAN-OUT TRAP

Chimneys used with coal-fired and oil-fired heating equipment should be equipped with clean-out traps. Access is provided in the ashpit (Fig. 3-4).

131

CHIMNEY DOWNDRAFT

Sometimes the air will not rise properly in a chimney, and will fall back into the fireplace or stove. This *downdraft* (or *backdraft)* condition results in poor combustion, smoke, and odors. It

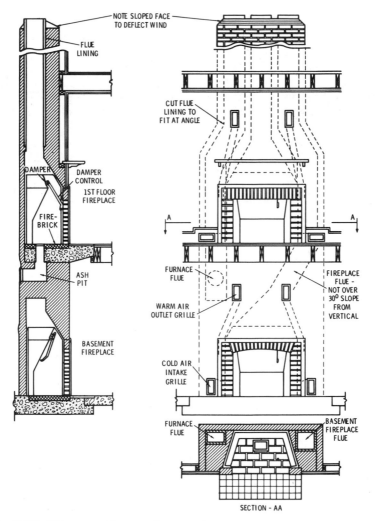

Fig. 3-22. Chimney with more than one flue.

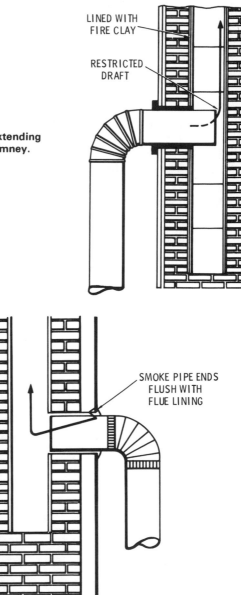

Fig. 3-23. Smoke pipe extending too far into chimney.

Fig. 3-24. Proper connection of smoke pipe to chimney flue.

133

is generally caused by either deflected air currents or chilled flue gases.

Air currents can be deflected down into a chimney by higher nearby objects, such as a portion of the structure, another building, a tree, or a hill. It is therefore important to build the chimney higher than any other part of the structure or any nearby objects. Because the deflected air entering the chimney has not passed through a hot fire, it will cool the air in the flue and weaken the draft. When the air becomes cooler, it also becomes heavier and falls back down into the chimney.

Chimneys built on the outside of a structure, particularly those exposed on three sides, must have walls at least 8 in. thick in order to prevent chilling the flue gases. Remember: flue gases

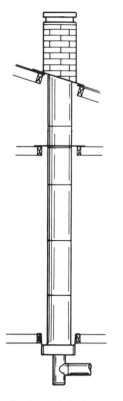

Fig. 3-25. Prefabricated metal chimney used with oil-, gas-, coal-, and wood-burning fireplaces.

must not be allowed to cool. The cooler the gas, the heavier it becomes.

PREFABRICATED METAL CHIMNEYS

Prefabricated (factory built) metal chimneys are commonly made of 24-gauge galvanized steel or 16-oz. copper (Figs. 3-25 and 3-26). These chimneys can be used with gas, oil, wood, or coal-burning appliances. The manufacturer's instructions should be carefully followed when installing one of these chimneys.

Freestanding fireplaces, such as those illustrated in Figs. 3-17,

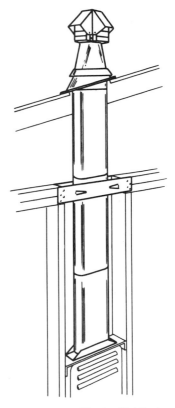

Fig. 3-26. Vent pipe design for wall heaters with inputs to 65,000 Btu.

Courtesy Metalbestos

135

3-18, and 3-19, are equipped with lengths of pipe designed to extend upward from the top of the unit for approximately 8 ft. (Fig. 3-27). These lengths of pipe are available with porcelain enamel surfaces that match the fireplace in color.

Check the local codes and regulations before installing a metal chimney. A single-wall metal pipe chimney has a tendency to chill rapidly, causing creosote to form on the interior surfaces. This is considered an extreme fire hazard.

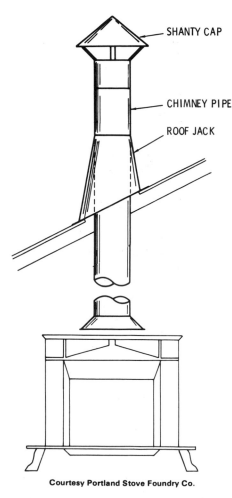

SHANTY CAP

CHIMNEY PIPE

ROOF JACK

Fig. 3-27. Shanty cap and pipe.

Courtesy Portland Stove Foundry Co.

TROUBLESHOOTING
FIREPLACES AND CHIMNEYS

The troubleshooting chart that follows lists the most common problems associated with the operation of a fireplace.

Symptom and Possible Cause   *Possible Remedy*

Persistent Smokiness

1. No damper.
2. Damper set too low.
3. Damper too narrow.

4. Damper set at back of fireplace opening
5. No smoke shelf.
6. No smoke chamber or poorly designed one.

7. Fireplace opening too large for flue size.
8. Insufficient combustion air.

9. Downdrafts occur.

10. Chimney clogged with debris.

1. Install a suitable damper.
2. Correct.
3. Extend damper all the way across opening.
4. Relocate damper to the front.
5. Install smoke shelf.
6. Install a suitable smoke chamber or make necessary modifications in old one.
7. Reduce the size of opening.
8. Open window a crack or provide other means of supplying combustion air.
9. Extend chimney higher; protect chimney opening with cap designed to deflect downdraft.
10. Clean chimney.

Fire Dies Out

1. Insufficient combustion air.

2. Clogged or dirty flue.

1. Open window a crack or provide other means of supplying combustion air.
2. Check flue, correct, and/or clean.

137

Symptom and Possible Cause   *Possible Remedy*

3. Closed damper.	3. Open damper.
4. Fuel logs too green.	4. Replace with suitable logs.
5. Fuel logs improperly arranged.	5. Rearrange and rebuild fire.

STOVES, RANGES, AND HEATERS

A *stove* is a device used for heating or for both heating and cooking purposes (Figs. 3-28 through 3-32). Heating is generally regarded as the primary function of a stove. The heat is delivered to the room by both radiation and convection.

Brick, tile, and masonry stoves first appeared in Europe as early as the fifteenth century. The first cast-iron stove was produced in Massachusetts in 1642, but it was a rather crude device by modern standards because it had no grates. Benjamin Franklin revolutionized the design of the stove with the introduction in 1742 of the model that bears his name (Fig. 3-33).

Sometimes the terms "stove" and "range" are used synonymously. This confusion, or blurring of the differences between the two, probably results from the incorporation of such features as an oven, cooking surface, and hot-water tank in the design of the stove.

A *range* is a cooking surface and an oven combined in a single unit. It differs from a stove by being *specifically* designed for cooking, although it can supply a certain amount of warmth to the kitchen.

Ranges can be divided into a number of different categories depending upon the type of fuel used. There are four basic categories of ranges:

1. Gas ranges.
2. Electric ranges.
3. Kerosene ranges.
4. Solid-fuel ranges.

Fig. 3-28. Wood-burning parlor stove.

Courtesy Portland Stove Foundry Co.

Courtesy Portland Stove Foundry Co.

Fig. 3-29. Atlantic box stove.

139

Gas ranges use both natural and manufactured gases. Kerosene ranges have been further developed to use either gasoline or fuel oil. The solid fuels used in ranges include wood, charcoal, coal (lignite, bituminous, anthracite), and coke.

Fig. 3-30. The Franklin stove with folding doors.

The ranges shown in Figs 3-34 an 3-35 use wood, coal, or oil. The fuel grates and ashes (from the ashpit) are removed through the front of the range. These ranges also contain a large removable copper tank, which is used to supply hot water for cooking and other purposes.

A *wood heater* is a device used specifically to provide heat to a room or a similarly limited space. These are wood-burning units that can also be equipped to burn coal, oil, or gas. These heaters are similar to stoves in their operating principle. They should not be confused with unit (space) heaters, which are suspended from

Fig. 3-31. Reproduction of the eighteenth-century cast-iron stove.

ceilings or walls and operate on a completely different principle (see Chapter 2, Radiators, Convectors, and Unit Heaters).

The wood-burning heater shown in Fig. 3-36 is equipped with a thermostat. The thermostat consists of a bimetallic helix coil that opens and closes a damper in response to temperature

141

Fig. 3-32. Reproduction of the 1812 Franklin stove.

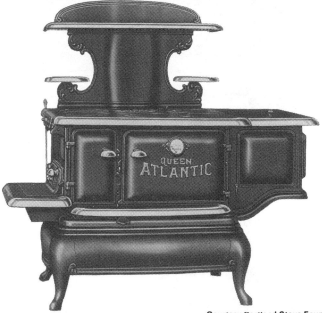

Fig. 3-33. Wood- and coal-burning range.

changes. The damper opens just enough to admit precisely the amount of combustion air necessary to maintain the heat at a comfortable level.

A mat or floor protector made of some sort of fireproof material should be placed under a stove, range, or heater to protect the floor or floor covering against live sparks or hot ashes. These units should also be placed as far from a wall or partition as is necessary to prevent heat damage.

Courtesy Portland Stove Foundry Co.

Fig. 3-34. Franklin cast-iron stove based on the 1742 design.

INSTALLATION INSTRUCTIONS

Before installing a new stove, make certain the chimney is large enough to accept the necessary smoke pipe. If another stove or furnace is using the same chimney flue, connect the stove above or below the other one *(never* at the same level). The area of a chimney flue should be about 25 percent larger than the area of the smoke pipe that enters the chimney.

The same design and construction features required for a chimney serving a fireplace also apply to one used with a stove. If there is no chimney, then one must be built.

Fig. 3-35. Cast-iron range used with coal, wood, or oil.

OPERATING INSTRUCTIONS

Stoves are similar to fireplaces in that they possess no power in themselves to force smoke up a chimney. The pipe and chimney

Courtesy Ashley Automatic Heater Co.

Fig. 3-36. Wood-burning heater equipped with a thermostat.

provide the draft, and if they are defective, the stove will not work satisfactorily.

The installation must be capable of providing sufficient air for the combustion process. At the same time, it must be able to expel smoke and other gases to the outdoors.

CHAPTER 4

Water Heaters and Other Appliances

A *water heater* is an appliance designed to heat water for purposes other than space or central heating. The hot water supplied by this appliance is used for such purposes as cooking, washing, and bathing. This water is generally referred to as *service* or *domestic hot water*.

There are two basic types of water heating systems. One is independent of the central or space heating system and has a separately fired water heater. The other water heating system depends upon the central heating boiler for its heat. The service hot water is heated when it circulates through a heat exchanger inserted in the hot-water or steam heating boiler. These heat exchangers are only used with boilers. In a warm-air heating system, service hot water is provided by an independent and separately fired water heater.

TYPES OF WATER HEATERS

Water heaters can be divided into several groups or classes. The most common classifications are based on the following criteria:

1. Size and intended usage.
2. Heating method.
3. Heat-control method.
4. Fuel type.
5. Flue location and design.
6. Recovery rate.

Water heaters are classified as either domestic or commercial water heaters on the basis of their size and intended usage. Domestic water heaters are those with input rates up to 75,000 Btu per hour. Water heaters regarded as commercial types have input rates in excess of 75,000 Btu per hour. Temperature is another of the criteria used to distinguish between domestic and commercial water heaters. For commercial usage, hot-water temperatures of 180°F or more are generally required. The hot water used in residences does not normally exceed 180°F. For domestic usage, 140°F is generally considered adequate.

Another method of classifying water heaters is by how the heat is applied. Direct-fired water heaters are those in which the water is heated by the direct combustion of the fuel. In indirect water heaters, the service water obtains its heat from steam or hot water and not directly from the combustion process. The advantages and disadvantages of both direct-fired and indirect water heaters are described elsewhere in this chapter.

The heat-control method in a water heater may be either automatic or manual (nonautomatic). All water heaters used in residences are now of the automatic type.

Gas, oil, coal, electricity, steam, or hot water can be used to heat the water in a water heater. Either steam or hot water can be used as the heat-conveying medium in indirect water heaters. Neither, of course, is a fuel. The fuels used to heat the domestic water supply are gas, oil, and coal. Gas is by far the most popular fuel. Oil is gaining some popularity, but it still falls far short of gas. The use of coal as a fuel for heating water is now found only in rare cases. Although electricity is not a fuel, in the strict sense of the word, it is generally used along with the three fossil fuels as an additional category for classification.

Quite often, water heaters are classified on the basis of flue location and design. This is a particularly useful criterion for classifying the various automatic storage-type water heaters.

Water heaters can also be classified on the basis of their recovery rate. Quick-recovery water heaters are capable of producing hot water at a more rapid recovery rate than the slow-recovery types. Quick-recovery heaters are often used in commercial structures where there is a constant demand for hot water.

Direct-Fired Water Heaters

A *direct-fired water heater* is one in which the water is heated by the direct combustion of a fuel such as gas, oil, or coal. The combustion flame directly impinges on a metal surface, which divides the flame from the hot water. This metal surface is quite often (but not always) the external wall of the hot-water storage tank. It serves as a convenient heat transfer surface.

A direct-fired water heater is easily distinguished from an indirect water heater (see below), which uses an intermediate heat conveying medium such as steam or hot water, and an electric water heater, which depends upon an immersed electric heating element for its heat.

Automatic Storage Water Heaters

The underfired automatic storage heater in which a single tank is used for both heating and storing the water is the most common water heater used in houses, apartments and small commercial buildings. These heaters generally have a 30-, 40-, 50-gal. storage capacity, although it is possible to purchase automatic storage heaters with capacities ranging from 70 gal. to several hundred gal.

Automatic storage heaters are classified according to the placement of the flue. Using flue placement as a basis for classification, *gas-fired* automatic storage water heaters can be divided into the following three basic types:

1. Internal or center flue.
2. External channel flue tank.
3. External flue and floating tank.

The internal or center flue gas-fired water heater (Fig. 4-1) is very economical to manufacture, but proper flue baffling is required for good efficiency.

The external channel flue gas-fired water heater provides a much larger heating surface than the center flue. As shown in Fig. 4-2, the entire bottom surface of the tank serves as the heating surface. Heating is from the bottom of the tank, which acts to increase efficiency.

An even larger heat transfer surface is provided by the external flue and floating tank water heater illustrated in Fig. 4-3. Both the full bottom and sides of the tank serve as heat transfer surfaces. The heat (and gases) pass around the storage tank and exit through the vent.

The tank and flue construction of oil-fired automatic storage water heaters is very similar to gas-fired types. The major difference between the two lies in the combustion chamber. In gas-fired water heaters, the burners are located inside the combustion chamber (Figs. 4-1, 4-2, and 4-3). In oil-fired water heaters, the oil burner is mounted externally, and the flames are shot into the combustion chamber (Fig. 4-4).

Additional information about automatic storage water heaters can be found in this chapter in the section "Gas-Fired Water Heaters."

Multicoil Water Heaters

Sometimes the size of a structure or its use will result in a greater demand for hot water than a conventional water heater can supply. When this is the situation, it is recommended that a *multicoil,* or *large-volume, water heater* be installed. These water heaters resemble instantaneous heaters externally, but differ by containing more than one heating coil. The separate heating coils are connected to manifolds operated by a thermostat inserted in the storage tank. Because multicoil water heaters have large storage tanks, they are suitable for use in restaurants, clubs, and structures of similar size and use.

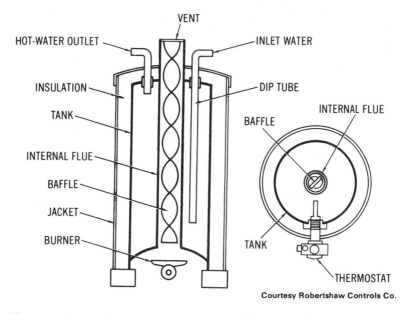

Fig. 4-1. Internal flue tank construction (gas-fired water heater).

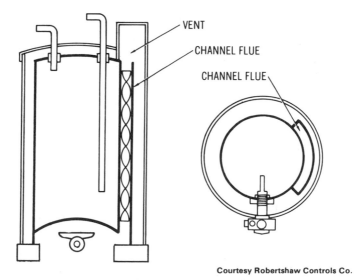

Fig. 4-2. External channel flue tank construction (gas-fired water heater).

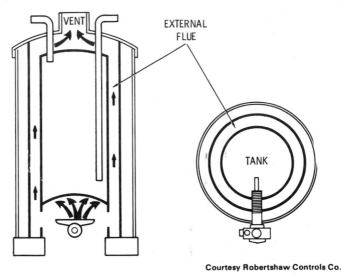

Fig. 4-3. Floating tank external flue construction (gas-fired water heater).

Multiflue Water Heaters

The *multiflue water heater* was developed for commercial uses to satisfy the need for greater heat transfer surface and to efficiently remove the higher volume of noncombustible gases resulting from the higher gas inputs. The diagram of a gas-fired multiflue water heater is shown in Fig. 4-5. Compare this with the diagram of the oil-fired multiflue water heater illustrated in Fig. 4-4. Each is essentially a vertical fire tube boiler enclosed in an insulated jacket.

Multiflue water heaters are characterized by a relatively high hot-water recovery rate. These heaters are used in conjunction with auxiliary storage and are frequently employed as booster heaters. It is common usage to refer to a multiflue water heater as a *booster heater*.

Instantaneous Water Heaters

Automatic instantaneous water heaters are self-contained units available in capacities ranging up to 445 gal. per hour (or up to 7.43 gal. per minute at 100° temperature rise) (Fig. 4-6).

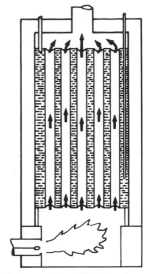

Fig. 4-4. Oil-fired multiflue design used in some high-capacity commercial water heaters.

Courtesy National Oil Fuel Institute

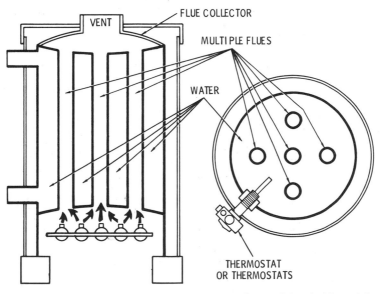

Courtesy Robertshaw Controls Co.

Fig. 4-5. Multiple-flue, multiple-burner commercial water heater.

153

Basically, an instantaneous heater consists of a large copper coil suspended over a series of burners. The copper coil, burners, valves, and thermostats are all enclosed in a protective steel casing. An instantaneous water heater does not have a hot-water storage tank; this feature distinguishes them from other types of commercial water heaters, such as the multicoil and multiflue.

The water to be heated circulates through the copper coil. The operating principle of these heaters is very simple and ideal for situations requiring intermittent use. Any pressure drop in the system (such as that caused by opening a faucet) provides the

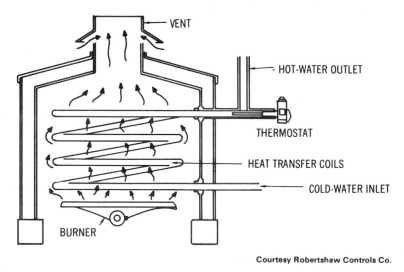

Courtesy Robertshaw Controls Co.

Fig. 4-6. Instantaneous water heater.

necessary power to force open the gas valve and allow gas to flow to the burners. The gas is ignited by the pilot, and the water in the coil is heated to the desired temperature. These heaters are ideally suited for those public buildings or sites that require periods of high demand (e.g., public washrooms, sports arenas, ball parks, stadiums, or recreational centers).

Indirect Water Heaters

An *indirect water heater* uses either steam or hot water to heat the water used in the domestic hot-water supply system.

In small, residential-type boilers, a water heater element consisting of straight copper tubes with U-bends or a coiled tube is located in the hottest portion of the boiler water (Figs. 4-7 and 4-8). It is positioned at the side of the boiler to create rapid natural circulation. Because there is no separate water storage tank, this type of unit is commonly called a *tankless water heater.*

In steam boilers, the copper tubes are generally placed below the water line in the boiler. The system is designed so that the water is heated after a single passage through the copper tubes.

Another type of indirect water heater utilizes a separate water heater outside the boiler (Fig 4-9). Hot water from the boiler flows into the heater and around copper coils containing the domestic hot-water supply before returning to the boiler. The water inside the coils is heated and returned to a hot-water storage tank, which is usually located at a level slightly higher than the heating boiler.

The water to be heated can also be circulated *around* tubes through which steam is circulated. The steam tubes are submerged in a steel tank as shown in Fig. 4-10. As in all methods of indirect heating, the domestic water supply is kept sealed off from the hot water or steam being used to heat it.

Indirect water heaters of the type illustrated in Fig. 4-11 are essentially space heating boilers with built-in coil bundles through which the hot water circulates. They differ from hot-water space heating boilers in the following two ways:

1. The coils are sized to absorb the *total* output of the boiler.
2. Water temperatures do not exceed 210°F.

These indirect water heaters may be used as instantaneous heaters. Because they can be used with any size storage tank, there is no real upper limit on their storage capacity. The same is true of their recovery rate.

The principal advantages of an indirect water heater are:

1. Longer operating life (the metal surfaces of the heater are not directly exposed to a flame).
2. Scale formation and corrosion in the secondary heat exchanger coils are minimized because of the relatively low operating temperatures.

155

3. No hot-water storage tank (with the accompanying circulator, controls, and piping) is required if the heater has been properly sized.

Quick-Recovery Heaters

This class of heater, as the name implies, has the ability to produce hot water at a more rapid rate than the slow-recovery type. It is frequently employed where repeated heavy requirement for hot water makes it essential to have a sufficient amount of water available upon demand.

Thermostats used are of the throttling or snap-action type. Primarily the throttling thermostat is the one in which the amount of gas valve opening is directly proportional to the temperature changes of water in the tank. In the snap-action thermostat the change from a completely open to a completely closed position of the valve, or vice versa, is accomplished by a snap action produced by a clicker diaphragm motivated by the temperature in the tank.

The essential difference between the slow- and quick-recovery heater lies in the amount of gas consumed by each unit. Thus, for example, a quick-recovery heater with a 25,000 Btu input will deliver 25 gal. of hot water per hour indefinitely, whereas a slow-recovery heater can never burn more than a certain relatively low and known amount of gas with an accompanying reduction in hot water delivery.

Slow-Recovery Heaters

The gas-fired slow-recovery water heater is designed to keep a supply of hot water in the storage tank and by means of a constantly burning gas flame deliver hot water continuously to this tank.

Thermostats employed can be of the graduating or snap-action type. With the graduating thermostat, the burner operates between a low and high flame—the low position of the flame serving as a pilot to keep the heater lighted and serving as a source of standby heat.

The slow-recovery automatic water heater is very economical since it can never burn more than a certain amount of gas, depending upon the regulation offered by the thermostat. Also, the small amount of heating surface keeps the standby loss at a minimum, making this type of heater very advantageous where economy is the primary consideration.

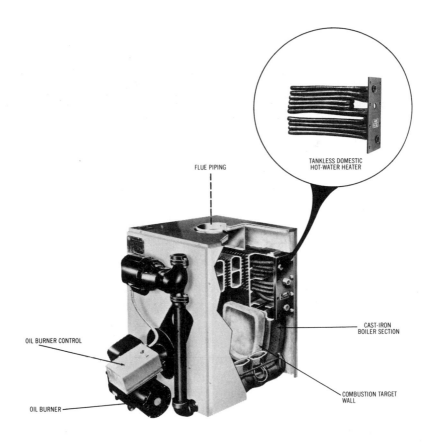

FLUE PIPING

TANKLESS DOMESTIC
HOT-WATER HEATER

CAST-IRON
BOILER SECTION

OIL BURNER CONTROL

COMBUSTION TARGET
WALL

OIL BURNER

Fig. 4-7. Oil-fired hydronic boiler with a straight-tube tankless water heater.

157

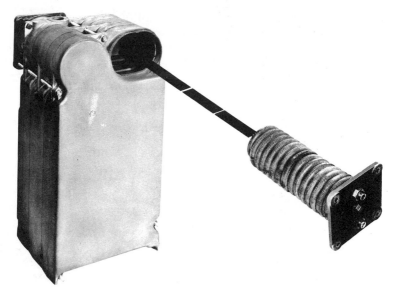

Courtesy H.B. Smith Co., Inc.

Fig. 4-8. Oil-fired hydronic boiler with a coiled-tube tankless water heater.

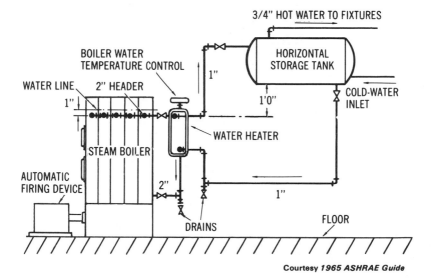

Courtesy *1965 ASHRAE Guide*

Fig. 4-9. Separate indirect water heater mounted outside the boiler.

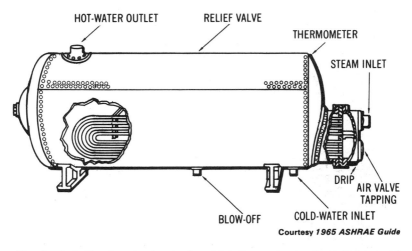

Fig. 4-10. Indirect water heater in which the water is circulated around steam-filled tubes.

WATER HEATER CONSTRUCTION DETAILS

The automatic controls, fuel-burning equipment (gas burners, oil burner, coal stoker, etc.), and venting system found on a water heater will depend upon the type of fuel used and certain other

Fig. 4-11. Indirect water heater with built-in coils designed to absorb the total heat output of the boiler.

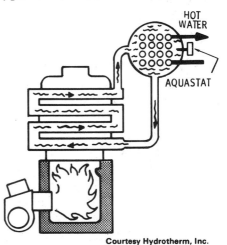

159

variables. These components are described in detail in appropriate sections of this chapter and elsewhere in the book.

The direct-fired automatic storage water heater is the most commonly used heater in residences. Certain construction details of these heaters remain essentially the same regardless of the type of controls or fuel-burning equipment.

Principal among these are:

1. Water storage tanks.
2. Tank fittings.
3. Dip tubes.
4. Safety relief valves.

WATER STORAGE TANKS

The tank of a water heater provides storage space for the hot water and also serves as a heat transfer surface. The design and construction of the tank must be strong enough to withstand at least 300 lb. per sq. in. without leakage or structural deformation.

Corrosion is a major problem in water heater storage tanks. One successful method of reducing corrosion has been to provide steel tanks with glass linings. Older water heater tanks will be found with copper or stone (portland cement) linings, but these are becoming increasingly rare.

Glass-lined tanks are also equipped with a sacrificial magnesium anode rod, as shown in Fig. 4-12. The corrosion process attacks the metal anode rod first, rather than the metal of the tank walls. The anode rod should be removed and replaced before decomposition is reached.

The operating life of a storage tank is directly related to the temperature of the stored water when it is over 140°F. The rule-of-thumb is that each 20° rise above 140°F will reduce tank life by 40 percent.

TANK FITTINGS

Figs. 4-13 and 4-14 illustrate the various fittings required on both domestic and commercial water heaters.

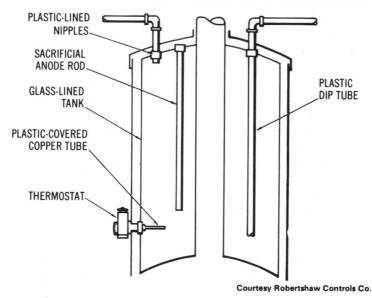

PLASTIC-LINED NIPPLES

SACRIFICIAL ANODE ROD

GLASS-LINED TANK

PLASTIC-COVERED COPPER TUBE

THERMOSTAT

PLASTIC DIP TUBE

Courtesy Robertshaw Controls Co.

Fig. 4-12. Sacrificial anode installation in a residential water heater tank.

Each tank should have fittings for hot-water outlet and cold-water inlet connections. These water connections generally consist of threaded spuds welded into openings in the tank.

A fitting is also provided in the top of most tanks for the insertion of the sacrificial anode. Other fittings provide for the use of immersion thermostats, immersion automatic gas shutoff devices, drain cocks, pressure- and temperature-relief valves, and dip tubes.

DIP TUBES

Some water heaters are designed so that the cold-water inlet is at the top of the tank. Because this is also the location of the hot-water supply outlet, there will be an excessive mixing of the cold water with the hot water unless provisions are made to keep the two separated. A *dip tube* is used for this purpose.

As shown in Fig. 4-15, a dip tube is an extension of the cold water supply pipe and is used to direct the cold water to the bottom of the tank. On older water heaters, dip tubes were made of metal. Now they are generally made from a high-density, temperature-resistant plastic. In all water heaters, the water at the top of the tank during cycling and intermittent standby conditions is always warmer than the water at the bottom of the tank. If the dip tube is too short, the cold water will mix with the water at the top of the tank and reduce the temperature of the hot-water supply to an unacceptable level. On the other hand, a dip tube that is too long will create an excessive water variation between the top of the tank and the lower thermostat control level. The dip tube must be of sufficient length to avoid both these conditions.

Each dip tube is provided with a small hole near the top of the tank, which functions as an antisiphon device. Sometimes it is possible for a malfunction to close off the cold-water supply

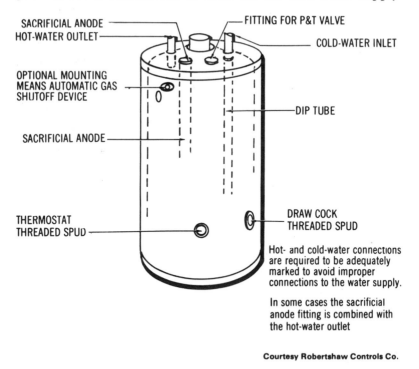

Courtesy Robertshaw Controls Co.

Fig. 4-13. Typical water heater fittings.

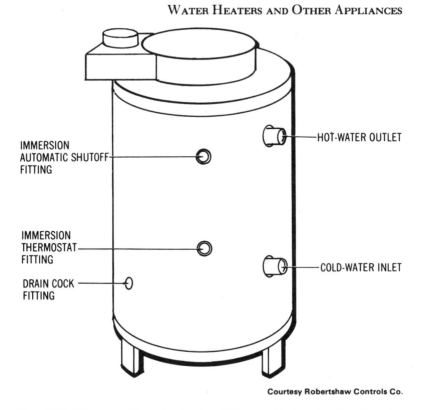

IMMERSION
AUTOMATIC SHUTOFF
FITTING

IMMERSION
THERMOSTAT
FITTING

DRAIN COCK
FITTING

HOT-WATER OUTLET

COLD-WATER INLET

Courtesy Robertshaw Controls Co.

Fig. 4-14. Commercial water heater fittings with low-level cold-water inlets.

while the hot water continues to be removed from the tank. Were it not for the antisiphon hole, the water would be drawn to the bottom of the dip tube, a dangerously low water level for the tank. The siphon action is broken when the water reaches the level of the antisiphon hole, and the hot water will not be drawn below this level.

RELIEF VALVES

A relief valve for a domestic hot-water supply heater is an emergency safety device. If properly installed, it allows water to escape and spill out when excessive pressure, dangerously high

163

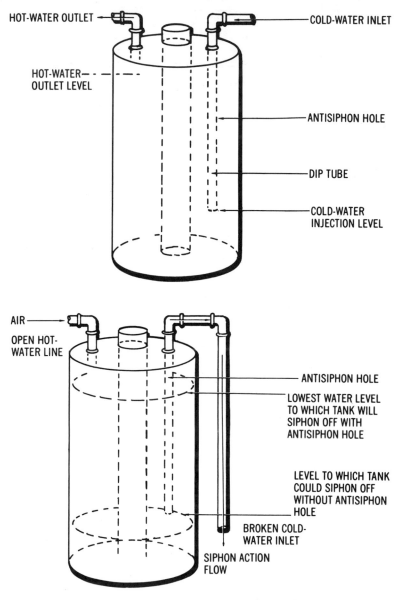

Fig. 4-15. Dip tube and antisiphon hole.

temperature, or both conditions are present in the water storage tank.

It is important to understand the relationship between excessive pressure and high temperature for the safe operation of a hot-water heater. Ignorance of this relationship may result in an explosion severe enough to cause tragic loss of life and extensive property damage. The two principal causes of hot-water storage tank explosions are: (1) water in the tank at an excessively high temperature; and (2) a physical weakness of the tank caused by a defect, age, corrosion, or general deterioration.

Water under pressure (greater than atmospheric pressure) can be heated above 212°F, and still not boil (Fig. 4-16). As shown in Fig. 4-16, the pressure in the tank will rise as it is heated because of thermal expansion. Such pressure cannot be relieved by backing into the main if the cold water is blocked by a check valve or other devices. If there is a failure in the heating-control device,

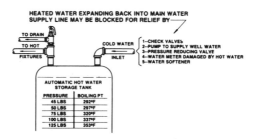

Courtesy A. W. Cash Valve Manufacturing Co.

Fig. 4-16. Boiling points of water at various pressures.

the water in the tank will continue to heat beyond 212°F, and the superheated water will immediately flash into steam when a rupture occurs in the storage tank. Such action instantaneously converts 1 cu. in. of water into 1 cu. ft. of steam with explosive force. This tremendous steam-flash explosive force can shoot water heaters, much like a jet-propelled rocket, through floors and roofs, burst foundations, destroy property, and cause serious or fatal injuries (Fig. 4-17).

Tests have shown that even though water pressure is raised above the normal safe tank limit, the worst that can happen is for

1 lb of nitroglycerine
2,000,000 ft lbs
of explosive energy

1 lb of water (flashed
into steam)
750,000 ft. lbs.
of explosive energy

Courtesy A. W. Cash Valve Manufacturing Co.

Fig. 4-17. Explosive power of superheated water.

the tank to rupture and cause water damage. When heat is applied, however, the tank becomes a potential hazard because of the steam-flash explosive possibilities. Protection against this danger can be obtained by the proper installation of suitable pressure- and temperature-relief valves. No water heater can operate safely without these safety-relief valves.

A pressure-relief valve is designed to prevent excessive pressure from developing in the hot-water heater storage tank (Fig. 4-18). This is exclusively a safety device and is not intended for use as a regulating valve to control or regulate the flow pressure. This type of valve starts to open at the set pressure and requires a certain percentage of overpressure to open fully. As the pressure drops, it starts to close and shuts at approximately the set pressure.

A temperature-relief valve is used to prevent excessively high water temperature from developing in the hot-water heater storage tank. A temperature-relief valve may be a separate unit or combined in the same housing with a pressure-relief valve to form a combination temperature- and pressure-relief valve (Fig. 4-19).

Water heater pressure- and temperature-relief valves must be installed in accordance with AGA, UL, or FHA standard safety requirements. Furthermore, the pressure- and temperature-relief valve should be constructed and located in conformance with current American National Standard (Z21.22) listing requirements. The manufacturer's recommendations for locating the valve on the storage tank should also be followed. *Never* reuse a relief valve if the water heater is being replaced. Always be sure to use the capacity relief valve recommended by the manufacturer for the installation.

A pressure- and temperature-relief valve is basically a pressure-relief valve with an added temperature-sensing element thermostat located at the inlet of the valve to prevent overheating or explosive dangers. The temperature-sensing element must be

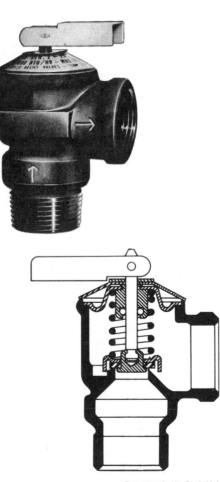

Fig. 4-18. Pressure-relief (only) valve used to protect hot-water supply systems from excessive pressure.

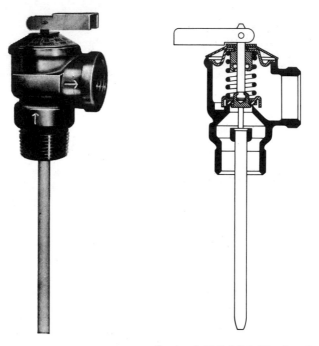

Fig. 4-19. Combination pressure- and temperature-relief valve.

immersed in the water within the top 6 in. of the tank (Figs. 4-20, 4-21, and 4-22). This location is required for the immersion element because the hottest water will occupy this portion of the tank.

The basic components of both a pressure-relief valve and a combination pressure- and temperature-relief valve are shown in Figs. 4-23 and 4-24. Note the position of the temperature-sensing element thermostat. When the water in the top of the tank approaches the danger zone (210°F), the thermostat expands and lifts the valve disc from the seat, allowing hot water to escape and cooler water to replace it in the tank. A decrease of less than 10° in the water temperature causes the thermostat to contract and allows the loading spring to reseat the valve, thus maintaining a supply of hot water at all times.

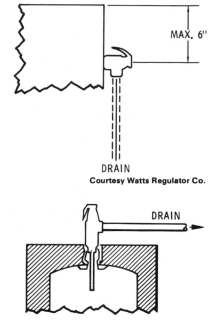

MAX. 6"

DRAIN

Courtesy Watts Regulator Co.

Fig. 4-20. Direct side tapping.

DRAIN

Fig. 4-21. Direct top tapping.

Courtesy Watts Regulator Co.

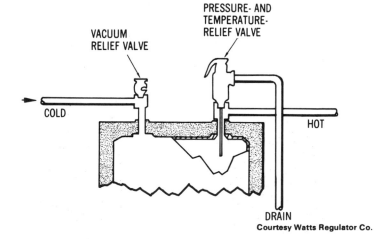

VACUUM
RELIEF VALVE

PRESSURE- AND
TEMPERATURE-
RELIEF VALVE

COLD

HOT

DRAIN
Courtesy Watts Regulator Co.

Fig. 4-22. Relief valve installed in hot-water discharge line.

169

If a combination pressure- and temperature-relief valve is used, it should be installed in a separate tapping in the top of the hot-water storage tank (Fig. 4-21). If a separate tapping is not available, then the valve should be installed in the hot-water discharge line to the fixture outlets at a point as close to the tank as possible (Fig. 4-22).

The relief valve should be installed in the upper end of a tee, the lower end of which is connected to the tapping in the top of the hot-water storage tank by means of a closed nipple. The hot-water supply line to the fixture outlets is then connected to the branch of the tee as shown in Fig. 4-22. If the relief valve is located in the hot-water supply line to the fixtures at a distance greater than 4 in. from the storage tank, excessive temperatures generated within the tank may result in serious damage before the excessive temperature is communicated through the water

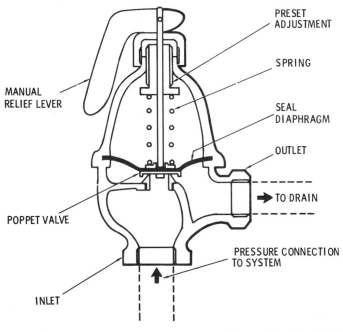

Fig. 4-23. Principal components of a pressure-relief valve.

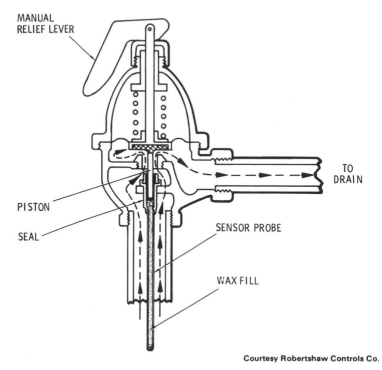

Fig. 4-24. Principal components of a pressure- and temperature-relief valve.

and piping to the relief valve. The possibility of this situation arising can be avoided by selecting a valve with a temperature-sensing element long enough to extend into the top 6 in. of the tank.

A temperature-relief valve must be installed so that the temperature-sensing element is in contact with hot water in the top 6 in. of the tank. This is a necessary safety precaution because of a condition called *temperature lag;* that is, the condition of temperature variation between the hottest water in the top of the tank and water temperature at varying distances away from the actual tank tapping (Fig. 4-25). When the average temperature in the top 6 to 8 in. is 210°F, there will be a considerable temperature lag (under no-flow conditions) at varying distances away from the tank. For example, a temperature of 191°F can be found

171

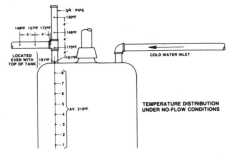

Fig. 4-25. Temperature distribution under no-flow conditions (temperature lag).

at a point even with the top of the tank. At 4 in. above the top of the tank, the temperature has dropped off to 170°F. Taking these conditions into account, then, it becomes clear why relief valves with extension-type temperature-sensing elements are recommended. The temperature-sensing element must reach down to the point at which the highest water temperature occurs.

Where *separate* pressure- and temperature-relief valves are used, the temperature-relief valve should be installed in the top of the hot-water storage tank or as close as possible to the tank in the hot-water supply line to the fixtures as previously described, and the pressure-relief valve should be installed in the cold-water supply line at a point as close to the hot-water storage tank as possible.

The rated capacity of temperature-relief valves (in Btuh) should equal or exceed the input capacity of the hot-water heater (also expressed in Btuh). Both the pressure and temperature steam ratings should be listed on the side of the valve (Fig. 4-26).

Typical piping connections for an automatic storage water heater are shown in Fig. 4-27. A two-temperature capability is provided by feeding the hot water directly to the appliances. A balancing valve is installed in the cold-water line to the tempering valve to compensate for pressure drop through the heater. Because the tempering valve cannot compensate for rapid pressure fluctuations, a pressure equalizing valve should be installed where the system is subjected to water-pressure fluctuations.

172

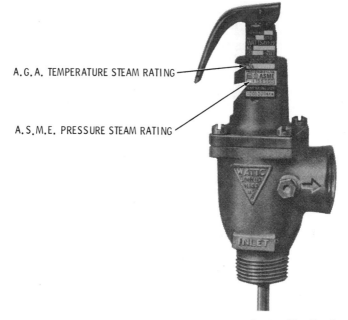

A.G.A. TEMPERATURE STEAM RATING

A.S.M.E. PRESSURE STEAM RATING

Courtesy Watts Regulator Co.

Fig. 4-26. Rating of a typical pressure- and temperature-relief valve.

Fig. 4-28 illustrates the piping connections for large-size instantaneous heat exchanger or converter heater applications. If either leg of the circulator is valved off from the heater, an ASME pressure-relief valve must be installed.

Tankless heater piping connections are shown in Fig. 4-29. As in other installations, a balancing valve has been installed in the cold-water line to the tempering valve to compensate for pressure drop.

VACUUM RELIEF VALVE

A vacuum relief valve is used to protect a hot-water supply system by preventing vacuum conditions that could drain the system by siphonage, burn out the water heater, or cause the storage tank to collapse (Fig. 4-30).

173

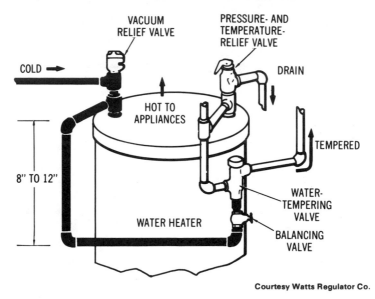

Fig. 4-27. Automatic storage water heater piping connections.

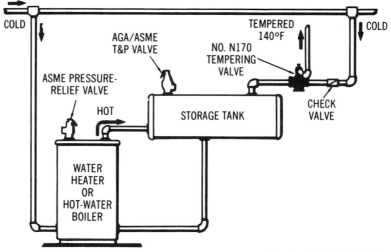

Fig. 4-28. Typical piping connections for large instantaneous heat exchanger or converter-type heater applications.

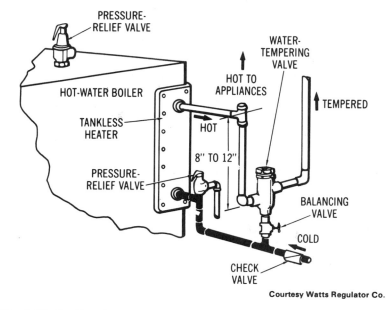

Fig. 4-29. Tankless heater piping connections.

The vacuum relief valve is installed in the cold-water supply line. It closes tightly under system pressure and opens quickly in case of emergency at less than ½-in. vacuum. When the valve opens, atmosphere is admitted and breaks the vacuum, prevent-

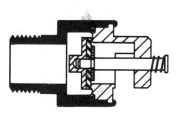

Fig. 4-30. Vacuum relief valve used to protect hot-water supply systems from internal vacuum conditions.

ing siphonage of the system and the possible collapse of the storage tank.

WATER-TEMPERING VALVES

A *water-tempering valve* (Fig. 4-31) is used in a hot-water supply system to provide domestic hot water at temperatures considerably lower than those of the water in the supply mains. These valves are especially recommended for larger hot-water supply systems requiring dependable control of the water temperature at the fixture outlets.

A water-tempering valve is not designed to compensate for system-pressure fluctuations and should never be used where more sophisticated pressure-equalized temperature controls are required to provide antiscald performance. Water-tempering valves are described in greater detail in Chapter 10 of Volume 2 (Steam and Hot-Water Line Controls).

GAS-FIRED WATER HEATERS

Gas-fired automatic storage water heaters are those in which the hot-water storage tank, the gas burner assembly, the combustion chamber and necessary insulation, and the automatic controls are combined in a single self-contained, prefabricated unit or package. Size limitations, resulting from the necessity of such heaters being readily portable, generally restrict the storage tank capacity to approximately 75 gal.

In this type of water heater, the heat of the gas flame is transmitted to the water by direct conduction through the tank bottom and flue surfaces. Some heaters have multiple central flues, while in other designs the hot exhaust gases pass between the outer surfaces of the tank and the insulating jacket. In either case, these areas become radiating surfaces serving to dissipate the heat of the stored hot water to the flue or chimney when the burner is off. This is particularly true if the flue or chimney has a good natural draft.

Fig. 4-31. Typical water-tempering valve.

Storage Capacity

The average ratio of hourly gas input to the storage capacity in gallons of water (for gas-fired automatic storage water heaters of the so-called rapid-recovery type) is such that the recovery (heating) capacity in gallons of water raised 100°F in temperature in one hour, in most instances, approximately equals the storage capacity of the tank in gallons.

Where the water must be raised 120°F in temperature, the recovery (heating) capacity in gallons per hour will be approximately 83 percent of the storage capacity in gallons.

Where the water must be raised 140°F in temperature, the recovery (heating) capacity in gallons per hour will be approximately 71 percent.

GAS BURNER ASSEMBLIES

The burners used in gas-fired water heaters must be provided with inlet gas orificing and some means of air intake. These are necessary to provide the required air-gas mixture for the flame.

177

Beyond these two basic requirements, gas burners will vary widely in both design and construction. These variations in design and construction are generally concerned with providing good flame pattern and ignition. Flame characteristics are affected not only by the design of the ports (raised, drilled, ribbon, slotted, or flush) but also by their number, distribution, depth, and spacing. The gas input rating is an important factor in determining the number, distribution, and size of the ports. Proper spacing is generally determined by observation. Some common types of gas burners used on water heaters are shown in Fig. 4-32.

The purpose of the gas orifice is to provide the proper input for the type of gas (e.g., natural, LP) and the normal range of gas pressures. The gas passes into the mixing tube of the burner where it mixes with the air. Air is generally admitted through adjustable air shutters located around the gas orifice. The design and arrangement of the burner ports control the burning characteristics and distribution of the flame.

The size of the ports and their distribution affects the flame characteristics. If the porting is too large (both individually and in their distribution), the flame may flash back to the burner orifice. On the other hand, blowing flames can result from porting that is too small. As can be readily understood, good flame patterns are in part determined by proper porting. The number and size of ports necessary to give proper flame characteristics must be calculated.

AUTOMATIC GAS CONTROLS

The principal automatic controls used to govern the operation of a *gas-fired* water heater are:

1. Thermostatic valve.
2. Automatic pilot valve.
3. Manual gas valves.
4. Main gas-pressure regulator.
5. Pilot-gas-pressure regulator.
6. Temperature- and pressure-relief valves.

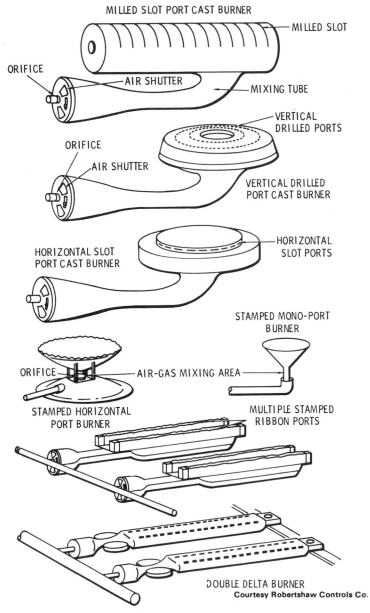

Fig. 4-32. Common types of gas burners used in water heaters.

179

7. Automatic gas shutoff device.
8. Pilot burner.

A *thermostatic valve* (Figs 4-33 and 4-34) is used to control the gas input to the burners in relation to the water temperature in the storage tank. Thermostatic water temperature controls are usually direct, snap-acting bimetallic devices that react to a drop in the temperature of the water in the storage tank. This drop in temperature causes a thermal element immersed in the stored hot

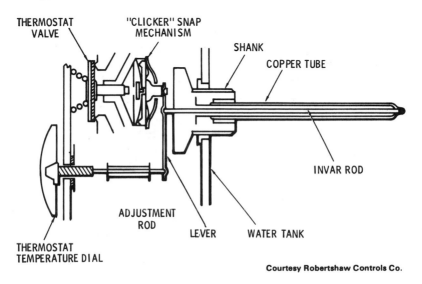

Courtesy Robertshaw Controls Co.

Fig. 4-33. Principal thermostatic valve components.

water to contract and, through mechanical linkage, to open the main gas valve on the unit. When the water in the tank reaches a selected, predetermined setting, the thermal element expands and closes the main gas valve. These thermostatic valves normally operate at a temperature differential of approximately 12°F. In other words, if set to shut off the gas to the main burner when the tank water temperature reaches 140°F, they will react to open the valve when the temperature drops to 128°F.

Another important control on a *gas-fired* water heater is the *automatic pilot valve*, which operates on the thermocouple principle (Fig. 4-35 and 4-36). This control automatically shuts off the

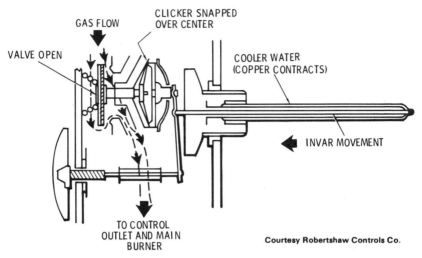

Fig. 4-34. Thermostatic valve operating principles.

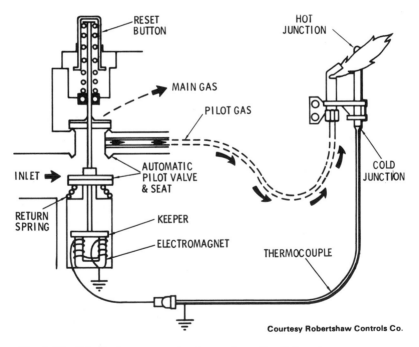

Fig. 4-35. Principal components of an automatic pilot system.

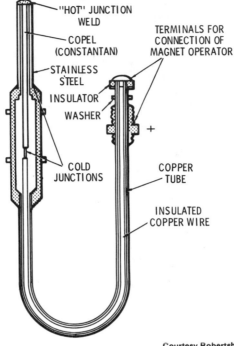

Courtesy Robertshaw Controls Co.

Fig. 4-36. Typical thermocouple construction.

gas when pilot outage or improper ignition conditions occur. When the electromagnet is deenergized, the magnet allows the return spring to close the automatic pilot valve and shut off the gas supply to *both* the main burner and the pilot. The 100 percent automatic pilot shutoff condition is shown in Fig. 4-37. As shown in Fig. 4-38, the reset button must be depressed while the pilot is being relit. If the flame is established, the pilot will continue to burn after the reset button is released (Fig. 4-39).

Manual gas valves (gas cocks) function as a backup safety system to the automatic pilot valve by providing manual control of the main burner and pilot burner gas supply (Fig. 4-40). The manual gas valve is also used with the automatic pilot valve to provide safe pilot lighting by ensuring that *only* pilot gas is flowing during the pilot lighting operation.

Both pilot-gas- and main-gas-pressure regulators are used on

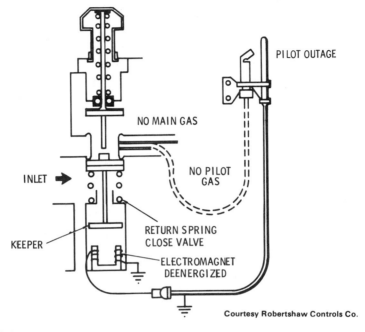

PILOT OUTAGE

NO MAIN GAS

NO PILOT GAS

INLET →

RETURN SPRING
CLOSE VALVE

KEEPER

ELECTROMAGNET
DEENERGIZED

Courtesy Robertshaw Controls Co.

Fig. 4-37. Automatic (100 percent) pilot shutoff condition.

gas-fired water heaters. The *pilot-gas-pressure regulator* is used to regulate the pressure of the gas flowing to the pilot. The *main-gas pressure regulator* performs the same function for gas flowing to the main burners. These gas-pressure controls have been mandatory on gas-fired water heaters since 1972.

Schematics of several types of pressure regulators used on water heaters are shown in Figs. 4-41, 4-42, and 4-43. These devices operate on the balanced pressure principle; that is to say, a main pressure diaphragm and a balancing diaphragm act to "balance out" or cancel the differences in inlet and outlet pressures caused by pressure variations.

Domestic gas water heater combination controls provide for main-gas-pressure regulation by incorporating a pressure regulator in the control. Independent pilot-gas-pressure regulation is optional (see "Combination Gas Controls" in this chapter).

The schematic of a Robertshaw gas-pilot-pressure regulator is shown in Fig. 4-41. This particular pressure regulator has the approximate diameter of a penny and is inserted downstream from the pilot filter.

The following three types of safety controls (individually or in combination) are also found on water heaters:

1. Automatic gas shutoff device.
2. Pressure-relief valve.
3. Temperature-relief valve.

An *automatic gas shutoff device* is designed to shut off *all* gas to the water heater when excessively high water temperature conditions occur. In most automatic gas shutoff systems, this device operates in conjunction with the automatic pilot valve in the pilot safety shutoff circuit. The automatic shutoff device is generally set to shut off the automatic pilot valve when the water temperature in the storage tank approaches 210°F.

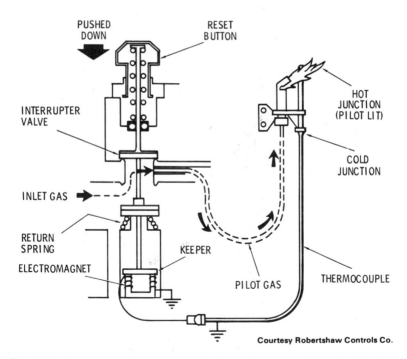

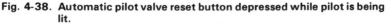

Fig. 4-38. Automatic pilot valve reset button depressed while pilot is being lit.

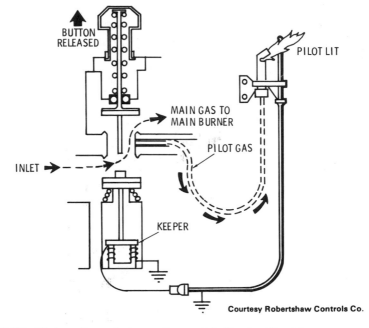

Fig. 4-39. Pilot continues to burn after reset button is released.

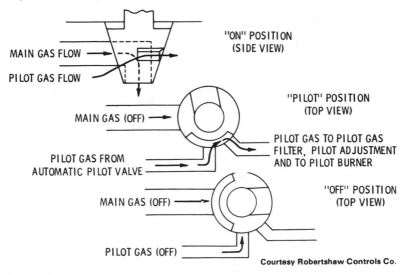

Fig. 4-40. Diagram of a gas-cock parting.

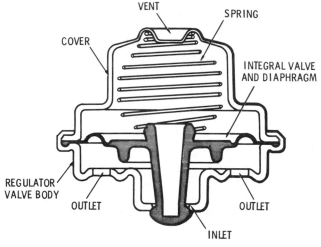

Courtesy Robertshaw Controls Co.

Fig. 4-41. Principal components of a pilot-gas-pressure regulator.

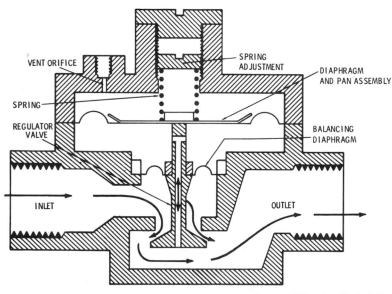

Courtesy Robertshaw Controls Co.

Fig. 4-42. Operating principles of a balanced pressure regulator.

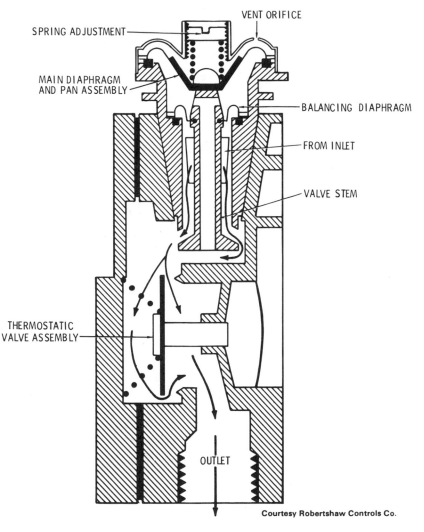

VENT ORIFICE

SPRING ADJUSTMENT

MAIN DIAPHRAGM
AND PAN ASSEMBLY

BALANCING DIAPHRAGM

FROM INLET

VALVE STEM

THERMOSTATIC
VALVE ASSEMBLY

OUTLET

Courtesy Robertshaw Controls Co.

Fig. 4-43. Diagram of a Robertshaw Unitrol R11OR series water heater control incorporating a balanced pressure regulator.

A typical automatic shutoff device, such as the one shown in Fig. 4-44, is mounted on the tank surface so as to sense the water temperature through the tank wall. These devices are normally closed electrical switches connected in series in the automatic

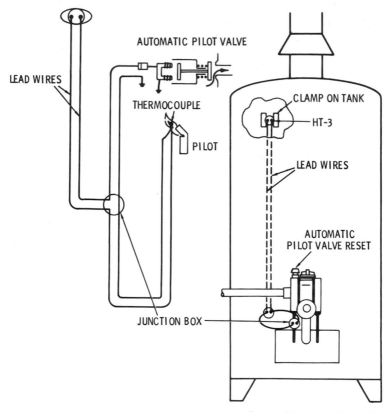

AUTOMATIC PILOT VALVE

LEAD WIRES

THERMOCOUPLE

PILOT

CLAMP ON TANK

HT-3

LEAD WIRES

AUTOMATIC
PILOT VALVE RESET

JUNCTION BOX

Courtesy Robertshaw Controls Co.

Fig. 4-44. Automatic gas shutoff device installation.

pilot millivolt circuit. When the switch reaches its preset temperature limit, it snaps open and deenergizes the electromagnet of the automatic pilot valve. This allows the closure spring of the automatic pilot valve to close the valve. As a result, all gas is shut off to all burners (including the pilot burner).

When the water temperature becomes too high, the volume of water in the tank tends to expand. A *temperature-relief valve* is designed to release a portion of the water and at the same time introduce cold water to reduce the temperature of the remaining water. The temperature-relief function must occur at or below 210°F on residential and commercial water heaters.

188

The purpose of a *pressure-relief valve* is to release a portion of the water from the heater or hot-water heating system when excessive pressure conditions occur. Pressure- and temperature-relief valves are used on *all* types of tank water heaters *regardless* of the fuel used to heat the water. See "Relief Valves" in this chapter.

The pilot burner gas supply is taken off ahead of the gas valve in the thermostatic control. Pilots are usually of the safety type, functioning to shut off the gas supply if the pilot burner flame is extinguished.

Most of the control functions described in the preceding paragraphs can be combined in a single unit or combination gas control (see below).

COMBINATION GAS CONTROL

A *combination gas control* (or *combination gas valve*) combines in a single unit all the automatic and manual control functions necessary to govern the operation of a gas-fired water heater.

A typical combination gas control is shown in Figs. 4-45 and 4-46. This particular unit contains a water heater thermostat (thermostatic valve), automatic pilot valve, automatic gas shutoff device, main-gas-pressure regulator, pilot-gas-pressure regulator, and manual valve (gas cock).

The installation instructions included with most combination gas controls are usually very complete and should cover the following points:

1. Disassembly and assembly instructions.
2. Automatic shutoff valve and magnet replacement.
3. Gas-cock lubrication.
4. Thermostatic valve cleaning instructions.
5. Pressure-regulator adjustment.
6. Thermostat calibration.
7. Lighting procedure.
8. Test procedure.

Most control manufacturers provide test kits to test the opera-

189

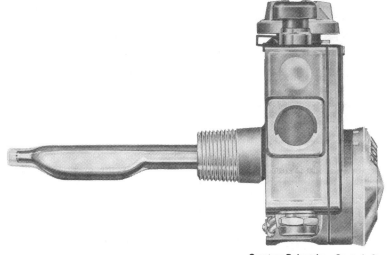

Fig. 4-45. Robertshaw Unitrol R110RTP combustion gas control.

tion of the thermocouple, thermomagnet, and automatic safety shutoff device. The test kit consists of a millivolt meter and an adapter for testing the thermomagnet (Fig. 4-47).

As shown in Fig. 4-48, the manual valve (gas cock) is used when lighting the pilot. The gas-cock dial (1) is turned to the *off* position and at least 5 minutes is allowed to pass. This should be sufficient time for any gas that has accumulated in the burner compartment to escape. The gas-cock dial (1) is then turned to the *start* position, and the set button (2) is depressed while the burner is being lit (Fig. 4-49). The standby flame is allowed to burn for approximately ½ minute before the reset button is released (Fig. 4-50). Unless there is a problem, the burner should stay lit after the reset button has been released. After releasing the reset button, turn the gas-cock dial (1) to the *on* position and turn the temperature dial (3) to the desired setting.

The combination control shown in Fig. 4-51 does not use a reset button in the lighting procedure. The upper dial is turned counterclockwise to the *pilot* position and held against the spring-loaded stop until the pilot burner lights. After the pilot burner burns for about 30 to 60 seconds, the upper dial is turned

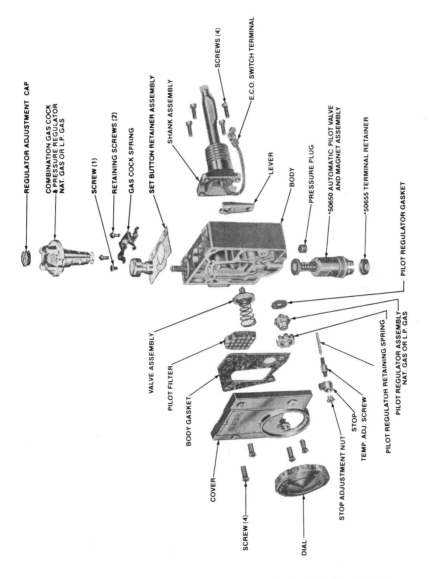

REGULATOR ADJUSTMENT CAP

COMBINATION GAS COCK & PRESSURE REGULATOR NAT. GAS OR L.P. GAS

SCREW (1)

RETAINING SCREWS (2)

GAS COCK SPRING

SET BUTTON RETAINER ASSEMBLY

SHANK ASSEMBLY

SCREWS (4)

E.C.O. SWITCH TERMINAL

LEVER

BODY

PRESSURE PLUG

*50650 AUTOMATIC PILOT VALVE AND MAGNET ASSEMBLY

*50655 TERMINAL RETAINER

PILOT REGULATOR GASKET

VALVE ASSEMBLY

PILOT FILTER

BODY GASKET

COVER

SCREW (4)

DIAL

STOP ADJUSTMENT NUT

TEMP. ADJ. SCREW

STOP

PILOT REGULATOR RETAINING SPRING

PILOT REGULATOR ASSEMBLY NAT. GAS OR L.P. GAS

Courtesy Robertshaw Controls Co.

Fig. 4-46. Exploded view of a Robertshaw Unitrol R11ORTP combination gas control.

191

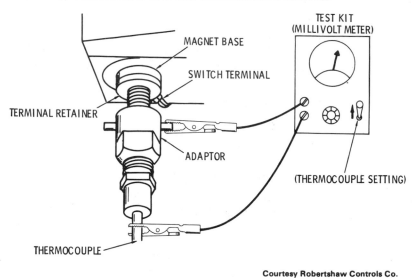

Fig. 4-47. Typical testing procedure with millivolt meter and adapter.

Fig. 4-48. Pilot lighting procedure.

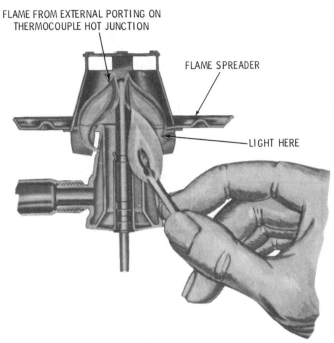

FLAME FROM EXTERNAL PORTING ON
THERMOCOUPLE HOT JUNCTION

FLAME SPREADER

LIGHT HERE

Courtesy Robertshaw Controls Co.

Fig. 4-49. Standby flame pattern.

clockwise to *on* for automatic control. The lower dial is then set
for the desired water temperature.

INSTALLATION AND OPERATION

The installation and operation of a gas-fired automatic storage
water heater involves little possibility of error if the instructions
of the manufacturer are strictly followed and due consideration is
given to the following factors:

1. Location.
2. Venting regulations.
3. Water heater venting system.
4. Size of flue pipe.
5. Runs of flue pipe.

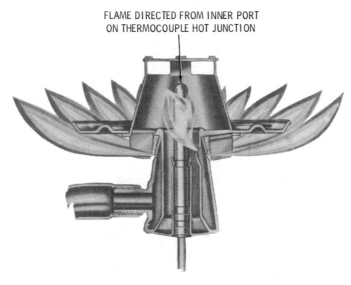

FLAME DIRECTED FROM INNER PORT
ON THERMOCOUPLE HOT JUNCTION

Courtesy Robertshaw Controls Co.

Fig. 4-50. Full input flame pattern.

6. Gas meter.
7. Gas supply line.
8. Hot-water circulation methods.
9. Safety-relief valves.
10. Building and safety code provisions.
11. Lighting and operating instructions.

Location

The heater should be located at a point convenient to flue or chimney and, if possible, at a point approximately equidistant from all hot-water outlets.

Venting Regulations

Many building codes prohibit connecting appliances to a common flue or chimney with coal- or oil-fired equipment. Where regulations do not require venting of gas-fired equipment to a separate flue and it is vented to a common flue with coal- or oil-fired units, the flue pipe of the gas-fired water heater or other appliance should be connected to the chimney at a point *above* the flue pipe from the coal- or oil-fired equipment.

When possible, a separate hole in the chimney should be used for the water heater flue. If this is not possible, join the flue from the water heater and the flue from the heating boiler (or furnace) with a Y connection *(never* a T connection) and install a separate draft regulator for each unit.

When the chimney cannot handle the combined input of both the water heater and the heating plant, wire the two so that they do not operate simultaneously.

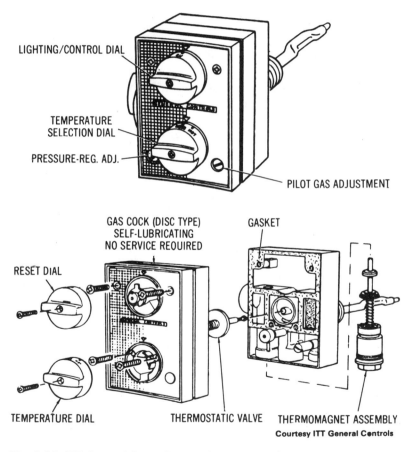

LIGHTING/CONTROL DIAL

TEMPERATURE SELECTION DIAL

PRESSURE-REG. ADJ.

PILOT GAS ADJUSTMENT

GAS COCK (DISC TYPE) SELF-LUBRICATING NO SERVICE REQUIRED

GASKET

RESET DIAL

TEMPERATURE DIAL

THERMOSTATIC VALVE

THERMOMAGNET ASSEMBLY

Courtesy ITT General Controls

Fig. 4-51. ITT General Controls water heater control.

Water Heater Venting Systems

The venting system of a fuel-fired water heater is designed to transfer the products of combustion to the outdoors. By transferring these potentially harmful gases outside the structure, the living spaces are maintained free of any possible air contamination. Electric water heaters do not require venting systems.

A typical venting system for a fuel-fired water heater consists of the following basic components:

1. Heater flues or heat exchangers.
2. Draft diverter.
3. Vent pipe connections.

The design and arrangement of a venting system should take advantage of the natural tendency of hot gases to rise. These flue gases are a waste by-product of the combustion process. Because they are hot gases, they are lighter than the surrounding ambient air, and they tend to rise in a vertical path. The venting system should provide essentially a vertical path to take advantage of this natural vertical flow of the gases. The vent pipes should be of sufficient diameter to carry the volume of gas without restricting its natural flow rate.

Excessive cooling of the flue gases can be avoided by providing a controlled mixture of dilutant air from the draft diverter. If the flue gases cool, condensation will occur, the gases will grow heavier, and it will be impossible to maintain draft. The same condition (i.e., excessive cooling) will occur if the vents are unusually long or high. This can be minimized by insulating the vent pipes.

Blowers are used on some high-input water heaters to compensate natural venting. The use of blowers is sometimes referred to as *power venting*. In most installations in which a blower is used, proper venting can generally be obtained from a smaller-diameter vent pipe. However, some provision should be made to automatically shut off the water heater in case of blower failure.

Vent pipe connections are generally made directly to the outside or through a chimney wall. When direct venting is the case, the vent pipe connects to a vent outlet hood.

Draft hoods are used on all water heaters which rely on natural vent action to eliminate contaminating flue gases. The draft hood should contain an inlet and outlet opening for the flue gases, and an air dilutant intake and relief opening to relieve downdraft conditions.

Examples of common types of draft hoods used on gas-fired water heaters are shown in Fig. 4-52. The vertical draft hood is usually the most efficient for venting.

A well-designed draft hood should be able to prevent excessive updraft in the burner compartment and momentary excessive downdraft conditions. Sometimes the flow rate of flue gases suddenly increases and results in an excessive updraft condition

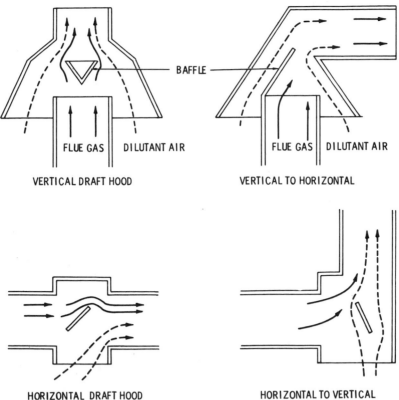

VERTICAL DRAFT HOOD

VERTICAL TO HORIZONTAL

HORIZONTAL DRAFT HOOD

HORIZONTAL TO VERTICAL

Courtesy Robertshaw Controls Co.

Fig. 4-52. Draft hoods commonly used on water heaters.

197

in the burner compartment. The amount of air flowing into the flue through the air dilutant intake is clearly insufficient to control the condition. The diverter should be designed to allow an increase in air intake during excessive updraft conditions so that the weight of the flue gases is increased and the flow rate slowed.

When downdraft conditions occur, the baffles and relief opening in the draft hood allow the downdraft to be relieved *outside* the flue so that the burner flame is unaffected. This expulsion of the combustion byproducts through the air intake is only a *momentary* condition.

Size of Flue Pipe

The size of the flue pipe should not be less than that specified by the manufacturer or that shown in available tables for the rated gas input.

Runs of Flue Pipe

Horizontal runs of the flue pipe should pitch upward toward the chimney connection and should run as directly as possible, avoiding unnecessary bends or elbows. The backdraft diverter supplied by the manufacturer or other approved draft hood of adequate size should be employed.

Gas Meter

The gas meter must be of adequate size or capacity to supply not only the requirements of the water heater but also the requirements of all other gas-fired equipment.

Gas Supply Line

The gas supply line to the water heater should be adequate in size to supply the full rated gas input of the heater at the available pressure, taking into consideration the pressure drop through the supply line. A separate gas supply line from the meter to the water heater should be employed if the existing line from the meter is too small to supply the combined requirements of the water heater and such other equipment or appliances as may be connected to it.

Safety-Relief Valves

Both a pressure-relief valve and a temperature-relief valve are used to ensure the safe operation of a water heater. The operating characteristics of these two safety-relief valves have already been described (see "Relief Valves" in this chapter).

HOT-WATER CIRCULATING METHODS

If a building circulation loop is employed to maintain circulation of the hot-water supply throughout the building for the purpose of making the hot water more readily available at each fixture, the return line from the circulating loop should be connected to the cold-water supply line of the heater. A swing-check valve *must* be installed in a horizontal section of the return line at a point as close to its connection to the cold-water supply line as possible to prevent possibility of backflow of cold water to the hot-water outlets of fixtures that may be connected to the circulation loop at a location which may be closer to the cold-water supply line than to the water heater.

If a check valve was not employed, and the pressure drop in the line between the cold-water supply line and the hot-water outlet was less than the pressure drop in the line between the heater and the hot-water outlet, cold water could flow to the hot-water outlet.

Where a circulating pump is employed to accelerate the circulation in a building hot-water supply loop, it should be installed in the return water line at a point as close to its connection to the cold-water supply line as possible. To eliminate the unnecessary wear and expense incidental to operating a circulating pump continuously, it should be controlled by a direct aquastat installed in the return circulating line at a point conveniently close to the pump.

It is suggested that this aquastat be adjusted to close the pump circuit when the water in the return line drops to (or below) 100° to 110°F and break the pump circuit when the water temperature at that point rises to approximately 120° to 130°F if the desired hot-water temperature at the fixture is approximately 130° to 140°F.

BUILDING AND SAFETY CODE PROVISIONS

The building and safety codes of certain states require the use of dip tubes in the hot-water storage tanks of automatic storage water heaters or recovery water heating systems, while other codes prohibit their use.

The apparent purpose of code provisions prohibiting the use of cold-water supply dip tubes in the tanks of underfired, storage water heaters is to prevent the possibility of developing dangerously excessive temperatures and pressures in the tank of the gas supply to a *manually* controlled heater (if not turned off) or the safety control of an automatic water heater that fails to function.

Under such circumstances, the water in the tank would drain to the levels of the holes drilled in the dip tubes close to the top of the dip tube before the siphon action would be broken.

The water remaining in the tank, being practically at zero pressure, would vaporize rapidly if automatic temperature and safety controls failed to function or the gas supply was not shut off. The steam thus generated in the remaining space in the upper portion in the tank would create a personal scalding hazard if communicated to the hot-water supply piping and outlets, or could attain sufficient pressure to rupture the tank or piping.

An equally hazardous condition would be created if the tank of a storage heater not equipped with a cold-water supply dip tube was completely drained.

Under this condition, if the gas supply to a manually controlled heater was not shut off, or the thermostatic and safety controls of an automatic heater failed to function with the gas valve in the thermostat in the open position, explosive pressures would be almost instantaneously developed if cold water was introduced into the empty heater tank.

To avoid such hazards, the gas supply to either a manually controlled or automatic storage water heater should always be shut off before draining the tank of the heater or the entire hot-water supply system.

Where the use of a cold-water dip tube installed in a tapping in the top and extending to a point close to the bottom of the hot water storage tank is prohibited, the cold-water supply line should be connected to a tapping as close to the bottom as possible.

If the cold water enters the tank at a point considerably above the bottom, the water below that point will be in a more or less static state, which may be conducive to more rapid deposition of lime or other scale on the tank bottom.

Lime or other scale on the tank bottom retards the transfer of heat to the water if the heater is of the underfired type, which impairs its efficiency.

LIGHTING AND OPERATING INSTRUCTIONS

Lighting and operating instructions are supplied by the manufacturer and should be read and thoroughly understood before attempting to adjust the rate of gas input and the thermostat or other automatic controls, particularly if the serviceman is not completely familiar with the design and construction of such controls.

Before leaving the premises, it is imperative that the serviceman hang the operating instructions at a point convenient to the heater where it will be available for ready reference whenever needed by the owner or operating personnel.

The return warranty registration card, if provided, should also be properly filled in by the serviceman and handed to the owner for mailing to the manufacturer. In many instances, warranty provisions are invalidated if the registration card is not returned to the manufacturer.

INSTALLATION AND MAINTENANCE CHECKLIST

The water temperature control in a residential water heater should be set as low as possible and still provide satisfactory hot water at the faucets. This adjustment for low-water temperature prolongs the life of the tank and prevents "stacking." The temperature dial is adjustable and should be used to meet varying conditions and requirements. For control maintenance, it is best to call a qualified serviceman. Instructions in the manufacturer's field information bulletin should be followed when servicing or repairing water heater controls. Fig. 4-53 provides a general checklist for water heater installation and maintenance.

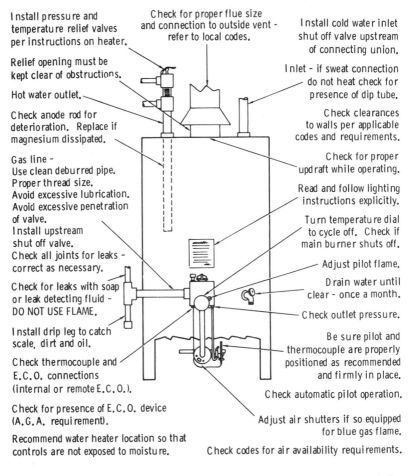

Install pressure and temperature relief valves per instructions on heater.

Relief opening must be kept clear of obstructions.

Hot water outlet.

Check anode rod for deterioration. Replace if magnesium dissipated.

Gas line –
Use clean deburred pipe.
Proper thread size.
Avoid excessive lubrication.
Avoid excessive penetration of valve.
Install upstream shut off valve.
Check all joints for leaks – correct as necessary.

Check for leaks with soap or leak detecting fluid – DO NOT USE FLAME.

Install drip leg to catch scale, dirt and oil.

Check thermocouple and E.C.O. connections (internal or remote E.C.O.).

Check for presence of E.C.O. device (A.G.A. requirement).

Recommend water heater location so that controls are not exposed to moisture.

Check for proper flue size and connection to outside vent – refer to local codes.

Install cold water inlet shut off valve upstream of connecting union.

Inlet – if sweat connection do not heat check for presence of dip tube.

Check clearances to walls per applicable codes and requirements.

Check for proper updraft while operating.

Read and follow lighting instructions explicitly.

Turn temperature dial to cycle off. Check if main burner shuts off.

Adjust pilot flame.

Drain water until clear - once a month.

Check outlet pressure.

Be sure pilot and thermocouple are properly positioned as recommended and firmly in place.

Check automatic pilot operation.

Adjust air shutters if so equipped for blue gas flame.

Check codes for air availability requirements.

Fig. 4-53. Water heater service guide.

TROUBLESHOOTING
GAS-FIRED WATER HEATERS

The following troubleshooting chart, courtesy Robertshaw Controls Company, includes many of the more common symptoms and possible causes of operating problems associated with gas-fired water heaters.

Symptom and Possible Cause *Possible Remedy*

Flame Too Large

1. Pressure regulator set too high.
2. Defective regulator.
3. Burner orifice too large.

1. Reset.
2. Replace.
3. Replace with correct size.

Noisy Flame

1. Too much primary air.
2. Noisy pilot.
3. Burr in orifice.

1. Adjust air shutters.
2. Reduce pilot gas.
3. Remove burr or replace orifice.

Yellow Tipped Flame

1. Too little primary air.
2. Clogged burner ports.
3. Misaligned orifices.
4. Clogged draft hood.

1. Adjust air shutters.
2. Clean ports.
3. Realign.
4. Clean.

Floating Flame

1. Blocked venting.
2. Insufficient primary or secondary air.

1. Clean.
2. Increase air supply; adjust air shutters.

Delayed Ignition

1. Improper pilot location.
2. Pilot flame too small.

1. Reposition pilot.
2. Check orifices; clean; increase pilot gas.

Symptom and Possible Cause: *Possible Remedy:*

3. Burner ports clogged near pilot.
3. Clean ports.

4. Low pressure.
4. Adjust pressure regulator.

Failure to Ignite

1. Gas off.
1. Open manual valve.

2. Thermostat out of calibration.
2. Recalibrate or repair.

3. Defective thermocouple and/or automatic pilot valve.
3. Check and replace.

Burner Will Not Turn Off

1. Thermostat set too high.
1. Lower setting.

2. Thermostat out of calibration.
2. Recalibrate or replace.

3. Dirt on thermostat valve seat.
3. Clean or replace.

4. Defective thermostat.
4. Replace thermostat.

Pilot Will Not Stay Lit

1. Too much primary air.
1. Adjust pilot shutter.

2. Dirt in pilot orifice.
2. Open orifice.

3. Too much draft.
3. Provide shielding or reduce draft.

4. Automatic pilot magnet valve defective.
4. Replace.

5. Loose thermocouple connection.
5. Tighten connection.

6. Defective thermocouple.
6. Replace.

7. Improper pilot-gas adjustment.
7. Adjust pilot gas.

8. Improper pilot orifice size.
8. Replace with correct size.

Symptom and Possible Cause: *Possible Remedy:*

Hot-Water Temperature Too High

1. Thermostat setting too high.
2. Leaking thermostat valve.
3. Pilot too high.
4. Thermostat out of calibration.

1. Set thermostat lower.
2. Clean valve or replace.
3. Adjust pilot lower.
4. Recalibrate or replace thermostat.

Hot-Water Temperature Too Low

1. Thermostat setting too low.
2. Thermostat out of calibration.
3. Undersized heater for hot-water demand.
4. Clogged burner.
5. Undersized burner orifice.
6. Gas pressure too low.

1. Set thermostat higher.
2. Recalibrate or replace thermostat.
3. Recommend larger heater.
4. Clean.
5. Reorifice to correct size.
6. Readjust regulator (if so equipped).

OIL-FIRED WATER HEATERS

Most of the oil-fired heaters used in residences and small buildings are of the external, or floating, tank design (Fig. 4-54). The products of the combustion process pass upward through the flue passages located between the suspended hot-water storage tank and the outer walls of the water heater. Multiflue and internal flue oil-fired water heaters, though less common, are also used (Fig. 4-4 and Fig. 4-55).

205

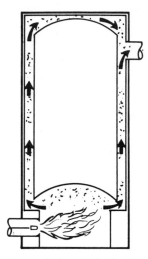

Fig. 4-54. External (floating) tank oil-fired water heater.

Courtesy National Oil Fuel Institute

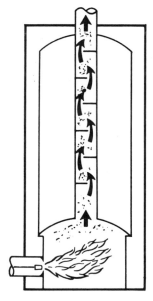

Fig. 4-55. Internal flue oil-fired water heater.

Courtesy National Oil Fuel Institute

As with gas-fired water heaters, an oil heater must have an adequate draft for proper combustion. Sometimes it is necessary to vent two oil-fired appliances (e.g., furnace and water heater) into the same flue. When this situation occurs, the two can be connected to the flue either with a Y connection or in such a way that both feed directly into the flue (Fig. 4-56). When two oil-fired appliances, such as a furnace and a water heater, are connected to a single vent, the controls can be wired to prevent simultaneous operations and give priority to the water heater (Fig. 4-57).

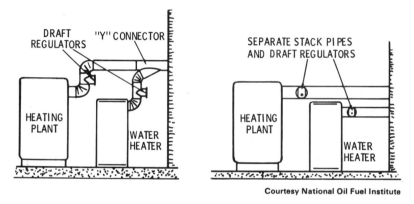

Fig. 4-56. Connecting the boiler and water heater flues to chimney.

Oil-fired water heater controls are similar to those used to control hot-water space heating boilers. They are also designed to regulate the temperature of the water in the water heater storage tank. For example, the primary control shown in Figs. 4-58 and 4-59 is used in conjunction with an immersion aquastat and a remote sensor (cadmium detection cell) to simultaneously regulate the water temperature and provide oil burner control. Both oil burner malfunctions and water temperatures that fall outside the rated temperature range of the heater will cause the primary control to start or stop the oil burner as conditions require.

Pressure- and temperature-relief valves are vital for the protection of the hot-water storage tank. These valves are described elsewhere in this chapter (see "Relief Valves").

When adjusting an oil-fired water heater, always adjust for a

207

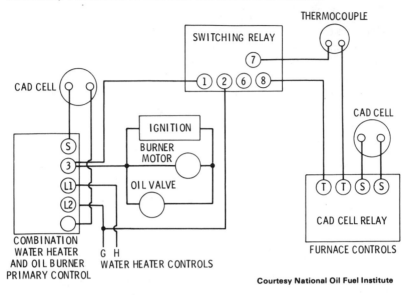

Courtesy National Oil Fuel Institute

Fig. 4-57. Wiring the water heater and heating plant so that they do not operate simultaneously.

smoke-free fire first, then make the necessary adjustment for efficient operation.

ELECTRIC WATER HEATERS

Most electric water heaters used in residences are the automatic storage type. Although some instantaneous heaters are in use, the high electric power input required make them uneconomical to operate when compared with fuel fired types.

An electric water heater generally consists of a vertical tank with a primary heating element or resistor inserted near the bottom of the tank. Some water heaters have a secondary heating element located in the upper one-fourth of the tank (Fig. 4-60). The number of heating elements used in the heater will depend on the size of the storage tank. Large-capacity storage tanks will require two heating elements.

The manual and thermostatic (automatic) controls are located inside the storage tank. A water heater thermostat is designed to automatically open or close the electrical circuit to the heating element(s) whenever the hot-water temperatures exceed or fall below the temperature range of the water heater. Depending on the size of the storage tank, the water heater will be equipped with either one or two thermostats.

The rated voltage for electric water heaters is 240 volts. The thermostats controlling the primary and secondary heating elements are generally set at 150°F.

Several types of heating units are available for electric water heaters, but the two most popular are probably the immersion element and the strap-on unit. The immersion element is inserted through an opening in the side of the tank. The strap-on unit is externally mounted on the surface of the tank.

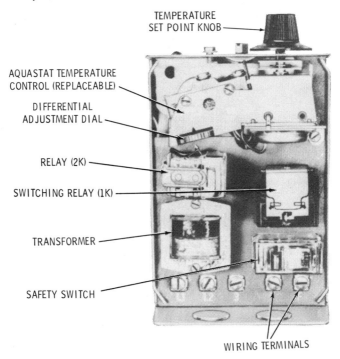

TEMPERATURE SET POINT KNOB

AQUASTAT TEMPERATURE CONTROL (REPLACEABLE)

DIFFERENTIAL ADJUSTMENT DIAL

RELAY (2K)

SWITCHING RELAY (1K)

TRANSFORMER

SAFETY SWITCH

WIRING TERMINALS

Courtesy Honeywell Tradeline Controls

Fig. 4-58. Honeywell R4166 combination water heater and oil burner primary control.

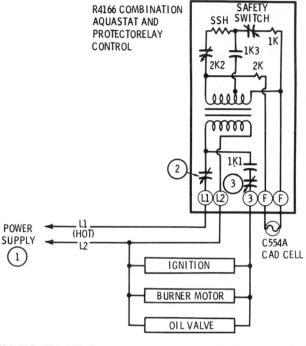

R4166 COMBINATION
AQUASTAT AND
PROTECTORELAY
CONTROL

SSH   SAFETY SWITCH

1K

1K3

2K2   2K

1K1

2

3

L1 L2   3 F F

POWER
SUPPLY

L1
(HOT)
L2

1

IGNITION

BURNER MOTOR

OIL VALVE

C554A
CAD CELL

① PROVIDE DISCONNECT MEANS AND OVERLOAD PROTECTION AS REQUIRED。

② AQUASTAT CONTROL BREAKS ON TEMPERATURE RISE TO SET POINT.

③ R4166B ONLY - HIGH LIMIT CONTROL BREAKS ON TEMPERATURE RISE TO
SET POINT.

Courtesy Honeywell Tradeline Controls

**Fig. 4-59. Internal diagram and typical wiring hookup of a Honeywell
R4166 combination water heater and oil burner primary control.**

A larger storage tank capacity is required for electric water
heaters than for fuel-fired types in order to compensate for the
limited recovery rate. As a result, initial equipment costs are
higher. Care should be taken not to oversize or undersize the
storage tank. Incorrect sizing will result in an inefficient water
heater.

Electric water heaters of the automatic storage type are generally available in storage tank capacities ranging from 30 to 140
gal. The electric power input requirement will range from 1600
to 7000 watts.

MANUAL WATER HEATERS

Manual water heaters, also referred to as *circulating tank* or *side arm heaters*, are of the conventional design with the gas burner and accompanying heating coils mounted on the side of the hot-water storage tank as shown in Fig. 4-61.

Manual water heaters are generally equipped with copper coils 16½ to 20 ft. in length and either ¾- or 1-in. outside diameter. The coils are usually made of copper tubing of No. 20 Stubbs gauge. Other designs are occasionally employed as, for example, the internal or underfired units intended to overcome liming in hard water areas.

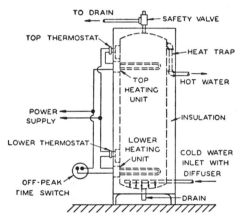

Courtesy *1965 ASHRAE Guide*

Fig. 4-60. Components of an electric-fired water heater.

These heaters usually have between 20,000 and 30,000 Btu capacity, although the maximum sizes run up to 85,000 Btu capacity. The smallest-size manual heater will deliver about 19 gal. of hot water per hour. This size heater is generally ample for most homes in which the conventional 30 gal. tank is used for storing the hot water.

The manual water heater was formerly widely employed because of a comparatively low first cost and economy in operation. As the name implies, manual water heaters are nonauto-

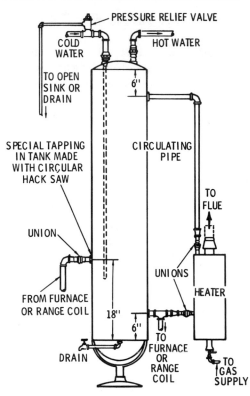

PRESSURE RELIEF VALVE

COLD WATER

HOT WATER

TO OPEN SINK OR DRAIN

6"

SPECIAL TAPPING IN TANK MADE WITH CIRCULAR HACK SAW

CIRCULATING PIPE

TO FLUE

UNION

UNIONS

FROM FURNACE OR RANGE COIL

HEATER

18"

6"

DRAIN

TO FURNACE OR RANGE COIL

TO GAS SUPPLY

Fig. 4-61. Connection of a manual gas water heater where interconnected with a waterback or range coil.

matic and supply hot water quickly by turning on and lighting the gas shortly before the warm water is required. Automatic water heaters have largely replaced the manual type because the former require little or no attention when in operation.

Assembly and Installation of Manual Water Heaters

A great many installations of manual water heaters have given poor service, not because of any fault in the heater, but mainly due to the improper method of connecting the heater to the storage tank. This lack of good service can be largely overcome if a few simple installation rules are followed.

All range boilers of recent manufacture have tappings provided in them for hot- and cold-water connections, two tappings in the side of the tank 6 in. from the top and bottom as shown in Fig. 4-61 to accommodate the circulating water connections, which should be made of ¾-in. or larger pipe. The bottom of the tank has a tapping for connection to the drain or blow-off which is used to drain water or sediment from the boiler. Circulating pipes between the heater and tank should be made as free from fittings and bends as possible.

The use of brass pipe on the hot-water circulating line from the

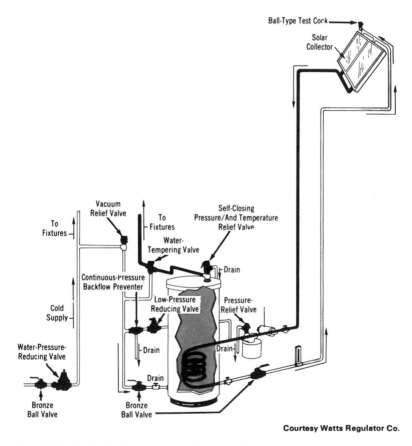

Fig. 4-62. Typical solar water heating installation.

heater to the tank is recommended, particularly for high temperature circulation.

When required, unions should be placed as close to the heater as possible on the hot-water and cold-water circulating lines.

The arrangement of hot- and cold-water tappings of 6 in. from the top and bottom of the tank allows free circulation, which results in relatively large volume storage without overheating. It also eliminates short circuiting of the water through the heater and provides ample sediment storage below the circulating line, thus preventing sediment from getting into the heating element.

Courtesy Watts Regulator Co.

Fig. 4-63. Thermostatic element for 210°F to 250°F service.

SOLAR WATER HEATERS

Many of the valves and heating controls used with gas, oil, and electric water heaters can also be used in solar water heating systems. These components are designed to protect and control a solar water heating installation from a single source. A typical system is shown in Fig. 4-62. A specially designed thermostatic element (for 210°F to 250°F service) in a thermal bypass control should be included in the system to divert high-temperature water to a cooling section (Fig. 4-63). This prevents structural damage to the solar panels and improves system efficiency.

CHAPTER 5

Heating
Swimming Pools

Heating systems have been added to both private and commercial swimming pools for a variety of reasons ranging from the simple desire to increase body comfort with warmer water temperatures to the more practical reason of extending the swimming season.

In many commercial buildings and larger residences that have hot-water heating systems, the central boiler provides the heat for space heating, domestic hot water, and pool water. In structures equipped with other heating systems (e.g., forced warm-air), a pool heater operating independently from the central heating unit is necessary to heat the pool water.

The water in a swimming pool may be heated in one of the three following ways:

1. Solar heating.
2. Radiant heating.
3. Recycling.

Solar heating as a method for heating pool water is strictly a do-it-yourself proposition. This is not the type of pool heating system that a contractor will install.

In a solar pool heating system, water pipes leading to the pool are left bare so that they can absorb solar radiation. Sometimes a reflector (e.g., aluminum coil) is placed under the pipe to intensify the radiated heat. Unfortunately, there is no storage capacity for the heat, and the system is inoperable when there is a cloud cover. Its major advantage is that no fuel cost is involved in its operation.

Pools can also be heated with a *radiant heating system,* but such a system *must* be installed when the pool is constructed, because it entails embedding copper tubing (commonly ½-in. diameter) in the pool itself. As shown in Fig. 5-1, the copper tubing is buried close to the surface in the walls and floor of the pool. The hot water is fed through a supply pipe from the boiler. The water returns to the boiler through a return line. Water temperature in the pipes is controlled by a thermostat, which is usually located on the return line.

Radiant heating systems for swimming pools are probably the most expensive types to install because they require a great deal more pipe than other heating systems. Furthermore, the construction feature of embedding the copper tubing in the walls and floor of the pool adds to the installation cost.

Most pool heating systems today are based on the *recycling* principle. The water is drawn from the pool by a pump, passed to the heat exchanger in the boiler or pool heater, heated to the desired temperature, and returned to the swimming pool.

When both a pool and its heating system are installed during the original construction of the house or building, it is possible (and advisable) to install a single boiler with more than one heat exchanger. As a result, the boiler will have the capacity to provide hot water for not only the pool, but also for space heating and domestic hot-water needs. Three such systems for commercial buildings are illustrated in Figs. 5-2, 5-3, and 5-4. In these systems, the primary boiler water is maintained at the desired space heating water temperature by the operating aquastat in the boiler. The temperature of the pool water is maintained by the opening and closing of a pool temperature-control valve located

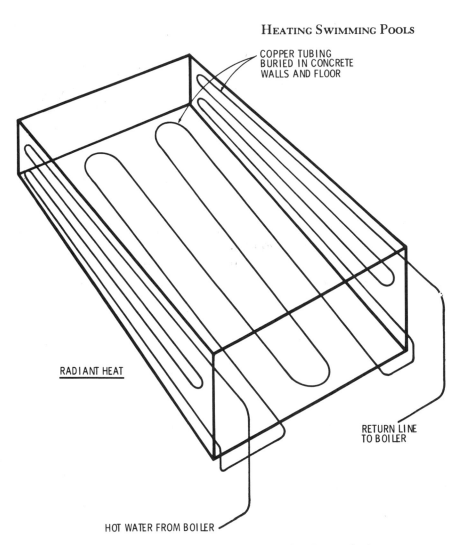

COPPER TUBING
BURIED IN CONCRETE
WALLS AND FLOOR

RADIANT HEAT

RETURN LINE
TO BOILER

HOT WATER FROM BOILER

Fig. 5-1. Heating a swimming pool by the radiant heating method.

in the circulation line between the filter and heater. This valve is controlled by a pool temperature-control aquastat, which senses the temperature of the water being returned from the pool.

Pool heaters can also be installed independently of space heating or domestic hot-water heating systems. This is usually the case when a heating unit is installed at an existing pool.

217

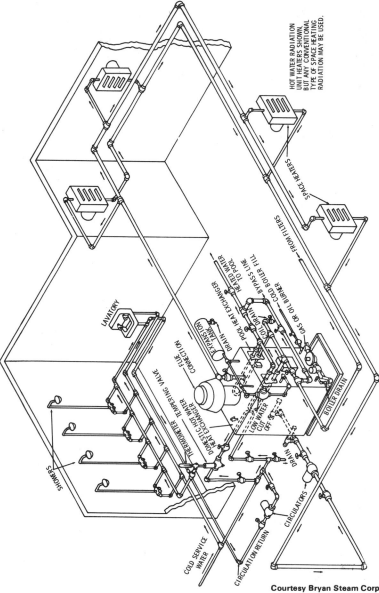

HOT WATER RADIATION UNIT HEATERS SHOWN, BUT ANY CONVENTIONAL TYPE OF SPACE HEATING RADIATION MAY BE USED.

SPACE HEATERS

FROM FILTERS

POOL HEAT EXCHANGER

HEATED WATER TO POOL

BY-PASS LINE

COLD BOILER FILL

GAS OR OIL BURNER

DRAIN

POOL DRAIN

EXPANSION TANK

LAVATORY

FLUE CONNECTION

THERMOMETER

TEMPERING VALVE

DOMESTIC HOT WATER HEAT EXCHANGER

LOW WATER CUT OFF

BOILER DRAIN

DRAIN

SHOWERS

CIRCULATORS

COLD SERVICE WATER

CIRCULATION RETURN

Courtesy Bryan Steam Corp.

Fig. 5-2. Typical multipurpose water heating system for pool heating, space heating, and hot-water service.

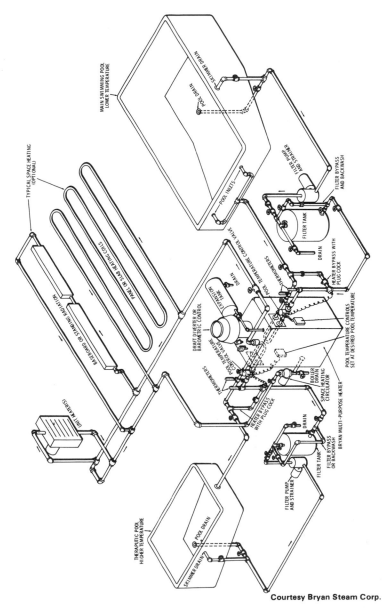

Courtesy Bryan Steam Corp.

Fig. 5-3. Typical multipurpose water heating system for two pools of different temperatures and space heating.

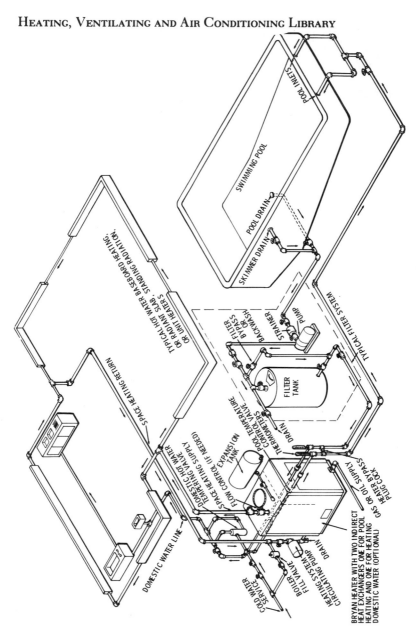

POOL INLETS

SWIMMING POOL

POOL DRAIN

SKIMMER DRAIN

TYPICAL HOT WATER BASEBOARD RADIATION,
OR RADIANT SLAB,
OR UNIT HEATERS

SPACE HEATING RETURN

FILTER
BYPASS
OR
BACKWASH

STRAINER

PUMP

TYPICAL FILTER SYSTEM

FILTER
TANK

DRAIN

POOL TEMPERATURE
CONTROLS

THERMOMETERS

DOMESTIC HOT WATER
TEMPERING VALVE

SPACE HEATING SUPPLY

FLOW CONTROL (IF NEEDED)

POOL EXPANSION
TANK

GAS OR OIL SUPPLY

HEATER BYPASS
PLUG COCK

DOMESTIC WATER LINE

COLD WATER
SERVICE

BOILER VALVE
FILL SYSTEM

HEATING SYSTEM DRAIN
CIRCULATING PUMP

BRYAN HEATER WITH TWO INDIRECT
HEAT EXCHANGERS ONE FOR POOL
HEATING AND ONE FOR HEATING
DOMESTIC WATER (OPTIONAL)

Fig. 5-4. Typical multipurpose water heating system for pool heating, space heating, and domestic hot water.

A pool heater operates on the same principle as the domestic hot-water heater, but the two differ considerably in their functions. The hot-water heater is required to heat only 30–40 gal. of water in an *enclosed* container. The pool heater must heat thousands of gallons of water with a surface *exposed* to the outside air. Because there is a high degree of heat loss from the large surface area of the water to the colder air above it, a pool heater uses a considerable amount of fuel to replace the lost heat and to maintain the pool water at the desired temperature. The fact that water has a high heat capacity contributes to the problem. As a result, it generally takes a pool heater 20 to 24 hours to warm the water to the desired temperature.

CLASSIFYING POOL HEATERS

Pool heaters can be classified in a number of different ways. If they are classified according to their basic operating principle, the following two types are recognized:

1. Direct-type pool heaters.
2. Indirect-type pool heaters.

In a *direct-type pool heater* (Fig. 5-5), the water from the pool passes through the heating unit in pipes which are heated directly by a gas or oil burner. In an *indirect-type pool heater* (Fig. 5-6), the pipes containing the pool water pass through a compartment in the heating unit, which also contains water. The water in the compartment of the heating unit is heated by a gas or oil burner

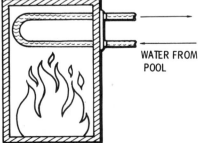

Fig. 5-5. Operating principle of a direct-type pool heater.

WATER FROM POOL

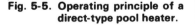

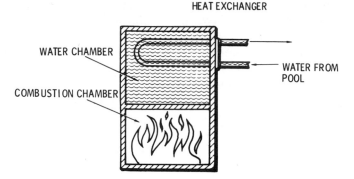

Fig. 5-6. Operating principle of an indirect-type pool heater.

located outside the water compartment. The heat of the water is then transferred to the water in the pipes from the pool; hence the name: *indirect* pool heater. Because the coils of the heat exchanger never come in direct contact with the combustion heat (as is the case with direct-type pool heaters), scale and corrosion are virtually eliminated.

Another means of classifying pool heaters is on the basis of the type of fuel they use. In addition to gas or oil, electric-fired pool heaters are also available. These, too, operate on the indirect heating principle.

Oil-Fired Indirect Pool Heaters

The oil-fired indirect pool heater illustrated in Fig. 5-7 is available in several input ratings ranging from 175,000 Btu/hr (1.25 gph firing rate) to 280,000 Btu/hr (2.00 gph firing rate) with No. 2 fuel oil. The corresponding *output* ratings range from 133,000 Btu/hr to 208,000 Btu/hr, which gives this pool heater the capacity to heat swimming pools containing 16,000 to 25,000 gal. of water.

These pool heaters have a 100 psi ASME pressure rating, which makes possible their direct connection to city water lines without the use of pressure-reducing valves. The heat exchanger consists of multiple-pass finned copper tubes, which are designed for low-pressure drop and which can be removed for cleaning.

Fig. 5-8 illustrates the recommended piping arrangement for

this particular pool heater. Note that the filter is installed on the line from the pool between the pump (circulator) and the shutoff valve. No provision for bypass piping is made between the pipes leading to and from the pool. An example of bypass piping in a pool heating installation for a similar oil-fired indirect heater is shown in Fig. 5-9.

The pool heater described above is suitable for small residential pools. A typical 25,000-gal. pool is approximately 18 ft. wide by 36 ft. long if a rectangular-shaped pool is used as an example. Pool heaters, whether oil-, gas-, or electric-fired, are available for almost any size pool. It is simply a matter of matching the pool

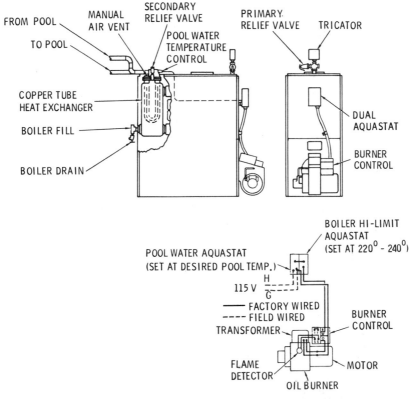

Courtesy Hydrotherm, Inc.

Fig. 5-7. Oil-fired indirect-type pool heater with wiring diagram.

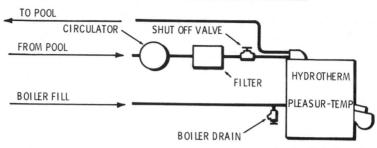

Courtesy Hydrotherm, Inc.

Fig. 5-8. Recommended piping arrangement for heater in Fig. 5-7.

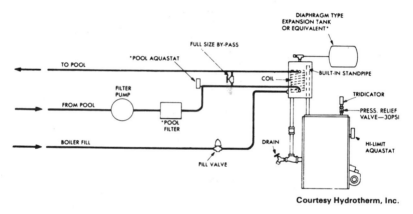

Courtesy Hydrotherm, Inc.

Fig. 5-9. Piping diagram with bypass arrangement.

size with the capacity of the pool heater. For example, Fig. 5-7 shows an indirect oil-fired pool heater used for commercial applications, which is capable of heating a 105,000-gal. swimming pool. It has an input rating of 1,155,000 Btu per hour with 8.25 gph firing rate. This pool heater actually represents the combination of three heating modules; each has its own oil burner and modulating aquastat.

Gas-Fired Indirect Pool Heaters

Gas-fired indirect pool heaters (Figs. 5-10 and 5-11) are available in the same Btu input/output ratings as the oil-fired types. They differ only in certain items of standard equipment that relate directly to the type of fuel used. For example, a gas-fired indirect pool heater will have an automatic safety pilot light, a gas-pressure regulator, an electric gas valve, a gas shutoff valve, a gas burner, and gas controls. The rest of the equipment (e.g., indirect heat exchanger, immersion aquastat, and boiler fill valve) will be identical to that found on the oil-fired pool heaters. Some manufacturers incorporate special design features on their models, such as the vertical draft diverter attached to the top of the boiler jacket on Bryan gas-fired indirect pool heaters (Fig. 5-11).

A typical wiring diagram for a gas-fired indirect pool heater is shown in Fig. 5-12. The control sequence is similar to the one that occurs in oil-fired pool heaters. The pool water control on the dual aquastat is set at the desired pool temperatures. The cold water from the pool starts the boiler, which continues to produce heat until the pool water temperature reaches the setting on the pool control. When this point is reached, the boiler shuts off. The boiler water temperature is controlled by the water-limit control.

Pool heaters are shipped with a complete set of instructions for installing and starting the boiler. If the instructions are missing, the manufacturer should be contacted for a replacement set. The following outline represents a summarized version of the instruc-

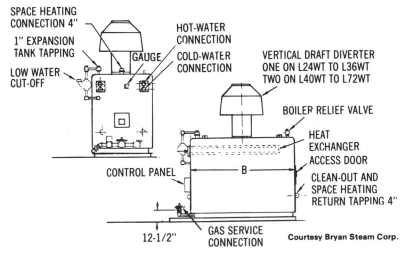

SPACE HEATING
CONNECTION 4"

1" EXPANSION
TANK TAPPING

LOW WATER
CUT-OFF

GAUGE

HOT-WATER
CONNECTION

COLD-WATER
CONNECTION

VERTICAL DRAFT DIVERTER
ONE ON L24WT TO L36WT
TWO ON L40WT TO L72WT

BOILER RELIEF VALVE

HEAT
EXCHANGER
ACCESS DOOR

CLEAN-OUT AND
SPACE HEATING
RETURN TAPPING 4"

CONTROL PANEL

B

12-1/2"

GAS SERVICE
CONNECTION

Courtesy Bryan Steam Corp.

Fig. 5-10. Gas-fired indirect pool heater.

Fig. 5-11. Bryan gas-fired indirect pool heater with vertical draft diverter.

tions for the Hydrotherm gas-fired indirect pool heater illustrated in Fig. 5-13. After unpacking the cartons and ascertaining to your satisfaction that all items are accounted for, you should proceed as follows:

1. Place the boiler in the desired position.
2. Screw tridicator assembly into the ¾-in. tapping on the top of the absorption unit.
3. Connect discharge of relief valve to outside of building or to a sump (do *not* plug the end of the discharge pipe).
4. Connect cold water supply to tee at boiler drain.
5. Open manual air vent on tee at secondary relief valve, and

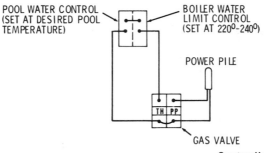

Courtesy Hydrotherm, Inc.

Fig. 5-12. Typical wiring diagram for a gas-fired indirect pool heater.

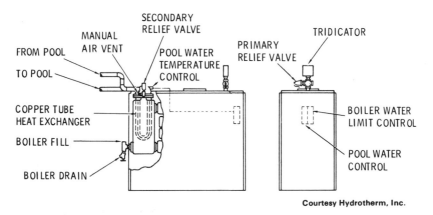

Courtesy Hydrotherm, Inc.

Fig. 5-13. Hydrotherm gas-fired indirect pool heater.

227

then open cold-water supply valve (the boiler should now be filling).

6. Close air vent when water flows freely.
7. Fill swimming pool in conventional manner.
8. Assemble boiler jacket around boiler (following instructions enclosed in jacket carton).
9. Connect gas line to boiler.
10. Light boiler according to lighting instruction plate on control panel.

Electric-Fired Indirect Pool Heaters

The input of an electric-fired indirect pool heater is measured in kilowatts. The pool heater shown in Fig. 5-14 is available in eleven different models ranging in input capacity from 15 kW to 300 kW, and a corresponding output of from 50,000 Btu per hour to 1,000,000 Btu per hour. These pool heaters are also available with 240 volts (1 phase) or 240/480 volts (3 phase). A wiring diagram for an electric-fired indirect pool heater is shown in Fig. 5-15.

Heat-Exchanger Pool Heaters

The heat-exchanger pool heater illustrated in Fig. 5-16 is designed for use in either steam or hot-water heating systems. It is essentially a steel shell enclosing the copper tubes of the heat exchanger. The hot water or steam is taken from the heating main and circulated *around* the copper tubes inside the steel shell. The pool water enters the heat exchanger in a pipe leading from the pool filter and circulates *inside* the copper tubes. As it circulates, it is heated by the hot water or steam circulating around the copper tubes. The heated water eventually leaves the heat exchanger through a second exit point and is returned to the pool.

Figs. 5-17 and 5-18 illustrate the modifications necessary for use with either steam or hot water as the heat source. Basic construction is essentially the same in both cases, with the pool water circulating through copper tubes (the heat exchanger) enclosed in a steel shell. You will probably notice that different types of steel are used in constructing the shell; 125 psi ASME steel for heat exchangers using hot water as a heat source, and 15 psi ASME steel for those using steam.

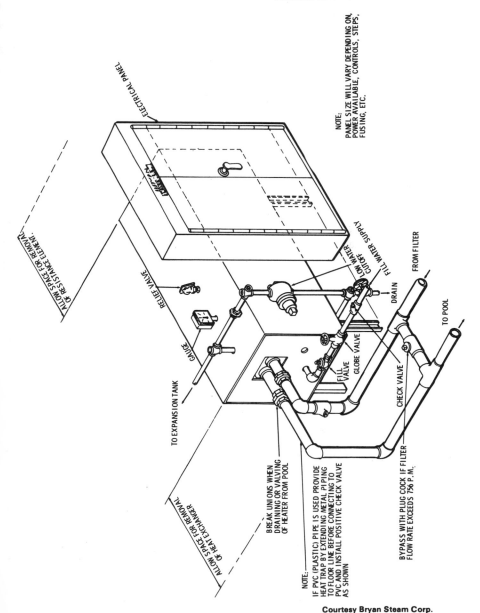

ELECTRICAL PANEL

NOTE:
PANEL SIZE WILL VARY DEPENDING ON,
POWER AVAILABLE, CONTROLS, STEPS,
FUSING, ETC.

ALLOW SPACE FOR REMOVAL
OF RESISTANCE ELEMENT.

RELIEF VALVE

GAUGE

TO EXPANSION TANK

LOW WATER CUTOFF

FILL WATER SUPPLY

FROM FILTER

DRAIN

TO POOL

GLOBE VALVE

FILL VALVE

CHECK VALVE

ALLOW SPACE FOR REMOVAL
OF HEAT EXCHANGER

BREAK UNIONS WHEN
DRAINING OR VALVING
OF HEATER FROM POOL

NOTE:
IF PVC (PLASTIC) PIPE IS USED PROVIDE
HEAT TRAP BY EXTENDING METAL PIPING
TO FLOOR LINE BEFORE CONNECTING TO
PVC AND INSTALL POSITIVE CHECK VALVE
AS SHOWN

BYPASS WITH PLUG COCK IF FILTER
FLOW RATE EXCEEDS 756 P.M.

Courtesy Bryan Steam Corp.

Fig. 5-14. Electric-fired pool heater.

229

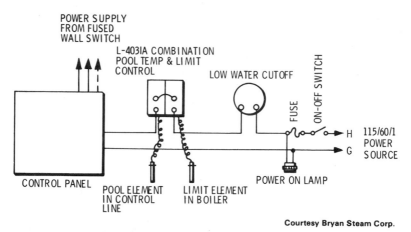

POWER SUPPLY
FROM FUSED
WALL SWITCH

L-403IA COMBINATION
POOL TEMP & LIMIT
CONTROL

LOW WATER CUTOFF

ON-OFF SWITCH

FUSE

CONTROL PANEL

POOL ELEMENT
IN CONTROL
LINE

LIMIT ELEMENT
IN BOILER

POWER ON LAMP

H

G

115/60/1
POWER
SOURCE

Courtesy Bryan Steam Corp.

Fig. 5-15. Wiring diagram for an electric-fired pool heater.

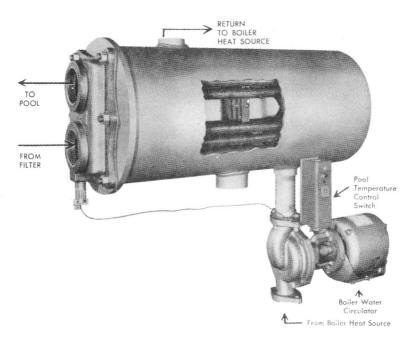

RETURN
TO BOILER
HEAT SOURCE

TO
POOL

FROM
FILTER

Pool
Temperature
Control
Switch

Boiler Water
Circulator

From Boiler Heat Source

Courtesy Bryan Steam Corp.

Fig. 5-16. Heat-exchanger pool heater.

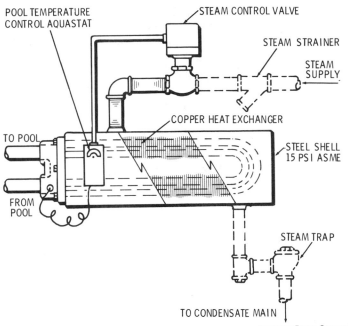

Fig. 5-17. Heat-exchanger pool heater using steam as a heat source.

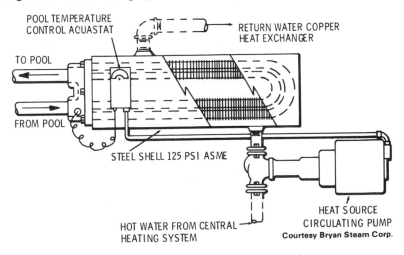

Fig. 5-18. Heat-exchanger pool heater using water as a heat source.

Both types of heat exchangers have temperature-control aquastats activated by the temperature of the pool water entering the heat exchanger. In heat exchangers that use water as a heat source, the aquastat is connected to a circulating pump (Fig. 5-18). If steam is the heat source, the aquastat is connected to a steam-control valve. Both the circulating pump and the steam-control valve control the *rate of flow* of the hot water or steam from the heating unit.

When steam is used as the heat source, the heat exchanger should also be equipped with a steam trap and a line running to the condensation main. The steam valve sizing is based on 2-psi-minimum steam pressure and O-psi return pressure. A steam strainer should be placed in the steam supply line just before it reaches the control valve.

Circulating pump sizing in units that use water as a heat source should be based upon a maximum temperature drop through the heat exchanger of 30°F, and a maximum head loss in piping between the hot-water boiler and the heater of 10 ft. H_2O.

The Btu per hour output for heat-exchanger pool heaters using water as a heat source ranges from 200,000 to 4,200,000 Btu per hour. The amount will depend upon water temperature (180°F or 210°F) and the size of the unit. Heaters using steam as a heat source are capable of generating an output of 400,000 to 3,600,000 Btu per hour. Here the difference depends upon the steam pressure used (2 lb. or 10 lb.) and the size of the unit.

SIZING THE POOL HEATER

The two principal methods used for sizing pool heaters are:

1. The surface-area method.
2. The time-rise method.

Table 5-1 contains all the necessary data for determining the required pool heater size with either method. The listed output ratings are for Bryan pool heaters.

The sizing data in Table 5-2 is based on heat loss from the surface of the heater with an assumed wind velocity of 3½ mph. This is the average wind velocity for a pool protected from direct wind exposure by trees, shrubs, fences, and buildings. An exposed pool will mean that the required pool heater capacity (i.e., the input ratings listed in Table 5-1) must be increased. Generally it is recommended that the input rating be increased by 1.25 for a 5-mph wind velocity and 2.0 for a 10-mph wind velocity.

THE SURFACE-AREA METHOD

The *surface-area method* of sizing a pool heater is based on the surface area (in square feet) of the pool and the temperature difference (in degrees Fahrenheit) between the desired water temperature and the average air temperature. The procedure is as follows:

1. Determine the *mean* average air temperature for the coldest month in which the pool is to be used.
2. Determine the desired pool water temperature.
3. Find the difference between the air temperature (Step 1) and the water temperature (Step 2).
4. Calculate the pool surface area in square feet.
5. In Table 5-1, find the surface area closest to your pool size. Move horizontally in a straight line across to the column that represents the temperature difference for your pool. The point at which the horizontal line and the temperature column intersect will be the required Btu per hour input rating for your pool heater.

The surface-area method can be illustrated with a simple example. Let's suppose that you have a small 15 ft. × 30 ft. pool with a 17,000 gal. capacity. You want to maintain a pool water temperature of 70°F (Step 2), and the mean average air temperature for the coldest month in which the pool is to be used is 50°F (Step 1). What will be the required input rating of your pool heater?

Your temperature difference required for sizing the heater is 20° (70°F − 50°F). The pool surface area is 450 sq. ft. (15 × 30 ft.). Moving horizontally across the top line in Table 5-1, you arrive at the 20° column and learn that the required input rating for a pool heater meeting these criteria is 150,000 Btu per hour.

Table 5-1. Sizing the Pool Heater. Ratings Given in 1,000 Btu

Pool Size Rectangular Width	Length	Pool Surface Area (Sq. Ft.)	SURFACE-AREA METHOD — Difference between Desired Water and Average Air Temperature — RECOMMENDED HEATER SIZED ON SURFACE AREA						TIME-RISE METHOD	
			10°	15°	20°	25°	30°	40°	Pool Gallonage (approx.)	RECOMMENDED HEATER Sized on Pool Capacity To Raise Temp. 1°F. Per Hr. (Approx.)
15	30	450	150	150	150	150	250	250	17,000	250
16	32	512				250			20,000	350
16	36	576						350	23,000	
18	36	648			250				25,000	450
18	40	720					350	450	30,000	
18	42	756							31,000	
20	40	800		250					34,000	
20	42	840							35,000	
20	45	900					450		37,000	650
20	50	1000	250		350			650	40,000	
25	50	1250		350	450	450	650	850	50,000	850
25	60	1500				650		900	62,000	900
30	60	1800		450	650		850	1200	79,000	1200
30	70	2100	350						84,000	
30	75	2250							92,000	
35	75	2625		650			1200	1500	107,000	1500
40	75	3000	450		850			1800	123,000	1800
42	75	3150			900				137,000	2100
42	80	3465		850			1500		143,000	2450

Table 5-1. Sizing the Pool Heater. Ratings Given in 1,000 Btu (Cont'd)

Pool Size Rectangular		Pool Surface Area (Sq. Ft.)	SURFACE-AREA METHOD						TIME-RISE METHOD	
			Difference between Desired Water and Average Air Temperature — RECOMMENDED HEATER SIZED ON SURFACE AREA						Pool Gallonage (approx.)	RECOMMENDED HEATER Sized on Pool Capacity To Raise Temp. 1°F. Per Hr. (Approx.)
Width	Length		10°	15°	20°	25°	30°	40°		
45	90	4050	650		1200			2100	184,000	3200
50	100	5000	850		1500	1500	2100	2700	217,000	3750
60	100	6000	900	1200	1800	1800	2450	3200	250,000	4300
60	110	6600	1200	1500	2100	2100	2700	3750	275,000	
65	120	7800	1500	2100	2450	2450	3200	4300	320,000	
75	130	9750	1800	2450	3200	3200	3750	4850	400,000	Two 2700
75	160	12,000	2100	2700	3750	4300	4850	Two 2700	490,000	Two 3200
80	175	14,000		3200	4300	4850	Two 2700	Two 3200	575,000	Two 3750
80	200	16,000				Two 2700	Two 3200	Two 4300	655,000	Two 4300

Courtesy Bryan Steam Corporation

235

Table 5-2. Wind Velocity and Recommended Pool Heater Capacity

Approximate Pool Gallonage	Wind Velocity	Recommended Pool Heater Capacity
17,000 Gal.	3 ½ mph	250,000 Btu
17,000 Gal.	5 mph	375,000 Btu
17,000 Gal.	10 mph	500,000 Btu

Another, less precise form of the surface-area method is to multiply the surface area of the pool by 15 by the temperature difference between the pool water and the air. Using the same data provided above, the following results are obtained:

$$450 \text{ sq.ft.} \times 15 \times 20° = 135,000 \text{ Btu per hour}$$

THE TIME-RISE METHOD

The first step in the time-rise method of sizing a pool heater is to determine the pool capacity in gallons of water (pool gallonage). This only needs to be an approximate figure and is commonly rounded off to the nearest thousand (e.g., 17,000, 20,000, 23,000). In Table 5-1, the extreme right-hand column lists pool heaters recommended for different pool capacities. The sizing in this chart is based on the number of Btu required to raise the pool temperature approximately 1°F per hour.

SIZING AN INDOOR POOL HEATER

If the swimming pool is located inside a heated building, the surface temperature of the water is naturally not affected by the colder outdoor air temperatures. A simple rule-of-thumb method for sizing an indoor pool heater is as follows:

1. Determine the surface area of the pool (e.g., 30 ft. × 40 ft. = 1200 sq. ft.).
2. Multiply the surface area by 125 Btu (1200 sq. ft. × 125 Btu = 150,000 Btu per hour input).

236

CHAPTER 6

Ventilation Principles

Ventilation is the process of moving air from one space to an entirely separate space and is primarily a matter of air volume. It should not be confused with *circulation*, which is the moving of air *around and within* a confined space. In contrast to ventilation, circulation is primarily a matter of air velocity. Ventilated air may or may not have been conditioned, and it may be supplied to or removed from the spaces by either natural or mechanical means.

Ventilation is a science with a majority of the people in the United States dependent upon its operation for their health and well-being. It developed slowly, as did other sciences, over a period of several thousand years.

Early in human history there was little need for ventilators because the majority of people lived in a warm climate where doors and windows were open most of the time. Those few who lived in colder climates had a hole in the roof to vent the smoke from the fire that kept them warm. This hole in the roof was the forerunner of our modern ventilators. The warm air from the fire rose through this hole, taking with it the majority of the smoke— and inducing a draft that constantly changed the air in the room. Furthermore, the houses were not tightly built, and ventilation took care of itself through chinks and cracks.

Castles, churches, and large public buildings were the only structures large enough to present any real problems in ventilation. Holes were cut in the roofs haphazardly to solve the problem. In some, large ducts or areaways were built to lead up to these holes. Of course, rain, sleet, and snow entered the building through these holes during every storm.

Hundreds of years passed without any further improvements in ventilation. When factories came into existence in the latter part of the nineteenth century, many people were brought together in one room. Little thought was given to ventilation, the only object being to keep the outside weather from entering the building. With these conditions prevailing, ventilation almost disappeared and was not again brought to public attention until the foul, polluted atmosphere in the factories began to seriously threaten the health (and thereby the productivity) of the workers.

Around 1875, people began to devise ways to get this foul air out through the roof. In most cases, this was only a hole covered by a flat or conical piece of metal. This did not keep out rain or snow when any wind was blowing.

In many cases the design of the covers hindered the ventilation, and covers did not work well unless temperature conditions were right. The air inside the room had to be much warmer than that outside before the ventilated air would move out through the passage provided.

Further attempts to exclude the weather resulted in the installation of stormbands and louvers. Unfortunately, most of these hampered the air passage until there was practically no ventilation.

Engineers, realizing the importance of ventilation in the rapidly growing factories, started to work on the problem and at the beginning of this century developed a scientific product. Today, roof ventilators used for domestic, commercial, and industrial purposes are as numerous as the ventilating problems which they were designed to solve.

THE MOTIVE FORCE

The force that moves the air in a room or building may be due to natural causes or mechanical means. In the first case, the ventilation is called *natural ventilation*. This kind of ventilation finds application in industrial plants, public buildings, schools, garages, dwellings, and farm buildings. The two natural forces available for moving air into, through, and out of buildings are: (1) induction, and (2) thermal effect. The inductive action is due to the wind force, whereas the thermal effect is due to difference in temperature inside and outside a building (this being in fact the same as *chimney effect*). The air movement may be caused by either of these forces acting alone or by a combination of the two, depending upon atmospheric conditions, building design, and location.

The nature of the ventilating results obtained by natural means will vary from time to time due to variation in the velocity and direction of the wind and the temperature difference.

The wind ventilating effect depends upon its velocity. In almost all localities, summer wind velocities are lower than those in the winter. There are relatively few places where the wind velocity falls below one-half of the average for many hours per month. Accordingly, if a natural ventilating system is proportioned for wind velocities of one-half the average seasonal velocity, it should prove satisfactory in almost every case.

When considering the use of natural wind forces for producing ventilation, three conditions must be considered:

1. The average wind velocity.
2. The prevailing wind direction.
3. Local wind interference by buildings, halls, or other obstructions.

INDUCTIVE ACTION OF THE WIND

When the wind blows without encountering any obstruction to change its direction, its movement may be represented by a series of parallel arrows, as in Fig. 6-1. The arrows indicate the

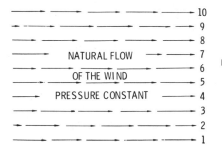

NATURAL FLOW

OF THE WIND

PRESSURE CONSTANT

Fig. 6-1. Natural flow of wind when there is no obstruction to change its directions.

direction of flow. Under such conditions the pressure may be considered as constant throughout the airstream.

If the airstream meets an obstruction of any kind, such as a house or ventilator, these parallel air lines will be pushed aside as in Fig. 6-2, crowding each other at points A and B, curving back on the sides of the obstruction and curving inward past points C and D to their original parallel positions.

A vacuum is formed here as indicated by the suction lines (i.e., the arrows that curve back and inward toward the space occupied by the vacuum). This vacuum (or reduction in pressure) is what causes inductive action.

A ventilator can be so constructed as to make this inductive action effective for ventilation. Fig. 6-3 shows the two essential parts of a simple ventilator: the head and connecting flue. The head is open at one end and closed at the other and in actual construction pivoted to rotate, guided by a vane so that the closed end always faces the wind.

In Fig. 6-3, as shown, the closed end forms an obstruction, which changes the direction of the wind expanding at the closed

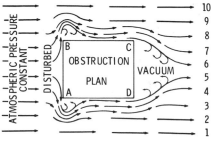

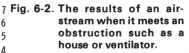

Fig. 6-2. The results of an airstream when it meets an obstruction such as a house or ventilator.

240

end and converging at the open end, producing a vacuum inside the head, which induces an upward flow of the air through the flue and out through the head. This is *inductive* action of the wind.

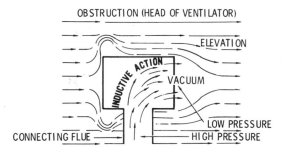

OBSTRUCTION (HEAD OF VENTILATOR)

ELEVATION

INDUCTIVE ACTION

VACUUM

LOW PRESSURE
HIGH PRESSURE

CONNECTING FLUE

Fig. 6-3. Two essential parts of a ventilator showing the results of wind action.

The *stack* or *flue effect* (Fig. 6-4) produced within a building when the outdoor temperature is lower than the indoor temperature is due to the difference in weight of the warm column of air within the building and the cooler air outside. The flow due to the stack (flue) effect is proportional to the square root of the draft head. The formula for determining the rate of flow is as follows:

$$Q = 9.4A\sqrt{h}(t\text{-}t_0)$$

Where:

Q = air flow, cubic feet per minute.

A = free area of inlets or outlets (assumed equal), square feet.

h = height from inlets to outlets, feet.

t = *average* temperature of indoor air in height h, Fahr. degrees.

t_0 = temperature of outdoor air, Fahr. degrees.

9.4 = constant of proportionality, including a value of of 65 percent for effectiveness of openings. This should be reduced to 50 percent (constant = 7.2) if conditions are not favorable.

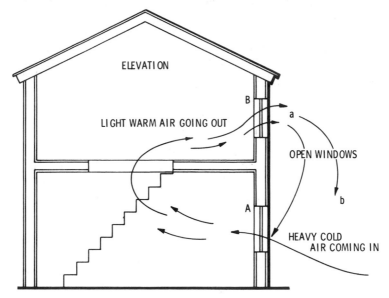

Fig. 6-4. Illustrating the stack effect in a two-story structure.

INDUCED DRAFT

A closed flue (stack) or chimney will induce a draft or "draw" air. In other words, it will cause the air to rise from the bottom level (or room level) to the top. Consider the schematic of a stack shown in Fig. 6-5. The air is cool at the bottom of the stack and hot at the top. Each unit of air in traversing the stack expands as the temperature increases and becomes lighter. Assuming that cube A, in Fig. 6-5, represents 1 lb. of air and that the initial volume undergoes first eight expansions and then sixteen, the corresponding weight of the initial volume (1 lb.) decreases to $\frac{1}{8}$ lb. and then to $\frac{1}{16}$ lb. Accordingly the sum of the weights of unit volume in ascending is $1 + \frac{1}{8} + \frac{1}{16}$ or $1\frac{3}{16}$ lb. Now, on the outside of the stack the volume and weight of each unit of air remain the same so that considering three units a, b, c of decreasing weights in the stack, there are three units of a', b', c' of constant weight outside the stack; the total weight outside the

242

stack being 3 lb. and only 1 ³⁄₁₆ lb. inside the stack. As a result of the downward force (3 lb.) outside the stack being greater than that in the stack, the heavy units a', b', c' push the lighter units a, b, c up and out of the stack, thus inducing a draft as indicated by the lever scales in Fig. 6-5. For purposes of explanation, induced draft may be considered virtually the same as thermal effect.

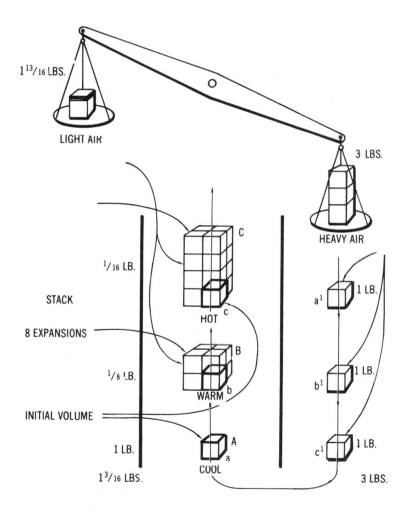

Fig. 6-5. The principle of induced draft.

COMBINED FORCE OF WIND EFFECT
AND THERMAL EFFECT

It should be noted that when the forces of wind effect and thermal effect are acting together (even when both forces are acting together without interference) the resulting air flow is not equal to the sum of the two estimated quantities. The flow through any opening is proportional to the square root of the sum of the heads acting upon the opening.

When the two heads are equal in value and the ventilating openings are operated so as to coordinate them, the total air flow through the building is about 10 percent greater than that produced by either head acting independently under conditions ideal to it. This percentage decreases rapidly as one head increases over the other, and the larger will predominate.

The wind velocity and direction, the outdoor temperature, or the indoor distribution cannot be predicted with certainty, and refinement in calculations is not justified; consequently, a simplified method can be used. This may be done by using the equations and calculating the flows produced by each force separately under conditions of openings best suited for coordination of the forces.

MECHANICAL VENTILATION

Mechanical ventilation is the process of supplying or removing air by mechanical means. In other words, it represents a form of *forced* ventilation. Some form of mechanical ventilation is necessary when the volume of air required to ventilate a space cannot be delivered adequately by natural means. As a result, it may be necessary to add a fan (or fans) to the ventilation system in order to obtain the required rate of air change.

Mechanical ventilation not only involves the use of fans but also ducts in some larger systems. The selection, installation, and operation of fans is described in Chapter 7 (Fan Selection and Operation). All aspects of duct sizing, the resistance to airflow, and other related aspects are described in considerable detail in Chapter 7 of Volume 2 (Ducts and Duct Systems).

AIR VENTILATION REQUIREMENTS

The volume of air required for proper ventilation is determined by the size of the space to be ventilated and the number of times per hour that the air in the space is to be changed. Table 6-1 gives some recommended rates of air change for various types of spaces. The volume of air required to ventilate a given space is determined by dividing the volume of the space by the number of minutes shown for that space in the "rate of change" column found in Table 6-1.

In many cases, existing local regulations or codes will govern the ventilating requirements. Some of these codes are based on a specified amount of air per person and others on the air required per foot of floor area. The table should thus serve only as a guide to average conditions; where local codes or regulations are involved, they should be taken into consideration.

If the number of persons occupying a given space is larger than would be normal for such a space, the air should be changed more often than shown in the table. It is recommended that an air exchange rate of 40 cu. ft. per minute per person be allowed for extremely crowded spaces.

If the cooling effect of rapid air movement is needed in localities having high temperatures or humidities, the number of air changes shown in Table 6-1 should be doubled.

ROOF VENTILATORS

The function of a roof ventilator is to provide a storm- and weatherproof air outlet. For maximum flow by induction, the ventilator should be located on that part of the roof where it will receive the full wind without interference.

One must exercise great care when installing ventilators. If the ventilators are installed with the vacuum region created by the wind passing over the building or in a light court, or on a low building between two buildings, their performance will be seriously influenced. Their normal ejector action, if any, may be completely lost.

The base of a ventilator should be a tapering cone shape. This

Table 6-1. Fresh Air Requirements

Type of Building or Room	Minimum Air Changes Per Hour	CFA Per Minute Per Occupant
Attic Spaces (For Cooling)	12–15	
Boiler Rooms...................................	15–20	
Churches, Auditoriums..........................	8	20–30
College Classrooms.............................		25–35
Dining Rooms (Hotel)	5	
Engine Rooms..................................	4–6	
Factory Buildings (Ordinary Manufacturing)	2–4	
Factory Buildings (Extreme Fumes or Moisture)....	10–15	
Foundries......................................	15–20	
Galvanizing Plants..............................	20–30*	
Garages (Repair)...............................	20–30	
Garages (Storage)	4–6	
Homes (Night Cooling)	9–17	
Hospitals (General)		40–50
Hospitals (Children's)		35–40
Hospitals (Contagious Diseases)		80–90
Kitchens (Hotel)	10–20	
Kitchens (Restaurant)	10–20	
Libraries (Public)..............................	4	
Laundries......................................	10–15	
Mills (Paper)	15–20*	
Mills (Textile—General Buildings)	4	
Mills (Textile—Dyehouses)	15–20*	
Offices (Public)	3	
Offices (Private)	4	
Pickling Plants	10–15†	
Pump Rooms...................................	5	
Restaurants....................................	8–12	
Schools (Grade)		15–25
Schools (High).................................		30–35
Shops (Machine)	5	
Shops (Paint)	15–20*	
Shops (Railroad)	5	
Shops (Woodworking)...........................	5	
Sub Stations (Electric)..........................	5–10	
Theatres		10–15
Turbine Rooms (Electric)	5–10	
Warehouses	2	
Waiting Rooms (Public).........................	4	

*Hoods should be installed over vats or machines.
†Unit heaters should be directed on vats to keep fumes superheated.

design provides the effect of a bell mouth nozzle, which gives considerably higher flow than that of a square entrance orifice.

Air inlet openings located at low levels in the building should be at least equal to, and preferably larger than, the combined throat areas of all roof ventilators.

The advantages of natural ventilation units are that they may be used to supplement power-driven supply fans, and under favorable conditions it may be possible to shut down the power-driven units.

TYPES OF ROOF VENTILATORS

Roof ventilators are manufactured in a variety of shapes and designs. Because of this variety, it is possible to classify roof ventilators under a number of broad categories, including:

1. Stationary head.
2. Revolving.
3. Turbine.
4. Ridge.
5. Siphonage.

Some of the examples illustrated in the paragraphs that follow are no longer manufactured, but they will be encountered on older buildings. For this reason, they are included in this chapter.

Stationary-Head Ventilators

More ventilators are of this type than any other, and when well-designed, it is usually considered the most efficient type of gravity ventilator. The chief advantages of stationary-head ventilators are:

1. Higher exhaustive capacity under all wind conditions.
2. No moving parts.
3. Quiet operation.
4. No upkeep.

A stationary head ventilator is illustrated in Fig. 6-6.

Revolving Ventilators

A typical revolving ventilator is shown in Fig. 6-7. This type of ventilator swings on a pivot, aided by the wind vane so that its open end points away from the wind. The inductive action of the wind, in passing around the head, draws air out of the building.

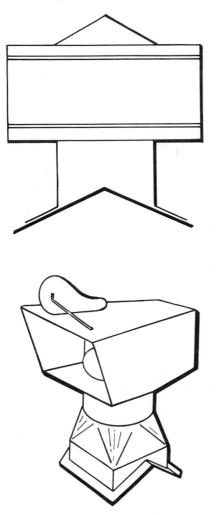

Fig. 6-6. Stationary-head ventilator.

Fig. 6-7. Revolving ventilator. The top portion pivots.

In general, this is not as efficient as the stationary-head ventilator, although it has general use.

A major disadvantage of the revolving ventilator is its low capacity in still air. Furthermore, if the pivot bearings stick so that the head cannot follow the wind, it fails to exhaust air (in some cases even admitting air, rain, or snow). The revolving ventilator also frequently becomes noisy, creaking when changing direction. Constant attention is required to overcome these disadvantages.

Turbine Ventilators

Fig. 6-8 represents one type of turbine ventilator. On its top is a series of vanes that, from the force of the wind, causes the entire top to rotate. Fastened to the top on the inside and placed in the air outlet openings is a series of propeller blades. The wind rotates this head, the blades drawing air up the shaft and exhausting it through the blade opening. The basic limitation of the turbine ventilator is that its blades draw the air outward as the top revolves. Its principal disadvantages include the following:

1. Low capacities in quiet air.
2. Lowered capacity if the pivot bearings do not turn freely.
3. The tendency to become noisy (as does any heavy rotating body that cannot accurately be kept in balance).
4. Required service attention in oiling, cleaning, and eventually replacing bearings.

Another type of turbine ventilator (Fig. 6-9) consists of a globe-shaped head composed of vane blades. The wind rotates this head, the blades drawing air up the shaft and exhausting it through the blade openings. The basic components of a globe-head turbine ventilator are shown in Fig. 6-10.

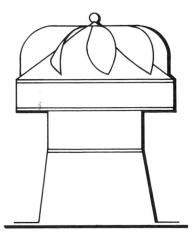

Fig. 6-8. Turbine ventilator. The blades draw air outward as the top revolves.

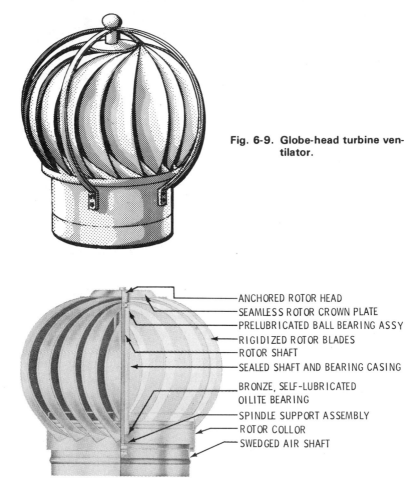

Fig. 6-9. Globe-head turbine ventilator.

ANCHORED ROTOR HEAD
SEAMLESS ROTOR CROWN PLATE
PRELUBRICATED BALL BEARING ASSY
RIGIDIZED ROTOR BLADES
ROTOR SHAFT
SEALED SHAFT AND BEARING CASING

BRONZE, SELF-LUBRICATED
OILITE BEARING
SPINDLE SUPPORT ASSEMBLY
ROTOR COLLOR
SWEDGED AIR SHAFT

Fig. 6-10. Construction details of a turbine ventilator.

Turbine ventilators with globe-shaped heads also have a number of disadvantages. For example, wind impact on the vanes allows outside air to enter the head on the windward side that must be exhausted, together with the air from the building on the leeward side. This decreases its efficiency inasmuch as the head must handle both volumes of air. Furthermore. this type of ventilator is apt not to be rainproof when the wind is not turning it. Like all revolving ventilators, its moving parts involve a service problem in oiling, keeping bearings free, and the replacement of worn parts.

Ridge Ventilators

Fig. 6-11 shows a sectional view of a ridge ventilator, which is installed along the entire roof ridge of the space to be ventilated. Basically, it consists of a valve in the top of the building, which lets the warm air out as it rises to the roof.

There are several advantages to using ridge ventilators in certain types of building construction. For example, it lets air out of the building at the highest point, and it allows an even air outlet distribution along the length of the structure. Then, too, it has a pleasing uniform appearance.

Its major disadvantage is that it does not take full advantage of wind action unless the wind direction is almost directly across it. This type of ventilator also involves a somewhat cumbersome damper and a rather difficult problem in building modernization.

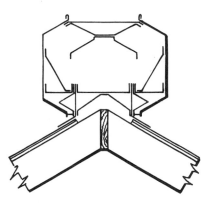

Fig. 6-11. Ridge ventilator.

Siphonage Ventilators

The siphonage ventilator (Fig. 6-12) acts by induction, drawing the building air upward. Wind causes a flow of air upward through a duct that is concentric to and parallel with the ventilator shaft extending from the building. Both of these air passages terminate in the ventilator head, and the upward flow of the outside air creates a siphonage action drawing the building air streams exhausting through the head. This ventilator is simply a stationary unit with an auxiliary air passageway to obtain the siphonage action.

The head of the siphonage ventilator must be designed to eject not only the building air, but also the siphon stream. This tends to retard air flow from the building due to the fact that egress from the head is usually somewhat restricted.

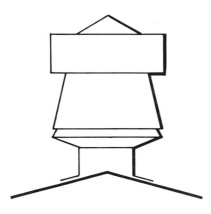

Fig. 6-12. Siphonage ventilator.

Fan Ventilators

A typical fan ventilator is shown in Fig. 6-13. The principal advantage of this type of ventilator is that the fan greatly increases the ventilating capacity when the fan is in operation. Fan ventilators also have a number of other advantages when the fan is not in use, and have great capacity when the fan is operated. Fan ventilators can be used on any building, regardless of access to wind flow, and can be spotted directly over any point at which ventilation is badly needed. These ventilators can also be used on ductwork flues to give greater capacities when friction losses are relatively large. It should be noted that a fan ventilator is a *forced* draft rather than an induced draft ventilator.

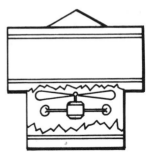

Fig. 6-13. Sectional view showing fan ventilator.

Fan ventilators (roof exhausters) are covered in considerable detail in Chapter 7 (Fan Selection and Operation).

COMPONENTS OF A ROOF VENTILATOR

The basic components of a typical roof ventilator are shown in Fig. 6-14 and may be listed as follows:

1. Base.
2. Barrel.
3. Top.
4. Windband.
5. Air Shaft.
6. Dampers.

The *base* serves as the connection between the roof and the ventilator proper. A trough is usually provided around the inside bottom edge of the base to collect any condensation occurring in the ventilator and draw it out to the roof.

As bases are made to fit either a round or square opening and to conform to the shape and slope of the roof, they are essentially tailor-made for each installation.

After leaving the base, the air passes through the *barrel* (or *neck*) of the ventilator, which is merely the lower section of the head. It then enters the head proper and is exhausted to the outside air through the openings between louvers. The *top*, of course, covers the opening in the roof.

The *windband* is the vertical band encircling the ventilator head. The *air shaft* is the entire passageway through the ventilator from the roof to the top of the barrel.

The *dampers* are mechanical devices for closing the air shaft. They assume a number of forms, and are referred to according to type as the *sliding sleeve, inverted cone, butterfly,* and *louver.*

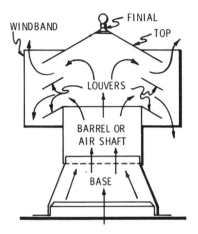

Fig. 6-14. Component parts of a roof ventilator.

MOTIVE FORCE TO CAUSE AIR CIRCULATION

Two agencies form the motive force to cause air circulation: (1) temperature differences inside and outside the building giving a chimney effect, and (2) inductive action caused by wind blowing against the ventilator.

CAPACITY OF VENTILATORS

Several factors must be taken into consideration in making a selection of the proper ventilator for any specific problem. They are:

1. Mean temperature difference.

2. Stack height (chimney effect).
3. Inductive action of the wind.
4. Area of opening in the ventilator.

The *mean temperature difference* refers to the average temperature difference between the air inside and outside the building (Fig. 6-15). If the ventilating problem is essentially a winter one, this difference may be as much as 40°F, while in summer, when doors and windows are open, this difference will probably drop to about 10°F.

Fig. 6-15. Mean temperature difference inside and outside.

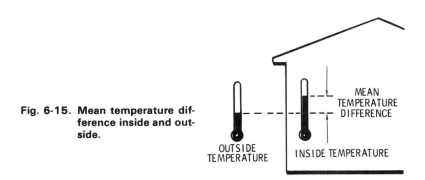

MEAN TEMPERATURE DIFFERENCE

OUTSIDE TEMPERATURE

INSIDE TEMPERATURE

Stack height (Fig. 6-16) is the height of the column of warm air that causes the chimney action of the ventilator. It is measured in feet from the floor of the building to the top of the ventilator barrel.

Fig. 6-16. Stack height upon which the chimney effect depends.

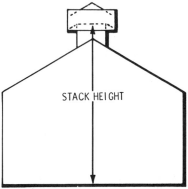
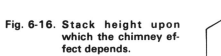

STACK HEIGHT

The *inductive action of the wind* (Fig. 6-17) is the effect of the wind as it passes over the ventilator in inducing a circulation out of the building.

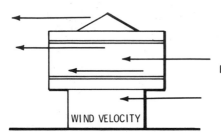

Fig. 6-17. The suction action of the wind as it passes over the ventilator.

The *area of the opening in the ventilator* (Fig. 6-18) is determined by the rated size of the ventilator, expressed as the inside diameter of the barrel in inches.

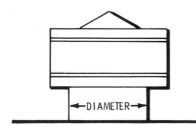

Fig. 6-18. Diameter of the base, which governs the capacity of gravity ventilation.

DESIGN AND PLACEMENT
OF INLET AIR OPENINGS

The purpose of a roof ventilator is to let air escape from the top of a building. This naturally means that a like amount of air must be admitted to the building to take the place of that exhausted. The nature, size, and location of these inlet openings is of importance in determining the effectiveness of the ventilating system.

Inlet air openings are frequently constructed in the form of louvered openings located near the floor line, but they can also often consist of the building windows. In general, there are three factors that should be considered when planning for an effective ventilating system:

1. The relationship between the area of the air intakes and the airshaft areas.
2. The distribution of inlet air openings.
3. The height at which air inlets are placed.

The area of the air intakes should at all times be twice that of the combined airshaft area of all ventilators (Fig. 6-19). By keeping this relationship, the full capacity of the ventilators will be realized, and the velocity of the incoming air will be low, thereby lessening the danger from drafts in the building. Any backdraft down the ventilator itself will also be eliminated.

These inlet air openings should also be so distributed as to allow the admission of fresh air to all parts of the building. They

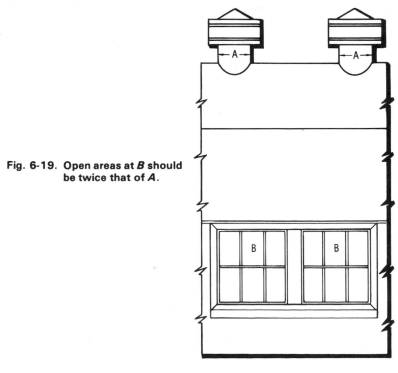

Fig. 6-19. Open areas at *B* should be twice that of *A*.

should be located as close to the floor as possible so as to bring the incoming fresh air into the breathing zone.

FRESH AIR REQUIREMENTS

Table 6-1 lists necessary air changes for various rooms and buildings, and thus offers a guide as to the amount of air required for efficient ventilation. This information should be used in connection with the Ventilator Capacity Tables in the proper selection of the number and size units required.

Where two figures are given for one type of building, use the smaller figure when conditions are normal and the larger figure when they are abnormal. Certain buildings are better figured on a CFM (cubic feet per minute) per occupant basis. Their requirements are given in this manner. See Chapter 7 (Fan Selection and Operation) for formulas for determining CFM.

VENTILATOR BASES

The ventilator base is the connection between the other elements of the ventilator and the roof. It must be designed so as to fit the contour of the roof and the opening on which it is mounted. Its design and construction is important, particularly in ventilator sizes and types where weight and wind resistance are high. Ventilator manufacturers provide a wide variety of different ventilator base sizes. In their literature, illustrated base designs (Fig. 6-20) are cross referenced with dimensional tables for the convenience of the buyer.

Gauges of metal are used that assure rigidity and strength. Reinforcement is added where needed. Flashing flanges are amply wide to assure rigid, weather-tight joints that will not leak when properly fastened and flashed to the roof. All seams are well riveted, soldered, or doped (the latter in asbestos protected steel) to assure storm-tight joints. All connections to roof and to next adjacent ventilator unit (the head in a gravity unit or the fan in a fan unit) are designed for 75-mph wind velocity as standard or 100 mph optional.

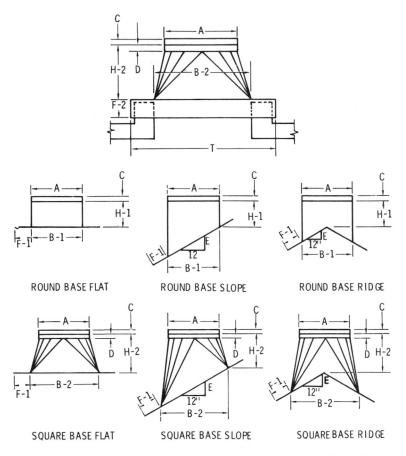

ROUND BASE FLAT   ROUND BASE SLOPE   ROUND BASE RIDGE

SQUARE BASE FLAT   SQUARE BASE SLOPE   SQUARE BASE RIDGE

Fig. 6-20. Standard ventilator bases. The letters refer to dimension on manufacturers' tables.

Since the outside sheet-metal surfaces of ventilators are exposed to atmospheric temperatures and the inside surfaces to room temperature, there is frequently a tendency for condensation to form on the inside surfaces. In order to minimize the possibility of condensation drip into the ventilated space, condensation gutters may be provided to drain the water to the outside of the ventilator. These condensation gutters are usually located at the extreme lower edge of the base where as little water as possible can get past them.

The round ventilator bases in Fig. 6-21 are standard but can be omitted if a duct is to be fitted into the lower end of the base with which the gutter would interfere. Examples of spare bases are shown in Fig. 6-22, but these are special. In either type, drains

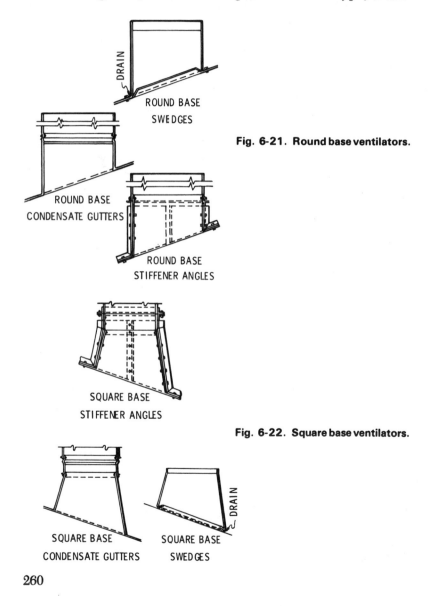

Fig. 6-21. Round base ventilators.

Fig. 6-22. Square base ventilators.

from the gutter to the roof are ample for drainage and to prevent clogging.

In ventilator sizes and types where weights are not great, bases are lapped into the next adjacent unit above. In order to properly position the units and to add stiffness, a swedge is provided in the base against which the next adjacent unit seats.

ANGLE RINGS

In larger ventilator sizes where added strength, rigidity, and ease of erection are important, angle rings have been used to connect the base to the next adjacent unit (ventilator head, fan section, or stack).

In certain instances a one-ring connection (Fig. 6-23) is used with the upper edge of the base lapping into the next adjacent unit, which rests upon the angle ring. The bolted connection is made through the sheet-metal lap. The lap connection is ample to assure proper bearing for connection screws or bolts. When angle rings are provided, swedges are not used.

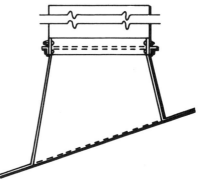

Fig. 6-23. One-ring connection.

Sometimes a two-ring connection (Fig. 6-24) is used, one ring at the top of the base, another at the bottom of the adjacent unit. The two rings form a flanged connection and are bolted together. In every instance where angle rings are used they are riveted to the bases to develop the full design strength of the assembly for velocities of 75 mph standard or 100 mph optional.

261

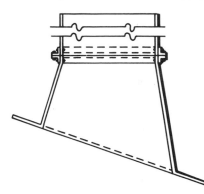

Fig. 6-24. Two-ring connection.

STIFFENER ANGLES

Stiffener angles are provided in conjunction with angle rings in certain applications where the sheet metal needs reinforcing. These lie vertically along the outside of the base surface at quarter points and extend from the angle ring at the the top to the outer edge of the roof flange at the bottom. They are riveted to the base sheet metal and welded to the angle ring. Bolted connections through portions of the stiffener that lie along the roof flange to the curb or framing members below provide rigid structural members through which load is carried to the roof structure.

Where applicable and with certain types of dampers, a small clip can be supplied and riveted to the base (Fig. 6-25). The damper chain drops through this clip and can be locked at any point to properly position the damper. This clip is useful when the damper is to be operated from a point directly below the ventilator since it avoids the use of pulleys and complicated chain arrangements.

PREFABRICATED ROOF CURBS

Prefabricated roof curbs (bases) for ventilators are available from a number of manufacturers. One of the advantages of the prefabricated type over those constructed at the site is the fact

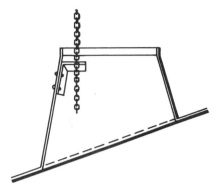

Fig. 6-25. Base chain clip.

that they are generally cheaper. This, of course, is due to their being mass produced. Although commonly designed for flat roof installation, prefabricated roof curbs can also be built for installation on the ridge or single slope of a roof. Figs. 6-26 and 6-27 illustrate some of the design features incorporated in a prefabricated roof curb.

VENTILATOR DAMPERS

Ventilator dampers (Fig. 6-28) are made in a variety of types from the single-disc butterfly to the multiblade louver, which has

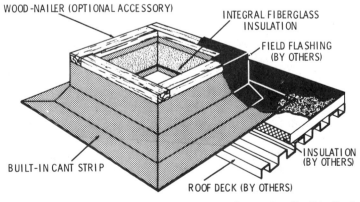

WOOD-NAILER (OPTIONAL ACCESSORY)

INTEGRAL FIBERGLASS INSULATION

FIELD FLASHING (BY OTHERS)

INSULATION (BY OTHERS)

ROOF DECK (BY OTHERS)

BUILT-IN CANT STRIP

Courtesy Penn Ventilator Co., Inc.

Fig. 6-26. Prefabricated roof curb designed for flat roof installation.

263

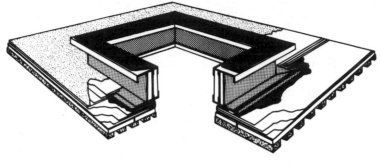

Fig. 6-27. Prefabricated roof curb featuring self-flashing design.

been designed to allow a flexible and reliable means of air movement control. A tight seal can be obtained through the use of a damper ring, which prevents the passage of air when the damper is closed.

Some dampers can be controlled only by hand chain; others may be remotely operated through the use of damper control motors of the electric or compressed air type.

The principal types of ventilator dampers are as follows:

1. Sliding sleeve.
2. Fire-retarding cone.
3. Single butterfly.
4. Divided butterfly.
5. Louver.

Louver Dampers

Fig. 6-28 shows the general appearance of a louver damper. It is considered the best type for airflow. Louver dampers offer little resistance to the passage of air and in a partially open position create little turbulence in the airstream. Multibladed, the blades on one half open up, those on the other half open down.

The center blade is double the width, and its edges are connected to adjacent blades by clips and bars. The spacing of the blades along the connecting bars have been closely studied in order that the blade edges shall seat tightly on the damper ring when closed. This damper ring also allows all the blade pivot bearings to be placed inside of the ventilator shaft, away from the weather. It also eliminates holes through the airshaft that might leak air and rain. Blades are regularly weighted so that the damper is automatic closing. Any type of control may be used.

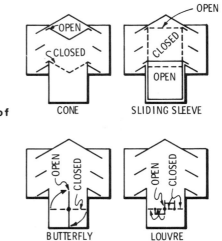

Fig. 6-28. Various forms of dampers.

Sliding Sleeve Dampers

The sliding sleeve damper is frequently used for gravity ventilators. In construction it closely fits the under circumference of the ventilator airshaft and operates vertically. It occupies almost no space in the airshaft and offers no resistance or turbulence to the airstream in any position. Its normal position is open, and with a fusible link it can be used as an automatic opening damper where there is need for such a control unit.

Sliding sleeve dampers have one principal advantage. Because of their vertical sides, there is no tendency for dust or dirt to accumulate.

The sliding sleeve damper is operated by chain only. Radial spiders keep the damper cylindrical to operating chain attaching to the hub of the spider, passing upward through a pulley fastened below the finial, then down past the spider hub, where a spring clip is located. The spring clip enables the damper to be positioned at any point between open and closed.

Sliding Cone Dampers

As shown in Fig. 6-28, a sliding cone damper consists of a cone with apex down and with a flared outer rim that seats on the upper edge of the airshaft when closed. This directs the airstream to the outlet openings of the ventilator, lessens turbulence in the ventilator head, and in consequence increases capacity.

From the apex of the damper the operating chain passes upward to a pulley located below the finial, then downward through a slot in the cone to a pulley located on the inside of the airshaft, then through the base to the building below.

When used in connection with the fusible link, this damper becomes an automatically closing, fire-retarding unit.

Butterfly Dampers

The two butterfly dampers used in roof ventilators are: (1) the single-disc damper, and (2) the divided-disc damper. They receive their name from the butterfly wing appearance of the metal disc used to open or close the air passage.

The single-disc butterfly damper seats tightly against the ring channel, preventing air leakage when closed. When operated by hand chain, these dampers are normally counterweighted to close, but can be supplied weighted to open when required.

The divided-disc butterfly damper is used for sizes that prevent the use of a single disc. The divided halves (or "wings") of the disc swing upward, pivoted on two rods whose ends bear in the damper ring channel on which damper edges seat tightly.

Butterfly dampers, single disc and divided, are the only dampers that can be successfully used in ventilators constructed of asbestos-protected steel. In this material the damper ring is omitted and the damper pivot rods are mounted on brackets secured to the airshaft wall.

266

METHOD OF CALCULATING NUMBER AND SIZE OF VENTILATORS REQUIRED

The number and size of ventilators required by a particular building can be calculated by taking the following steps:

1. Determine the gross volume of the interior of the building in cubic feet by multiplying its length by its width and by its height.

2. Determine the number of air changes per hour required for the building in question. This can be found by referring to the tables given in the "Typical Installation" section.

3. Find the total number of cubic feet of air per hour necessary for the ventilators to exhaust by multiplying the number of changes per hour by the volume of each change, or 1×2.

4. Reduce this to cubic feet per minute by dividing by 60.

5. The effective range of a gravity ventilator is about 10 to 15 ft. in the direction of the length of the building and 15 to 25 ft. in the direction of the width of the building, where air is being admitted on the sides. And so the next step is to space the units along the roof, using these restrictions, determining in this manner the number of ventilators that will be required.

   If the building is divided into bays, one unit is usually placed in each bay, providing the bays are not over 20 ft. long. If there are no bays, one ventilator every 20 ft. should be sufficient for buildings up to 50 ft. in width. In wider buildings, they should be so arranged as to maintain approximately the 20-ft. spacing. If there are any spots where there is urgent need of ventilation, such as a machine or vat emitting fumes or dust, they should be cared for by ventilators placed above them. Such a layout as suggested will give the number of ventilators required for the building.

6. Next determine the exhaustive capacity required of each ventilator by dividing the total amount of air per minute to be exhausted by the number of ventilators to be used, or $4/5$. This is the capacity for which you should look in the Capacity Tables obtained from the manufacturers.

267

7. Determine the conditions under which the ventilators will operate, that is, stack height, mean temperature difference, and wind velocity. These factors, their use, and their probable values have already been explained.

8. Turn to Manufacturers' Capacity Tables and under the proper columns of Temperature Difference, Stack Height, and Wind Velocity, as determined in Step 7, find a capacity in CFM that will fulfill the requisite as determined in Step 6. Whichever this capacity is listed under is the size to be selected.

VENTILATOR CALCULATION EXAMPLES

Assume a building used as an automobile storage garage 50 ft. wide by 150 ft. long by 20 ft. high to the eaves, 26 ft. to the ridge of the roof.

1. Volume of space from the eaves down is:
   $50 \times 20 \times 150 = 150,000$ cu. ft.
   Volume of space from the eaves up is:
   $50 \times 3 \times 150 = 22,500$ cu. ft.
   Total Volume = 172,500 cu. ft.

2. From the table of air change requirements you will find that storage garages require four changes per hour for proper ventilation. Total air to be exhausted is then $172,500 \times 4 = 690,000$ cu. ft. per hour, or $690,000 \div 60 = 11,500$ CFM.

3. Spacing ventilators along the roof as indicated in Fig. 6-29, with the rules given in mind, the resultant number of ventilators would be seven, spaced 15 ft. from each end and 20 ft. apart. Each ventilator would need to exhaust $11,500 \div 7 = 1643$ CFM.

4. Suppose we have a mean temperature difference of 10°F, an average wind velocity of 5 mph, and a stack height of 26 ft. (floor to roof ridge).

5. Turning to the Capacity Tables under that of the 30-in. cone damper unit in the proper columns for the factors as determined in Step 4, we find that the capacity is 1750 CFM,

which exceeds our requisite of 1643 CFM as obtained in Step 3. Even though this is a trifle higher than necessary, it should be used, inasmuch as the next size smaller would be too far under the required capacity. This can be seen by referring to the 24-in. unit, the capacity of which is 1165 CFM.

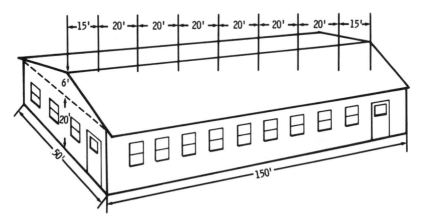

Fig. 6-29. Typical building illustrating method of calculating number and size of ventilators required.

We have therefore determined that to properly ventilate this building, seven 30-in. cone damper ventilators will be necessary, located on the ridge of the roof and spaced 20 ft. apart and 15 ft. from each end.

AIR LEAKAGE

Air leakage is the passage of air in and out of various cracks or openings in buildings. It is also sometimes referred to as *infiltration*.

Air leaking *into* a building may be caused by wind pressure or by differences in temperature inside and outside of the building. In the former case, the wind builds up a pressure on one or two sides of a building, causing air to leak into the building. As shown in Fig. 6-2, the action of the wind on the opposite side or sides produces a vacuum that draws air out of the building. Thus in Fig. 6-4, a plan view of a single-room building is shown having a window and a door on one side and a window on the opposite side. The details are greatly exaggerated so that you can see the cracks.

Note that when the wind hits the A side of the building, its momentum (dynamic inertia) builds up a pressure higher than inside the building, which causes the air to leak through any cracks present, as indicated by the arrows.

As the wind traverses the length of the building, the air currents as they continue past the side C converge and produce a vacuum along side C by induction. Because the pressure on the outside of C is lower than inside the building, air leaks out as indicated by the arrows.

Air leakage due to temperature difference or thermal effect is usually referred to as *stack* or *chimney effect*. Air leakage due to cold air outside and warm air inside takes place when the building contains cracks or openings at different levels. This results in the cold and heavy air entering at low level and pushing the warm and light air out at high levels, the same as draft taking place in a chimney.

Thus in Fig. 6-4, assume a two-story building having a window open on each floor. Evidently when the temperature inside the building is higher than outside, the heavy cold air from outside will enter the building through window A, and push the warm light air through window B, as indicated by arrow a; and as it cools, will increase in weight and circulate downward, as indicated by arrow b.

Although not appreciable in low buildings, this air leakage is considerable in high buildings unless sealing between various floors and rooms is adequate.

A reasonable amount of air leakage is actually beneficial to health. Any attempt to seal a building drum tight will cause the inside air to become stale and putrid. Emphasis should be placed on the reduction of heat transmission rather than an absolute elimination of air leakage.

The application of storm sash to poorly filtered windows will generally result in a reduction of air leakage of up to 50 percent. An equal effect can be obtained by properly installed weather stripping.

GARAGE VENTILATION

The importance of garage ventilation cannot be overestimated because of the ever-present danger of carbon monoxide poisoning. During warm weather there is usually adequate ventilation because doors and windows kept open. However, in cold and very cold weather people close up openings tight as a drum, with considerable danger. Nobody can breathe the resulting carbon monoxide concentration long without being knocked out; hence the importance of proper ventilation in cold weather, regardless of physical comfort.

Where it is impractical to operate an adequate natural ventilation system, a mechanical system should be used that will provide for either the supply of 1 cu. ft. of air per minute from out of doors for each square foot of floor area, or for removing the same amount and discharging it to the outside as a means of flushing the garage.

The following points should be carefully reviewed when considering a ventilating system capable or removing carbon monoxide from an enclosed area:

1. Upward ventilation results in a lower concentration of carbon monoxide at the breathing line and a lower temperature above the breathing line than does downward ventilation, for the same rate of carbon monoxide production, air change and the same temperature at the 30-in. level.
2. A lower rate of air change and a smaller heating load are required with upward than with downward ventilation.
3. In the average case, upward ventilation results in a lower concentration of carbon monoxide in the occupied portion of a garage than is had with complete mixing of the exhaust gases and the air supplied. However, the variations in concentration from point to point, together with the possible failure of the advantages of upward ventilation to accrue, suggest the basing of garage ventilation on complete mixing and an air change sufficient to dilute the exhaust gases to the allowable concentration of carbon monoxide.
4. The rate of carbon monoxide production by an idling car is shown to vary from 25 to 50 cu. ft. per hour, with an average rate of 35 cu. ft. per hour.

5. An air change of 350,000 cu. ft. per hour per idling car is required to keep the carbon monoxide concentration down to one part in 10,000 parts of air.

VENTILATION OF KITCHENS

In estimating the requirements for the ventilation of kitchens there are two methods to be considered. These are:

1. It is customary to allow a complete change of air every 2 minutes.
2. In many cases it is desirable to have all the extracted air leave via hoods or canopies located over ranges, steam tables, urns, dish washers, etc.

The first method applies only to average conditions, and modification from this average should be made depending upon the kitchen size and the heat and vapor-producing equipment.

In the second method, the air volume should be calculated from the hood entrance velocity rather than the air-change method.

Light cooking requires an entrance velocity of only 50 ft. per minute, while severe conditions may run to 150 ft. per minute or higher.

The size of a hood will depend upon its dimensions. For example, a hood 3 ft. by 8 ft. would have an area of 24 sq. ft. using an average velocity of 100 feet.

Where quiet operation is essential, the blower should be selected on the basis of a low outlet velocity. This will also result in lower operating costs.

If space is limited and noise is not a factor, smaller units with higher outlet velocities may be necessary. This may result in a lower initial cost.

GENERAL VENTILATION RULES

The American Society of Heating and Ventilating Engineers offers the following recommendations for designing and installing an adequate natural ventilation system:

1. Inlet openings in the building should be well distributed and should be located on the windward side near the bottom, while outlet openings are located on the leeward side near the top. Outside air will then be supplied to the zone to be ventilated.
2. Inlet openings should not be obstructed by buildings, trees, sign boards, etc., outside, nor by partitions inside.
3. Greatest flow per square foot of total openings is obtained by using inlet and outlet openings of nearly equal areas.
4. In the design of window-ventilated buildings, where the direction of the wind is constant and dependable, the orientation of the building together with amount and grouping of ventilation openings can be readily arranged to take full advantage of the force of the wind. Where the wind's direction is variable, the openings should be arranged in sidewalls and monitors so that, as far as possible, there will be approximately equal areas on all sides. Thus, no matter what the wind's direction, there will always be some openings directly exposed to the pressure force and others to a suction force, and effective movement through the building will be assured.
5. Direct short cuts between openings on two sides at a high level may clear the air at that level without producing any appreciable ventilation at the level of occupancy.
6. In order that temperature difference may produce a motive force, there must be vertical distance between openings. That is, if there are a number of openings available in a building, but all are at the same level, there will be no motive head produced by temperature difference, no matter how great that difference might be.
7. In order that the forces of temperature difference may operate to maximum advantage, the vertical distance between inlet and outlet openings should be as great as possible. Openings in the vicinity of the neutral zone are less effective for ventilation.
8. In the use of monitors, windows on the windward side should usually be kept closed, because if they are open, the inflow tendency of the wind counteracts the outflow tendency of temperature difference. Openings on the leeward

side of the monitor result in cooperation of wind and temperature difference.

9. In an industrial building where furnaces that give off heat and fumes are to be installed, it is better to locate them in the end of the building exposed to the prevailing wind. The strong suction effect of the wind at the roof near the windward end will then cooperate with temperature difference, to provide for the most active and satisfactory removal of the heat and gas-laden air.

10. In case it is impossible to locate furnaces in the windward end, that part of the building in which they are to be located should be built higher than the rest so that the wind, in splashing, will create a suction. The additional height also increases the effect of temperature difference to cooperate with the wind.

11. The intensity of suction or the vacuum produced by the jump of the wind is greatest just back of the building face. The area of suction does not vary with the wind velocity, but the flow due to suction is directly proportional to wind velocity.

12. Openings much larger than the calculated areas are sometimes desirable, especially when changes in occupancy are possible, or to provide for extremely hot days. In the former case, free openings should be located at the level of occupancy for psychological reasons.

13. In single-story industrial buildings, particularly those covering large areas, natural ventilation must be accomplished by taking air in and out of the roof openings. Openings in the pressure zones can be used for inflow and openings in the suction zone, or openings in zones of less pressure, can be used for outflow. The ventilation is accomplished by the manipulation of openings to get airflow through the zones to be ventilated.

CHAPTER 7

Fan Selection and Operation

Both ventilation and air circulation utilize fans to move the air. Ventilation is concerned with the moving of a *volume* of air from one space to another. It does not involve the weight of air but the volume of air in cubic feet moved per minute (CFM). The circulation of air, on the other hand, is concerned with the velocity at which air moves around a confined space and is expressed in feet per minute (FPM).

This chapter is concerned primarily with introducing the reader to the problems of fan selection and operation. Several sections of this chapter provide detailed instructions for fan sizing. Because the selection of a fan and the design of a duct system are mutually dependent, Chapter 7 of Volume 2 (Ducts and Duct Systems) should also be consulted.

CODES AND STANDARDS

Always consult local codes and standards before designing or attempting to install a fan system. Other sources of information on codes and standards pertaining to fans and fan systems are:

1. The Air Moving and Conditioning Association (AMCA).
2. The National Association of Fan Manufacturers (NAFM).
3. The American Society of Heating, Refrigerating and Air-Conditioning Engineers (ASHRAE).

DEFINITIONS

A number of terms and definitions largely related to fan selection and operation should be examined and learned for a clearer understanding of the materials in this chapter. Most of the terms and definitions contained in this section are provided by the Air Moving and Conditioning Association.

Air Horsepower (AHP)—The work done in moving a given volume (or weight) of air at a given speed. Air horsepower is also referred to as the Morse power output of a fan.

Area (A)—The square feet of any plane surface or cross section.

Area of Duct—The product of the height and width of the duct multiplied by the air velocity equals the cubic feet of air per minute flowing through the duct.

Brake Horsepower (BHP)—The work done by an electric motor in driving the fan, and measured as horsepower delivered to the fan shaft. In belt-drive units, the total workload is equal to the workload of the electric motor plus the drive losses from belts and pulleys. The brake horsepower is always a higher number than air horsepower (AHP). Brake horsepower is also referred to as the *horsepower input* of the fan.

Cubic Feet Per Minute (CFM)—The physical volume of air moved by a fan per minute expressed as fan outlet conditions.

276

Density—The actual weight of air in pounds per cubic foot (.075 at 70°F and 29.92 in. barometric pressure).

Fan Inlet Area—The inside area of the inlet collar.

Fan Outlet Area—The inside area of the fan outlet.

Mechanical Efficiency (ME)—A decimal number or a percentage representing the ratio between air horsepower (AHP) divided by brake horsepower (BHP) of a fan. It will always be less than 1.000 or 100 percent, and may be expressed as:

$$ME = \frac{AHP}{BHP}$$

Outlet Velocity (OV)—The outlet velocity at fan measured in feet per minute.

Revolutions Per Minute (RPM)—The speed at which a fan or motor turns. It is expressed in revolutions per minute.

Standard Air—Air at 70°F and 29.92 in. barometric pressure weighing .075 lb. per cubic foot.

Static Efficiency (SE)—The static efficiency of a fan is the mechanical efficiency multiplied by the ratio of static pressure to the total pressure.

Static Pressure (SP)—The static pressure of a fan is the total pressure diminished by the fan velocity pressure. It is measured in inches of water (*see* Velocity Pressure).

Tip Speed (TP)—Also referred to as the peripheral velocity of wheel. It is determined by multiplying the circumference of the wheel by the RPM.

$$TP = \frac{\times \text{ wheel diameter in feet } \times RPM}{12}$$

$$\times \quad \text{wheel diameter in feet } \times RPM$$

The tip speed should not exceed 3300 RPM if a quiet operation is to be obtained.

Total Pressure (TP)—Any fan produces a total pressure (TP), which is the sum of the static pressure (SP) and the velocity pressure (VP). Total pressure represents the rise of pressure from fan inlet to fan outlet.

Velocity—The speed in feet per minute (FPM) at which air is moving at any location (e.g., through a duct, inlet damper, outlet damper, or fan discharge point). When the performance data for air handling equipment are given in feet per minute (FPM), conversion to cubic feet per minute can be made by multiplying the FPM by the duct area. Example:

Air velocity	= 1000 FPM
Duct size	= 8 in. × 20 in. = 160 sq. in.
Duct area	= 160 ÷ 144 = 1.11 sq. ft.
Air flow	= 1000 FPM × 1.11 sq. ft. = 1110 CFM

Velocity Pressure (VP)—Velocity pressure results only when air is in motion, and is measured in inches of water: 1-in. water gauge corresponds to 4005 FPM (standard air) velocity. The formula for determining velocity pressure is:

$$VP = \left[\frac{\text{Air Velocity}}{4005}\right]^2$$

TYPES OF FANS

The various mechanical devices used to move the air in heating, ventilating, and air conditioning installations are known as fans, blowers, exhausts, or propellers.

Every fan is equipped with an impeller, which forces (impels) the airflow. The manner in which air flows through the impeller provides the basis for the following two general classifications of fans:

1. Centrifugal fans.
2. Axial-flow fans.

In a *centrifugal* (or *radial flow) fan*(Fig. 7-1), the air flows radially (i.e., diverging from the center) through the impeller,

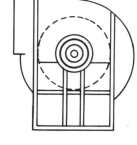

Fig. 7-1. Centrifugal fan principles.

which is mounted in a scroll-type housing. Centrifugal fans are further subdivided into a number of different types depending upon several design variations, such as the forward or backward inclination of the blade.

An *axial-flow fan* is mounted within a cylinder or ring, and the air flows axially (i.e., parallel to the main axis) through the impeller. Depending upon the design of the enclosure and impeller axial-flow fans can be subdivided into the following types:

1. Tubeaxial fans.
2. Vaneaxial fans.
3. Propeller fans.

A *tubeaxial fan* consists of an axial-flow wheel within a cylinder (Fig. 7-2). These fans are available in a number of different types depending upon the design and construction of the impeller blades.

A *vaneaxial fan* also consists of an axial-flow wheel, but differs

Fig. 7-2. Tubeaxial fan principles.

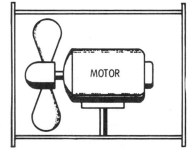

MOTOR

from a tubeaxial fan by using a set of vanes to guide the airflow and increase efficiency (Fig. 7-3).

A *propeller fan* consists of a propeller or disc wheel within a ring casing or plate. These fans are by far the simplest in construction, and operate best against low resistance (Fig. 7-4).

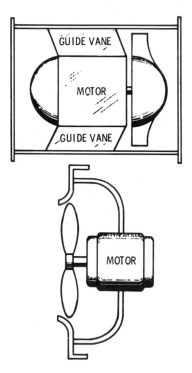

Fig. 7-3. Vaneaxial fan principles.

Fig. 7-4. Propeller fan.

FURNACE BLOWERS

The blower used in a forced warm-air furnace is similar to the centrifugal fan used in ducts and other types of applications. Most blowers are designed with a belt drive, although some are equipped with a direct drive to the motor. Furnace blowers are described in considerable detail in the several chapters dealing with specific types of furnaces. See, for example, Chapter 11 of Volume 1 (Gas-Fired Furnaces).

BASIC FAN LAWS

The performance of fans and their relationship to the ventilation system are governed by definite principles of fluid dynamics. An understanding of these principles is useful to anyone designing a ventilation system because they make possible the prediction of effects resulting from altered operating conditions. The principles (and formulas) associated with fan and ventilation system engineering are referred to collectively as *basic fan laws.*

The basic fan laws used in calculating fan performance depend upon the fact that the mechanical efficiency (ME) of a fan remains constant throughout its useful range of operating speeds (i.e., the fan RPM). They also apply only to fans that are geometrically similar.

A current edition of the *ASHRAE Guide* will contain detailed explanations of the principles and formulas associated with basic fan laws. A typical example is the production of fan speed (RPM), static pressure (SP), and horsepower when the volume of air moved by the fan is varied. The following three principles and formulas are involved:

1. Fan speed delivery will vary directly as the CFM ratio:

$$\text{New RPM} = \text{Old RPM} \times \left[\frac{\text{New CFM}}{\text{Old CFM}}\right]$$

2. Fan (and system) pressures will vary directly as the square of the RPM ratio:

$$\text{New SP (or TP or VP)} = \left[\frac{\text{New RPM}}{\text{Old RPM}}\right]^2 \times \text{Old SP (or TP or VP)}$$

3. Brake horsepower (BHP) on the fan motor (or air horsepower of the fan) will vary directly as the *cube* of the RPM ratio:

$$\text{New BHP (or AHP)} = \left[\frac{\text{New RPM}}{\text{Old RPM}}\right]^3 \times \text{Old BHP (or AHP)}$$

Example: A centrifugal fan delivers 10,000 CFM at a static pressure of 1.0 in. when operating at a speed of 600 RPM and

281

requires an input of 3 hp. If a 12,000 CFM is desired in the same installation, what will be the new fan speed (RPM), static pressure (SP), and horsepower (BHP) input? The three formulas listed above can be applied as follows:

1. New RPM $= 600 \times \left[\dfrac{12,000}{10,000}\right]$

$$600 \times 1.2 = 720$$

2. New SP $= \left[\dfrac{720}{600}\right]^2 \times 1$

$$1.44 \times 1 = 1.44$$

3. New BHP $= \left[\dfrac{720}{600}\right]^3 \times 3$

$$1.7 \times 3 = 5.1$$

The following three formulas also may prove useful in making fan calculations:

1. A (area) × V (velocity) = CFM
2. CFM ÷ V = A
3. CFM ÷ A = V

SERIES AND PARALLEL FAN OPERATION

Two separate and independent fans can be operated either in series or in parallel (Fig. 7-5). When two fans are operated in series, the CFM is *not* doubled. Instead, the total airflow is limited to the CFM capacity of one fan alone. Series operation is seldom desirable except when it is necessary to maintain the following conditions:

1. Constant pressure.
2. Zero pressure.
3. Constant vacuum.

When fans are operated in parallel, they produce a total airflow equal to the sum of their individual CFM capacities. Parallel fan operation is necessary when a single fan is incapable of moving the total volume of air required or when airflow distribution is a factor.

FAN PERFORMANCE CURVES

Fan performance curves are provided by fan manufacturers to graphically illustrate the relationship of total pressure, static pressure, power input, mechanical efficiency, and static efficiency to actual volume, for the desired range of volumes at constant speed and air density. A typical performance curve for a forward-curved blade centrifugal fan is shown in Fig. 7-6.

GENERAL VENTILATION

General ventilation involves the moving of a volume of air from one space to an entirely separate space, where concern for a concentrated source of heat or contamination is not a factor. In

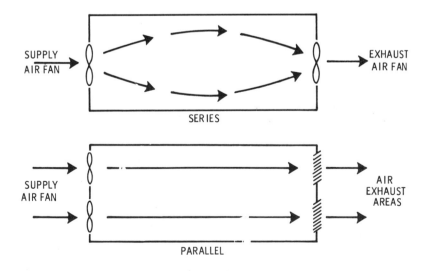

Fig. 7-5. Series and parallel fan operations.

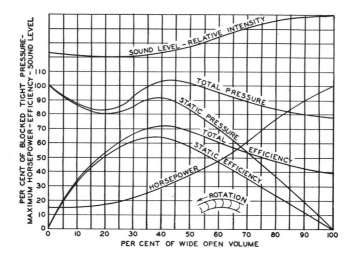

Fig. 7-6. Typical performance curves for a forward-curved blade centrifugal fan.

this respect, it differs from *local ventilation,* which is used primarily to control atmospheric contamination or excessive heat at its source (see below).

In general ventilation, the specific volume of air to be moved is measured in cubic feet per minute (CFM). The two principal methods of determining the required CFM are:

1. Air-change method.
2. Heat removal method.

Determining CFM by the Air-Change Method

In order to determine the required CFM for a structure of space by the air-change method, the following data are necessary:

1. The total cubic feet of air space in the structure or space.
2. The required number of air changes necessary to give satisfactory ventilation.

The total cubic feet of air space is easily determined by multiplying the dimensions of the structure or space. For example, a room 12 ft. long and 10 ft. wide with an 8-ft. ceiling would have 960 cu. ft. of air space (12 ft. × 10 ft. × 8 ft. = 960 cu. ft.).

The required number of air changes necessary to give satisfactory ventilation will depend on a variety of factors, including: (1) use, (2) number of people, (3) geographic location, and (4) height of ceiling.

Usually local health department codes will specify the required number of air changes for various installations. When there are no code requirements, the data given in Table 7-1 are recommended.

Once the necessary data have been obtained, the following formulas can be used to determine the CFM:

$$CFM = \frac{\text{Building Volume in Cubic Feet}}{\text{Minutes Air Change}}$$

Let's use the space shown in Fig. 7-7 to illustrate how the air-change method is used to determine CFM. First, let's assume that the space is being used as a bakery. In Table 7-1 you will note that a 2- to 3-minute air-change range is recommended for a bakery. The fact that a range is given is important because the number selected for the air-change method will depend on several variables. For example, a higher number is used when the structure or space is located in a warm climate, or when the ceiling is a particularly low one, or when there is a large number of people using a relatively small space. Comfort cooling, on the other hand, may be obtained by using the lowest figure in each stated range.

For the sake of our example, let's assume that the bakery has *not* been designed for comfort cooling. Furthermore, the ceilings are higher than average (15 feet), and the structure is located in a warm climate. With this information, the CFM can be determined in the following manner:

1. 100 ft. × 30 ft. × 15 ft. = 45,000 cu. ft. of air space (Fig. 7-7).
2. 3-minute required air change (Table 7-1).
3. CFM = $\frac{45,000}{3}$ = 15,000

Table 7-1. Average Air Changes Required Per Minute For Good Ventilation

	Minutes Per Change
Assembly Halls	2-10
Auditoriums	2-10
Bakeries	2-3
Banks	3-10
Barns	10-20
Bars	2-5
Beauty Parlors	2-5
Boiler Rooms	1-5
Bowling Alleys	2-10
Churches	5-10
Clubs	2-10
Dairies	2-5
Dance Halls	2-10
Dining Rooms	3-10
Dry Cleaners	1-5
Engine Rooms	1-3
Factories	2-5
Forge Shops	2-5
Foundries	1-5
Garages	2-10
Generator Rooms	2-5
Gymnasiums	2-10
Kitchens-Hospitals	2-5
Kitchens-Resident	2-5
Kitchens-Restaurant	1-3
Laboratories	1-5
Laundries	1-3
Markets	2-10
Offices	2-10
Packing Houses	2-5
Plating Rooms	1-5
Pool Rooms	2-5
Projection Rooms	1-3
Recreation Rooms	2-10
Residences	2-5
Sales Rooms	2-10
Theaters	2-8
Toilets	2-5
Transformer Rooms	1-5
Warehouses	2-10

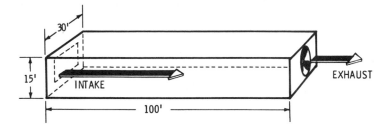

Fig. 7-7. Bakery building dimensions.

Thus, 15,000CFM are required to change the air in the bakery every 3 minutes. Assuming a 300 FPM intake velocity, 50 sq. ft. of free air intake are needed.

Determining CFM by the Heat Removal Method

The heat removal method is useful for determining CFM in installations where the ventilation of sensible heat is required.

In order to determine CFM by this method, you need to know the total Btu per minute, the average outdoor temperature, and the desired inside temperature. This information is then used in the following formula:

$$CFM = \frac{\text{Total Btu per Minute}}{.0175 \times \text{Temp. Rise} \degree F}$$

Note that the CFM determined by the heat removal method deals primarily with sensible heat, not with radiant heat. The CFM obtained from the formula above indicates the amount of air that needs to be passed through a structure or space in order to maintain the desired inside temperature.

DETERMINING AIR INTAKE

Adequate air intake area should be provided where fans are used to exhaust the air. The same holds true for fans used to *supply* air to a room (i.e., adequate air exhaust area should be provided). The size of the air intake (or air exhaust) area depends upon the velocity (FPM) of the entering or existing air and the total CFM required by the structure or space. This may be expressed by the following formula:

$$A = \frac{CFM}{FPM}$$

Where:

A = square feet of free intake (or exhaust) area.
CFM = cubic feet per minute.
FPM = feet per minute.

The bakery described above requires 15,000 CFM. Assuming a 300 FPM intake velocity, 50 sq. ft. of free air intake area are required.

$$Area = \frac{15,000\ CFM}{300\ FPM} = 50$$

Doors and windows are suitable air intake areas if they are located close enough to the floor and provide a full sweep through the area to be ventilated. When you have determined the total free air intake area by the formula illustrated above, *deduct* the area for doors and windows that function as passageways for air. Fixed or adjustable louvers can be installed over the other intake areas.

SCREEN EFFICIENCY

It is frequently necessary to cover an air intake (or exhaust) area with a bird or insect screen. These screens reduce the free intake area, but the amount of the reduction will depend upon the type of screen used. In Fig. 7-8, the net (effective) free area for each of three screens is shown as a percentage. The small holes required by an insect screen reduce the net free area to approximately 50 percent. The ½-inch mesh screen, on the other hand, provides a net free area of 90 percent.

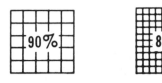

Fig. 7-8. Net free air of various screens.

The reduction of the free intake area by screens can be compensated for by using a larger overall area.

STATIC PRESSURE

The *static pressure* of a fan may be defined as the total pressure diminished by the fan velocity pressure.

Calculating the total external static pressure of a system is important to the selection of a fan or blower because it must be capable of handling the required volume of air (in terms of CFM) against this pressure.

The total external static pressure is determined by adding the static pressures of any of the air handling components in a system capable of offering resistance to the flow of air. A 10 percent allowance of the sum of these static pressures is added to obtain the total external static pressure.

The static pressures (i.e., friction losses in inches of water) used in determining the *total* external static pressure of an air handling system will include:

1. Entrance loss.
2. Friction loss through filters.
3. Friction loss through tempering coils.
4. Friction loss through air washer.
5. Duct system resistance.
6. Supply grille resistance.

Most friction losses can be obtained from data tables provided by manufacturers; however, duct loss is based on the longest run of duct, and this will vary from one installation to another.

In determining the length of duct, start at the point where the air enters the system and include all ducts in the main supply duct to the end of the system. An example of this type of calculation is show in Table 7-2. The sum (406 ft.) is then multiplied by the resistance for the ducts. For example, if the resistance is found to be 0.1 in. per 100 ft., the static pressure for the *total* run of duct will be 0.406 in. (i.e., 406 ft ÷ 100 ft. = 4.06 ft. × 0.1 in. = 0.406 in.).

Table 7-2. Total Length of Duct for an Installation

	Equivalent length of straight duct
1 goose neck at roof (double elbow 18″ × 63″)	39 ft.
28 feet straight duct to basement	28
1 elbow (18″ × 63″) to transition piece	20
40 feet straight duct	40
1 elbow (63″ × 18″)	68
60 feet straight duct	60
1 elbow (63″ × 18″)	68
15 feet straight duct	15
1 elbow (63″ × 18″)	68
	406

LOCAL VENTILATION

Local ventilation is used to control atmospheric contamination or excessive heat at its source with a minimum of airflow and power consumption. It is *not* used to move a volume of air from one space to another for human comfort.

Air velocity is an important factor in local ventilation. Air must move fast enough past the contaminant source to capture fumes, grease, dust, paint spray, and other materials and carry them into an exhaust hood.

Both the capture velocity of the air at the contaminant source and the velocity at the discharge duct must be considered when designing a localized ventilation system. It is important to remember that capture velocities will differ depending upon the contaminant. Table 7-3 lists the capture velocities for contami-

Table 7-3. Capture Velocities for Various Types of Booths

Process	Type of Hood	Capture Velocity (FPM)
Aluminum Furnace	Enclosed Hood, Open One Side	150–200 FPM
	Canopy or Island Hood	200–250 FPM
Brass Furnace	Enclosed Hood, Open One Side	200–250 FPM
	Canopy Hood	250–300 FPM
Chemical Lab.	Enclosed Hood, Front Opening	100–150 FPM
Degreasing	Canopy Hood	150 FPM
	Slotted Sides, 2"–4" Slots	1500–2000 FPM
Electric Welding	Open Front Booth	100–150 FPM
	Portable Hood, Open Face	200–250 FPM
Foundry Shake-Out	Open Front Booth	150–200 FPM
Kitchen Ranges	Canopy Hoods	125–150 FPM
Paint Spraying	Open Front Booth	100–175 FPM
Paper Drying Machine	Canopy Hood	250–300 FPM
Pickling Tanks	Canopy Hood	200–250 FPM
Plating Tanks	Canopy Hood	225–250 FPM
	Slotted Sides	250 CFM per Ft. of Tank Surface
Steam Tanks	Canopy Hood	125–175 FPM
Soldering Booth	Enclosed Booth, Open One Side	150–200 FPM

Courtesy Hayes-Albion Corporation

nants found in a variety of booths (i.e., enclosures designed to isolate areas used for special purposes).

The selection of a suitable fan for a local ventilation system requires knowledge of the required fan CFM capacity and the static pressure (SP) at which the fan must work. Once these facts are known, you have all the necessary information required for sizing the fan from the manufacturer's rating tables.

The required fan CFM capacity is determined by multiplying the open face area of any booth by the capture (face) velocity (FPM) of the air at the source of contamination.

$$CFM = \text{Face Area (sq. ft.)} \times \text{Face Velocity (FPM)}$$

The total open face area of any booth is determined by its physical size and the required access to the work area. If a booth is designed with several open face areas, all of them must be calculated and added together.

Capture (face) velocities (FPM) can be determined for various types of booths from data available from manufacturers and other sources. An example of such data is illustrated in Table 7-3.

The precise calculation of static pressure (SP) is not necessary for the sizing of a fan (or fans) in a local ventilation installation. An approximate calculation of static pressure will usually suffice, and the following steps are suggested for making such a calculation:

1. Assume the losses in the exhaust hood itself to be .05 in. to .10 in. SP.
2. Size the cross-sectional area of the main duct to ensure 1400–2000 FPM velocity.

$$\text{Duct Area (sq. ft.)} = \frac{\text{CFM}}{\text{FPM}}$$

3. Keep all ductwork as short and straight as possible and avoid elbows and sharp turns (Fig. 7-9).

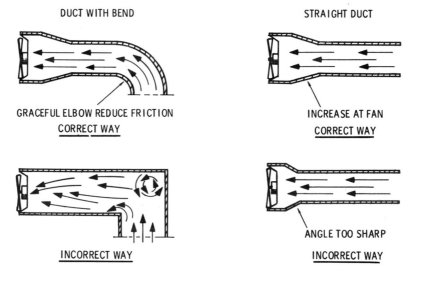

Fig. 7-9. Various designs to avoid when designing a duct system.

4. Determine the total straight duct length.
5. Add the equivalent straight duct length for each turn (Table 7-4), and add it to Step 4.
6. Multiply the sum of Steps 4 and 5 by .0025 in. to determine an approximate static pressure for the ductwork.

Table 7-4. Equivalent Lengths of Straight Pipe

Dia. of Pipe	90° Elbow * Centerline Radius		
	1.5D	2.0D	2.5D
3"	5	3	3
4"	6	4	4
5"	9	6	5
6"	12	7	6
7"	13	9	7
8"	15	10	8
10"	20	14	11
12"	25	17	14
14"	30	21	17
16"	30	24	20
18"	41	28	23
20"	46	32	26
24"	57	40	32
30"	74	51	41
36"	93	64	52
40"	105	72	59
48"	130	89	73

*For 60° elbows—x.67
*For 45° elbows—x.5

Courtesy Hayes-Albion Corporation

7. The static pressure for filters usually can be obtained from the filter manufacturer. If this is not possible, assume .25 in. SP for clean filters and .50 in. for dirty filters.
8. Add the static pressures from Steps 1, 6, and 7 to obtain the approximate total static pressure for the installation.
9. Use the total static pressure for the installation and the required CFM capacity for sizing the fan.

Some words of caution. Always check local building and safety codes for regulations pertaining to hazardous conditions. For extremely critical installations, such as those dealing with acids, poisons, or toxic fumes, consult the fan manufacturer for engineering analysis. It is *strongly* recommended that you do not attempt to make these calculations yourself. Finally, *never* under-size a fan. If you have any doubts about the correct size or horse-power, then select the next size larger.

EXHAUST-HOOD
DESIGN RECOMMENDATIONS

A properly designed exhaust hood (Fig. 7-10) is an important part of a local ventilation system. The following recommendations are offered as design guidelines:

1. Use the shortest duct run possible.
2. Avoid the use of elbows and transitions.
3. Size the exhaust hood to provide a minimum 100 FPM face velocity (150 FPM for island-type work).

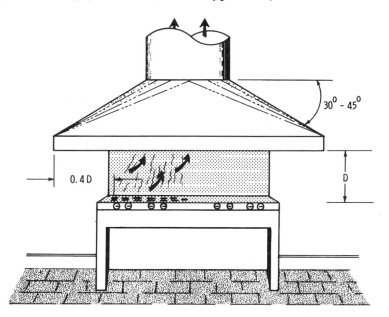

Fig. 7-10. Recommended hood design.

4. Provide sufficient hood overhang on all sides to overlap work area.
5. Use more than one exhaust fan on very large hoods.
6. Use as many individual hood and duct systems as possible (i.e., try to avoid grouping hoods together on the same duct system).
7. Use filters where required. Velocities over filters should be sized in accordance with filter manufacturer's recommendations.
8. Provide make-up air units.

FAN MOTORS

Most fans used used in heating, ventilating, and air conditioning installations are powered by electric motors. Because of the small size of many of these fans, the majority are equipped with direct-connected motors. A V-belt drive arrangement is used with larger fans, particularly centrifugal fans used in forced warm-air furnaces or the larger ventilating units found in commercial and industrial installations.

It is the general rule to select a fan one size larger than the fan requirements for the installation. This is particularly so in installations that may require the movement of large volumes of air for short intervals.

The advantage of fans equipped with a belt-drive arrangement is that adjustments can be made for different speeds. It is simply a matter of changing the pulley size. A belt drive is especially desirable when the required horsepower requirement is in doubt.

A fan wheel directly connected to the motor shaft is the best arrangement, but this is only feasible with the smaller centrifugal fans and propeller fans under 60 in. in diameter.

A direct-connected fan is generally driven by a single-phase AC motor of the split-phase, capacitor, or shaded-pole type. The capacitor motor is recommended when there are current limitations. Its major advantage is its greater efficiency electrically. The major disadvantage of such motors is that they are usually designed to operate at only one speed. A damper arrangement can be used to throttle the air when it becomes necessary to vary the air volume or pressure of the fan.

The variation of pressure and air volume in larger fan installations (e.g., mechanical draft fans) can be accomplished by means of a constant-speed, direct-connected motor equipped with movable guide vanes in the fan inlet.

The National Electrical Manufacturer's Association (NEMA) has recently revised motor voltage designations to conform to system voltages now present throughout the country. Single-phase motor voltages should now be specified as 115 volts or 230 volts instead of the 110-volt or 220-volt designations formerly in effect. Polyphase voltages should be expressed as 208 volts or 230/460 volts instead of the 220/440 voltages formerly in effect. Motors for special voltages such as 177, 480, or 575 volts are available from many fan manufacturers on special order. Fan motors have been designed to operate satisfactorily over the range of plus or minus 10 percent of the nameplate voltage ratings. *Always* check the fan motor nameplate voltage ratings before installing the motor.

Tables 7-5 and 7-6 list nominal full-load ampere ratings for single phase and three-phase motor voltages. The amperes given are approximate values only and represent averages compiled from tables of leading motor manufacturers. Compare these with the specific amperages listed for Airmaster fans in Table 7-7.

Overload relay heaters should not be selected solely on the basis of the data listed in Tables 7-5 and 7-6. Heaters must be selected in accordance with the actual motor current as shown on the nameplate. It is also important that ambient temperatures of the area in which the motor control is located be taken into consideration when making heater selections. Ambient compensated overload relays are available for abnormal temperature conditions.

Typical connection diagrams for two-speed, three-phase motors are illustrated in Figs. 7-11 and 7-12. Two-speed motors are available for most fan lines. Because *all* two-speed motors are always single-voltage, it is necessary to specify the available line voltage phase and frequency when ordering.

Table 7-5. Nominal Full-Load Ampere Ratings
for Single-Phase Motors

HP	RPM	FULL-LOAD CURRENT	
		115V	**230V**
$^1/_{25}$	1550	1.0	.5
$^1/_{25}$	1050	1.0	.5
$^1/_{12}$	1725	2.0	1.0
	1140	2.4	1.2
	860	3.2	1.6
$^1/_{10}$	1550	2.4	1.2
$^1/_8$	1725	2.8	1.4
	1140	3.4	1.7
	860	4.0	2.0
$^1/_6$	1725	3.2	1.6
	1140	3.84	1.92
	860	4.5	2.25
$^1/_4$	1725	4.6	2.3
	1140	6.15	3.07
	860	7.5	3.75
$^1/_3$	1725	5.2	2.6
	1140	6.25	3.13
	860	7.35	3.67
$^1/_2$	1725	7.4	3.7
	1140	9.15	4.57
	860	12.8	6.4
$^3/_4$	1725	10.2	5.1
	1140	12.5	6.25
	860	15.1	7.55
1	1725	13.0	6.5
	1140	15.1	7.55
	860	15.9	7.95

Courtesy Penn Ventilator Company, Inc.

Table 7-6. Nominal Full-Load Ampere Ratings
for Three-Phase Motors

| HP | RPM | FULL-LOAD CURRENT | |
		230V	460V
	1725	.95	.48
¼	1140	1.4	.7
	860	1.6	.8
	1725	1.19	.6
⅓	1140	1.59	.8
	860	1.8	.9
	1725	1.72	.86
½	1140	2.15	1.08
	860	2.38	1.19
	1725	2.46	1.23
¾	1140	2.92	1.46
	860	3.26	1.63
	1725	3.19	1.6
1	1140	3.7	1.85
	860	4.12	2.06
	1725	4.61	2.31
1½	1140	5.18	2.59
	860	5.75	2.88
	1725	5.98	2.99
2	1140	6.50	3.25
	860	7.28	3.64
	1725	8.70	4.35
3	1140	9.25	4.62
	860	10.3	5.15
	1725	14.0	7.0
5	1140	14.6	7.3
	860	16.2	8.1
	1725	20.3	10.2
7½	1140	20.9	10.5
	860	23.0	11.5

Courtesy Penn Ventilator Company, Inc.

Examples of fans equipped with the belt-drive arrangement and direct-connected motors are shown in Figs. 7-13 and 7-14.

TROUBLESHOOTING FAN MOTORS

Fan motors mounted in the airstream are cooled by a portion of the air drawn around them. This acts to hold motor temperatures down and makes it possible for the motor to run continuously at substantial brake horsepower (BHP) overloads without exceeding its rated temperature rise.

Table 7-7. Ampere Rating Table for Airmaster Fans

HP	RPM Syn. Speed	3-Phase, 60 Cycle A-C 220 Volts	3-Phase, 60 Cycle A-C 440 Volts	3-Phase, 60 Cycle A-C 550 Volts	Single-Phase A-C 110 Volts	Single-Phase A-C 220 Volts
$1/30$	1050	—	—	—	1.90	—
$1/25$	1550	—	—	—	2.00	—
$1/20$	1550	—	—	—	2.30	—
	1140	—	—	—	2.40	—
	1050	—	—	—	3.10	—
$1/15$	1550	—	—	—	3.20	—
$1/12$	1550	—	—	—	3.80	—
	860	—	—	—	2.80	—
$1/8$	1800	—	—	—	2.80	1.40
	1200	—	—	—	3.40	1.70
	860	—	—	—	5.40	2.70
$1/6$	1800	—	—	—	3.00	1.50
	1200	—	—	—	3.60	1.80
	860	—	—	—	5.60	2.80
$1/4$	1800	0.96	0.48	0.38	4.12	2.06
	1200	1.16	0.58	0.46	5.50	2.75
	900	1.45	0.73	0.58	6.50	3.25
$1/3$	1800	1.16	0.58	0.47	5.00	2.50
	1200	1.43	0.72	0.58	6.00	3.00
	900	1.75	0.88	0.71	8.40	4.20
$1/2$	1800	1.68	0.84	0.67	7.16	3.58
	1200	2.07	1.04	0.83	10.00	5.02
	900	2.90	1.45	1.16	—	—
$3/4$	1800	2.33	1.17	0.93	9.86	4.94
	1200	2.85	1.43	1.14	11.90	5.96
	900	3.45	1.73	1.38	—	—
1	1800	3.05	1.53	1.22	10.60	5.28
	1200	3.54	1.77	1.42	12.30	6.12
	900	3.74	1.87	1.50	12.90	6.48
$1\frac{1}{2}$	1800	4.28	2.14	1.71	14.80	7.40
	1200	4.85	2.43	1.94	16.80	8.40
	900	5.81	2.91	2.32	20.00	10.10
2	1800	5.76	2.88	2.30	20.00	10.00
	1200	6.35	3.18	2.54	22.00	11.00
	900	7.21	3.61	2.88	25.00	12.50
3	1800	8.29	4.14	3.32	28.80	14.30
	1200	8.92	4.46	3.56	30.80	15.40
	900	10.20	5.09	4.08	35.40	17.60
5	1800	13.20	6.60	5.28	45.60	22.80
	1200	13.10	7.05	5.64	48.80	24.40
	900	15.60	7.80	6.24	54.00	27.00
$7\frac{1}{2}$	1800	19.30	9.70	7.72	67.00	33.40
	1200	20.30	10.20	8.12	70.20	35.20
	900	23.80	11.90	9.51	82.40	41.20

Courtesy Hayes-Albion Corporation

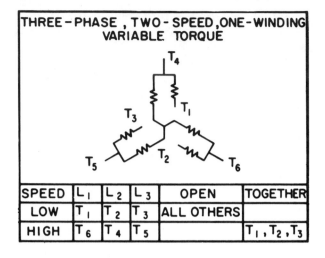

SPEED	L_1	L_2	L_3	OPEN	TOGETHER
LOW	T_1	T_2	T_3	ALL OTHERS	
HIGH	T_6	T_4	T_5		T_1, T_2, T_3

Courtesy Penn Ventilator Co., Inc.

Fig. 7-11. Connection diagram for three-phase, two-speed, one-winding variable-torque motors.

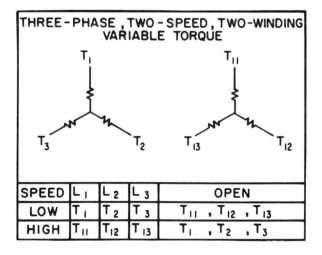

SPEED	L_1	L_2	L_3	OPEN	
LOW	T_1	T_2	T_3	T_{11}, T_{12}, T_{13}	
HIGH	T_{11}	T_{12}	T_{13}	T_1, T_2, T_3	

Courtesy Penn Ventilator Co., Inc.

Fig. 7-12. Connection diagram for three-phase, two-speed, two-winding variable-torque motors.

300

Fig. 7-13. Belt-driven propeller fan assembly.

Courtesy Hayes-Albion Corp.

The actual brake horsepower load has little relation to its nameplate horsepower as long as it remains at or below its rated maximum temperatures. It is *temperature rise* that is crucial to the breakdown or burning out of a fan motor (even though it may be physically underloaded).

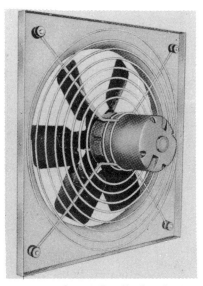

Fig. 7-14. Fan with direct-connected motor.

Courtesy Penn Ventilator Co., Inc.

301

A fan motor will normally run at a temperature too hot to hold a hand against it, but this will still be at or below the manufacturer's temperature rise limit as stated on the nameplate. When a fan motor does overheat or burn out, it is usually for one of the following reasons:

1. Defective motor.
2. Line voltage too high or too low for the rated motor voltage (more than plus or minus 10 percent voltage deviation is considered excessive).
3. The belts on belt-driven fans may be too tight or too loose, causing slippage and consequent loss of the cooling effect of air passing around the motor.
4. Improper pulleys on belt-driven fans will result in too high a fan RPM, which causes overloading.
5. Backward- and forward-curved wheel centrifugal fans or roof ventilators running backward will guarantee an overload condition. Remember that all centrifugal fans blow in one direction only, regardless of rotation; propeller fans that run backward blow backward.
6. Propeller fans may be starved for air as a result of insufficient intake (or outlet) area. The fans are literally starved for air, which causes the static pressure to rise and the brake horsepower load on the motor to increase. The air that flows around the motor is also reduced, causing the motor to overheat. Attic fans are frequently damaged due to inadequate outlet area.

FAN SELECTION

The following information is generally required for the selection of a suitable fan:

1. Volume of air required (CFM).
2. Static pressure (SP).
3. Type of application or service.
4. Maximum tolerable noise level.
5. Nature of load and available drive.
6. Ambient and airstream temperature.

7. Mounting arrangement of the system.
8. Available line voltage.

The *volume of air required* refers to the volume of air that must be moved by the fan to meet the needs of the building or space. It is expressed in cubic feet per minute and is determined by dividing the total cubic feet of air space by the required number of air changes necessary to give proper ventilation.

The *static pressure* of a fan may be defined as the total pressure diminished by the fan velocity. In other words, it is the resistance offered by the system (ducts, air intakes, etc.) to the flow of air. After the duct sizes have been determined, it is necessary to calculate the static pressure of the system so that the proper fan can be selected which will handle the desired volume of air (i.e., the required CFM) against the static pressure of the system. The various fan manufacturers provide tables indicating the operating characteristics of various-size fans against a wide range of static pressures. These tables list static pressures for different sizes of various fans.

The *type of application* (or service) is often an important consideration in what kind of fan is used in an installation. For example, a duct system will offer sufficient resistance to require a centrifugal, tubeaxial, or vaneaxial fan. A propeller fan is usually recommended for an installation without a duct system. Other factors, such as the volume of air that must be moved, the allowable noise level, the air temperature, use for general or local ventilation, and cost, are also important considerations in fan selection.

The *maximum tolerable noise level* is the highest acceptable noise level associated with air exchange equipment. The fan should be of suitable size and capacity to obtain a reasonable operating speed without overworking.

The *nature of load and available drive* is an important factor in controlling the noise level. High-speed motors are usually quieter than low-speed ones. Either belt- or direct-drive units are used in fan installations, and a high-speed motor connected to the fan with a V-belt offers the quietest operation.

Ambient or airstream temperature. The dry-bulb temperature of either ambient air or exhaust-stream air is a determining factor in selecting a suitable fan. Most fans operate satisfactorily at temperatures up to about 104°F (40°C). Special fans that can operate at higher temperatures are also available. For example, standard belt-driven tubeaxial fans are usable for temperatures up to 200°F (where the motor is out of the airstream).

The *mounting arrangement of the system* is directly determined by the application or service of the fan. Certain types of fans will prove to be more suitable than others, depending upon the kind of installations. Fan manufacturers often offer useful recommendations for mounting arrangements.

The *available line voltage* will determine the size and type of fan motor most suitable for the installation. Motor voltage designations conform to the following system of voltages now used throughout the country: 115 volts, 230 volts, and 460 volts. Motors for special voltages (i.e., 117, 480, or 575 volts) are available on special order.

Fan manufacturers provide information and assistance in selecting the most suitable fan or fans for your installation. Remember that ventilation requirements vary under different climatic conditions, and it is impossible to provide exact rules for determining the variables of local climate and topography (Table 7-8). Allowances must be made for these climatic variables.

The following suggestions are offered only as a general guide to the selection of a fan and should not be construed as applying in every situation.

1. Use a ½ HP, ⅓ HP, or ¼ HP 860 RPM direct-drive fan on three-phase motor voltages whenever possible to eliminate the possibility of single-phase magnetic hum.
2. A belt-driven fan is less expensive, less noisy, more flexible, and more adaptable to capacity change than the direct-drive type.
3. Prolonged motor life can be expected of direct-driven fans using other than shaded-pole motors. For that reason, 1550 RPM and 1050 RPM motors should be avoided when very heavy duty and/or extremely long motor life is required.
4. Use a propeller fan when operation offers little or no resistance, or when there is no duct system.

5. Use a centrifugal or axial-flow fan when a duct system is involved.
6. Never try to force air through ducts smaller than the area of the fan.

Table 7-8. Mimimum Fan Capacity (CFM) for Various Sections of the Country

Approx. Volume of House	Minimum Fan Capacity Needed For Satisfactory Results (CFM)		
(Cu. Ft.)	North	Central	South
3,000	1,000	2,000	3,000
4,000	1,320	2,640	4,000
5,000	1,650	3,300	5,000
6,000	2,000	4,000	6,000
7,000	2,310	4,620	7,000
8,000	2,540	5,280	8,000
9,000	3,000	6,000	9,000
10,000	3,330	6,660	10,000
11,000	3,630	7,260	11,000
12,000	4,000	8,000	12,000
13,000	4,290	8,580	13,000
14,000	4,620	9,240	14,000
15,000	5,000	10,000	15,000
16,000	5,280	10,560	16,000
17,000	5,610	11,220	17,000
18,000	6,000	12,000	18,000
19,000	6,270	12,540	19,000
20,000	6,660	13,320	20,000
21,000	7,000	14,000	21,000
22,000	7,260	14,520	22,000

(Fan size designations shown in table: North 24″, 30″; Central 24″, 36″, 42″; South 24″, 30″, 36″, 42″, 48″)

FAN INSTALLATION

The following recommendations are offered as guidelines for proper fan installation.

1. Install the fan and air intake openings at opposite ends of the enclosure so that the intake air passes lengthwise through the area being ventilated (Fig. 7-15).

305

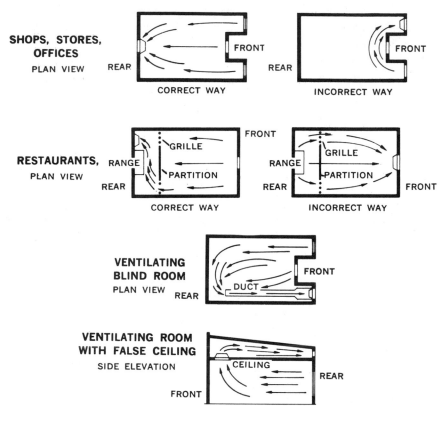

Fig. 7-15. Locations of fans and air intake openings.

2. When possible, install exhaust fans or air outlets on the lee-ward side so that the air leaves with the prevailing winds (Fig. 7-16).
3. When possible, install ventilation (supply) fans or air intakes on the windward side so that the entering air utilizes pressure produced by prevailing winds (Fig. 7-17).
4. Provide a net intake area *at least* 30 percent greater than the exhaust fan opening.

5. When filters are used, increase the net intake area to allow minimum pressure loss from resistance of the filter.
6. Steam, heat, or odors should be exhausted by fans using totally enclosed motors mounted near the ceiling. The air intakes should be located near the floor (Fig. 7-18).
7. An explosion-proof motor with a spark-proof fan should be used when the exhaust air is hazardous.
8. Spring-mount fans and connect them to the wall opening by a canvas boot when extremely quiet operation is required.

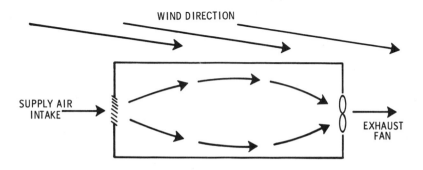

Fig. 7-16. **Exhaust fan installed on the leeward side, away from prevailing winds.**

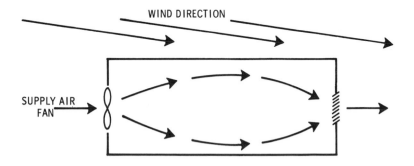

Fig. 7-17. **Supply air fan installed on the windward side, with the prevailing winds.**

FUME AND HEAT REMOVAL

AUTOMOBILE PAINT SHOP

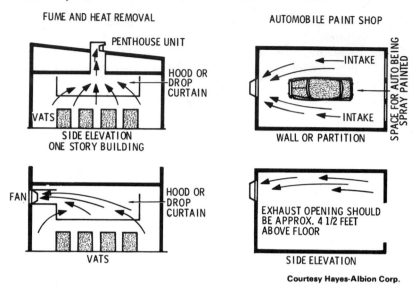

Courtesy Hayes-Albion Corp.

Fig. 7-18. Recommended locations for exhaust openings.

FAN INSTALLATION CHECKLIST

A properly installed fan motor (unless defective) will operate efficiently and will never overheat or burn out. To insure proper installation, the following guidelines are suggested:

1. Check the line current to be sure it is not more than a plus or minus 10 percent voltage deviation.
2. Check belt tension for looseness after the first or second day of operation.
3. Check to be sure the fan is running in the right direction.
4. Inspect and lubricate fans subject to heavy usage. Do this after the first 15,000 hours or five years service (whichever comes first).
5. Use open drip-proof motors where fan motors are installed outside the airstream.
6. Use enclosed, nonventilated, air-over motors where fan motors are mounted in the airstream to reduce obstruction by dirt, grease, or other contaminants.

AIR VOLUME CONTROL

Sometimes it becomes necessary to vary the air volume handled by a fan. The following methods can be used for this purpose:

1. Dampers placed in the duct system.
2. Changing the pulley on the fan motor.
3. Changing the pulley on the fan.
4. Using variable-speed pulleys.
5. Using variable-speed motors.
6. Using variable inlet vanes on fans.
7. Reduction or increase of speed through power control.

NOISE CONTROL

In order to decrease the noise associated with the air exchange equipment, the following recommendations should be observed:

1. The equipment should be located a reasonable distance from important rooms.
2. The fans should be of proper size and capacity to obtain reasonable operating speed.
3. The equipment should be mounted on resilient bases made from a sound dampening material.
4. When possible, the quieter high-speed AC motors with belt-driven fans should be used.

FAN APPLICATIONS

Many different fans are used in ventilation and air circulation systems, and these fans are classified on the basis of certain design and operating characteristics. Fan applications, on the other hand, are classified by the specific function they serve in the ventilation or air circulation system. Among the numerous fan applications used for ventilation and air circulation are the following:

1. Attic fans.
2. Exhaust fans.
3. Circulating fans.
4. Kitchen fans.
5. Cooling tower fans.
6. Unitary system fans.

Attic fans are generally propeller types used during the summer to draw the cool night air through the structure and discharge it from the attic. The air can be discharged through windows or grilles, or directly through an attic exhaust fan. The air is circulated, not conditioned.

Exhaust fans are used in local ventilation to remove contaminants at their source, or as a general means of discharging air from a space (e.g., attic exhaust fans, wall fans, bathroom fans). Hood exhaust fans used with a duct system are generally centrifugal types. Because wall fans operate against very low resistance or no resistance at all, they are most commonly propeller types.

Circulating fans and *kitchen fans* are propeller types used for air circulation purposes. *Cooling tower fans* are also generally propeller types, although centrifugal fans have also been used in installations requiring a forced draft.

Unitary system fans are centrifugal or propeller fans used in unit ventilators, unit heaters, unit humidifiers, and similar types of equipment. A propeller fan is used in these units when no ductwork exists.

ATTIC VENTILATING FANS

A noninsulated attic on a hot summer day will experience temperatures as high as 115° to 130°F. This accumulated heat penetrates down through the ceiling and raises the temperatures in the rooms below, making living and sleeping areas almost unbearable. After sunset, the outside temperature sinks to pleasant cool levels, whereas the indoor air (especially in the upstairs sleeping areas) remains sultry with temperature readings that can reach as high as 85°F. This condition results from the fact that a house emits accumulated heat at a very slow rate.

This condition can be alleviated considerably by installing an attic exhaust fan. Such a fan placed in the attic will cool the air by replacement; that is, it will ventilate rather than condition.

Because of the low static pressures involved (usually less than .125 in.), disc or propeller fans are generally recommended for attic installation. It is recommended that these fans have quiet operating characteristics and sufficient capacity to give at least thirty air changes per hour.

The two general types of attic fans in common use in houses and other buildings are:

1. Boxed-in fans.
2. Centrifugal fans.

The *boxed-in* type attic fan is installed within the attic in a box or suitable housing, located directly over a ceiling grille, or in a bulkhead enclosing an attic stair. This type of fan can be connected to individual room grilles by means of a duct system.

In operation, outside cool air entering through the windows in the downstairs rooms is discharged into the attic space and escapes to the outside through louvers, dormer windows, or screened openings under the eaves. A general arrangement showing a typical installation of this type of fan is illustrated in Fig. 7-19.

The installation of a multiblade centrifugal attic fan is shown in Fig. 7-20. At the inlet side, the fan is connected to exhaust ducts leading to grilles, which are placed in the ceiling of the two bedrooms. The air exchange is accomplished by admitting fresh air through open windows and then up through the suction side of the fan, and is finally discharged through louvers.

The fan shown in Fig. 7-21 is of the centrifugal curved-blade type, mounted on a light angle-iron frame, which supports the fan wheel, shaft, and bearings, with the motor that supplies the motive power to the fan through a belt drive.

The air inlet in this installation is placed close to a circular opening which is cut in an airtight board partition, which serves to divide the attic space into a suction and discharge chamber. The air is admitted through open windows and doors, then drawn up the attic stairway through the fan into the discharge chamber, from which it flows through the attic open window.

311

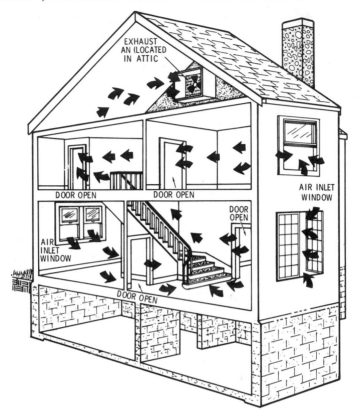

Fig. 7-19. Attic ventilation system.

For best results, the outlet area for an attic fan should be 1½ times the area of the fan. Satisfactory results will be obtained as long as the area is *at least* equal to the blade area of the fan. Recommended dimensions for attic fan exhaust outlets are given in Table 7-9.

Tables 7-10 and 7-11 provide data for square- and triangular-type louvers used with various fan diameters. These tables include the net free areas for 1-in. mesh wire screening. Remember: In order to keep insects and other foreign objects out of the attic, the exhaust air outlets should be covered with ½-in. or 1-in. wire mesh screen. The fan should be "boxed-in" on the

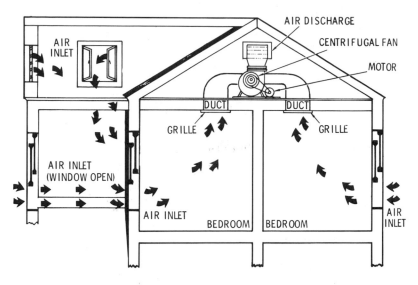

Fig. 7-20. Multiblade centrifugal attic fan connected to exhaust ducts.

intake side with conventional fly screening. This screening has only 50 percent free area; therefore, the "boxing" must have a total area twice that of the inlet opening.

The location of the attic fan always depends upon the design of the structure. In a house having a suitably sized window in an attic end wall or dormer, the best results (and fewest construction problems) can usually be obtained by mounting the fan directly against the window (Fig. 7-22) . An automatic shutter should be installed outside the window for window-mounted fans. If the window is opened and the fan is mounted inside the opening, louvers should be installed a few inches in front of the fan to keep rain out of the attic.

In one-floor houses, the ceiling openings can be installed in any convenient *central* location (Fig. 7-23). In houses of two or more stories, the opening is generally located in the ceiling of the top-floor hallway. Again, a central location is usually best. The ceiling opening and its accompanying grille or shutter must be of sufficient size to avoid excessive resistance to airflow, permitting the airstream to pass through at a moderate, quiet velocity.

313

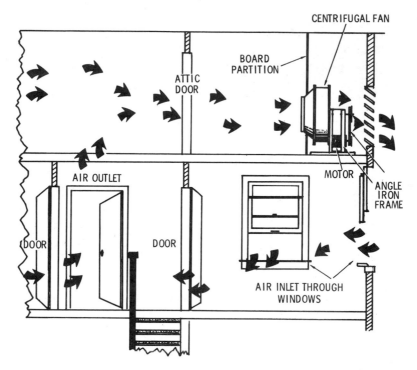

Fig. 7-21. Centrifugal curved-blade fan.

Table 7-9. Recommended Dimensions for Attic Fan Exhaust Outlets

Recommended Dimensions of Attic Fan Exhaust Outlet		
Fan Diameter	Air Delivery Range CFM	*Free Outlet Area Needed (Sq. Ft.)
24″	3500/5000	4.70
30″	4500/8500	7.35
36″	8000/12000	10.06
42″	10000/15500	14.40
48″	12000/20000	18.85

*1.5 times fan area

Courtesy Hayes-Albion Corporation

Table 7-10. Data for Square-Type Louvers

Fan Diameter	Minimum Size of Square Outlet (inches)		
	Metal Shutters		Wood Slats
	Automatic (90% Open Area)	Fixed (70% Open Area)	Fixed (60% Open Area)
24″	26 × 26	32 × 32	34 × 34
30″	32 × 32	39 × 39	42 × 42
36″	38 × 38	45 × 45	49 × 49
42″	44 × 44	54 × 54	60 × 60
48″	50 × 50	62 × 62	68 × 68

Courtesy Hayes-Albion Corporation

Table 7-11. Data for Triangular-Type Louvers

Fan Diameter	*Height of triangular louvers (for different roof pitches)			
	$^5/_{12}$ Pitch One Louver	$^6/_{12}$ Pitch One Louver	$^7/_{12}$ Pitch One Louver	$^8/_{12}$ Pitch One Louver
24″	2′ 0″	2′2″	2′4″	2′6″
30″	2′ 6″	2′9″	3′0″	3′3″
36″	3′ 0″	3′3″	3′6″	3′9″
42″	3′ 3″	3′9″	4′1″	4′4″
48″	3′10″	4′3″	4′7″	4′9″

*Heights given are for one triangular louvered opening only; when two openings are used reduce heights by approximately 80%.

Courtesy Hayes-Albion Corporation

KITCHEN EXHAUST SYSTEMS

In estimating the requirements for ventilating kitchens, it is customary to allow a complete change of air every 2 minutes. In many cases, it is also desirable to have all the extracted air leave

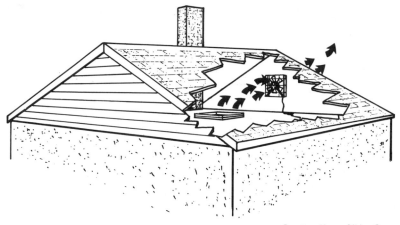

Fig. 7-22. Fan mounted at the attic window with the ceiling opening in central hallway.

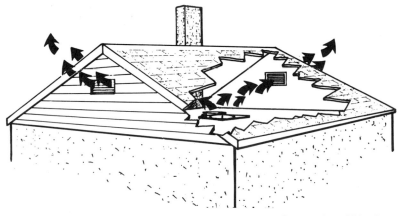

Fig. 7-23. Fan mounted in ceiling hallway.

via hoods or canopies located over ranges, steam tables, dishwashers, and similar sources of localized heat and contaminants.

Allowing for a complete change of air every 2 minutes only applies to average conditions, and modifications from this average should be made on the basis of the kitchen size and the type of heat and vapor-producing equipment.

An entrance velocity at the hood opening of 100 ft. per minute is considered satisfactory as an allowance for average conditions. For very light cooking, an entrance velocity of only 50 ft. per minute is usually sufficient. Heavy cooking may require an entrance velocity of 150 ft. per minute or higher.

Exhaust hoods are usually located overhead in the majority of kitchen exhaust systems. They should be placed directly over the heat and vapor producing equipment and approximately 80 in. from the floor line to allow sufficient head clearance.

An overhead exhaust hood should be larger in horizontal area than the source of the heat or fumes. When located not over 2 ft. above the range, the hood should be 6 in. larger in all directions that the overall dimensions of the range when the distance exceeds 2 ft. Thus, a range 2 ft. by 7 ft. with a clearance of 2 ft would require a hood 3 ft. by 8 ft. Such a hood would have an area of 24 sq. ft. Using an average entrance velocity at hood of 100 FPM, the volume of air to be handled would be 2400 CFM.

The area of the branch duct leading from the hood should be made $\frac{1}{16}$ of the hood area (i.e., 24 sq. ft. ÷ 16) or 1.5 sq. ft. With the hood located 4 ft. above the range, the dimensions would be 4 ft. by 9 ft. with a branch duct area of 2.25 sq. ft.

If a supply system is required, the amount of exhaust air should be greater than the volume of supply air to prevent undesirable cooking odors spreading to adjoining rooms. The supply air is usually figured on the basis of 75 percent of the exhaust air.

BATHROOM EXHAUST SYSTEMS

An air change every 3 minutes, or 20 complete changes per hour, is desirable for bathroom ventilation. Systems of this type should be entirely different and separated from other ventilating systems. Bathrooms located on the inside of a structure require ducts to exhaust air to the outside.

Compact unit blowers are especially recommended for use in bathroom exhaust systems. Their compact design requires a minimum of space, and they are capable of operating against the resistance of the system.

AIR CIRCULATORS

An air circulator is a mechanical device (most commonly a propeller fan) used to move the air around within a confined space. It is concerned primarily with air velocity and not with the movement of a volume of air, as is the case with ventilation.

Producing air movement for the cooling of people is the most common use of fans used as air circulators. A human body will reject excess heat by evaporation, convection, and radiation. Air circulators will assist in the removal of excess heat rejected by both evaporation and convection. They will not affect heat loss or gain resulting from radiation; therefore, they are of little or no value for reducing the effect of heat from hot furnaces, stoves, and similar heat sources.

Perspiration is an example of body heat loss by evaporation. Increasing the rate of air movement across the skin will also increase the rate of evaporation. This results in a more rapid loss of body heat, which produces a cooling sensation. Air circulators contribute to the acceleration of heat loss by evaporation. This phenomenon is referred to as *evaporative cooling*.

CHAPTER 8

Air Conditioning Principles

Air conditioning may be defined as the simultaneous control of all or at least the first three of those factors affecting both the physical and chemical conditions of the atmosphere within any structure. These factors include the following:

1. Temperature.
2. Humidity.
3. Air movement.
4. Air distribution.
5. Dust.
6. Bacteria.
7. Odors.
8. Toxic gases.

It should be evident at this point that it is incorrect to regard air conditioning as simply the *cooling* of air. This is only one aspect

319

of temperature control and does not involve the other seven factors affecting the atmosphere of the space being air conditioned. Conditioning the air of a space means to change it in whatever way necessary to improve the comfort of those living or working there. This may mean warming air to a livable temperature and holding it there; cooling the air; adding or subtracting moisture; filtering out contaminants such as dust, bacteria, and toxic gases; and maintaining a proper distribution and movement of the air. In general, air conditioning includes the following processes:

1. Heating.
2. Cooling.
3. Humidifying.
4. Cleaning and filtering.
5. Circulating.

Each of these processes contributes in some way to the improvement of atmospheric conditions that effect the comfort and health of the individual. For example, air nearly always contains certain impurities, such as ammonia, sulphurous acid, and carbon dioxide. The latter is a product of exhalation from the lungs and the combustion process in internal combustion engines. It is so universally present (about in the same proportions everywhere, except where concentrated by some local conditions) that it may be regarded as a normal constituent of the air. Air conditioning is an efficient means of eliminating carbon dioxide from the air. The same is true of other impurities. Some dry strainer filters used in air conditioning systems are capable of removing 99.98 percent of radioactive dust from the air.

The purpose of this chapter is to examine the factors affecting the physical and chemical conditions of the atmosphere, and to determine the effect they may have upon the basic requirements of air conditioning.

PROPERTIES OF AIR

Air is a gas consisting of a mechanical mixture of 23.3 percent oxygen (by weight), 75.5 percent nitrogen, and 1.32 percent argon. Carbonic acid is present to the extent of about .03 or .04 percent of the volume. Obscure constituents include krypton (.01 percent) with small amounts of other gases. The percentages by volume are 21 percent oxygen, 78.06 percent nitrogen, and .94 percent argon.

AIR CIRCULATION

The effectiveness of any air conditioning system depends as much upon the proper distribution of the air as upon the efficiency of the conditioning equipment itself. It may be said that an air conditioning installation is no better than its duct system. To be effective, the conditioned air must be uniformly distributed over the entire area of the enclosure, especially in closed or dry rooms.

SPECIFIC, LATENT, AND SENSIBLE HEAT

The *specific heat* of a substance represents the number of Btu's required to raise the temperature of one pound of that substance one degree Fahrenheit. The specific heat of water (1.00) is taken as the standard because of its large heat capacity. In comparison, air has a specific heat of 0.238. It should be noted that the more water a substance contains, the higher its specific heat will be.

Latent heat is the quantity of heat that disappears or becomes concealed in a body while producing some change in it other than a rise of temperature. Changing a liquid to a gas, or a gas to a liquid, are both activities involving latent heat. In the former process (i.e., evaporation), an absorption of heat accompanies the change of physical state.

Sensible heat is the part of heat that provides temperature change and that can be measured by a thermometer. It is referred to as such because it can be "sensed" by instruments or the touch. *Temperature* is the state of a body with respect to sensible heat. That is to say, the degree of heat or cold.

AIR PRESSURE

Atmospheric air is air at the pressure of the standard atmosphere. *Standard atmosphere* is considered to be air at a pressure of 29.921 in. of mercury, which is equal to 14.696 pounds per square inch (usually written 14.7 psi). In any air conditioning system, atmospheric air is regarded as being air at the atmospheric pressure at the point of installation.

Standard atmospheric pressure at sea level is 29.921 in. of mercury. Since most pressure gauges indicate gauge pressure or pounds per square inch, a barometer reading can be converted into gauge pressure by multiplying inches of mercury by 0.49116 psi. Thus, a barometer reading of 29.921 in. is equivalent to a gauge pressure of 14.696 psi (29.921 × 0.49116 = 14.696, or 14.7, psi).

So 0.49116 psi is a constant value for 1 in. of mercury, and is determined by dividing the pressure in pounds per square inch by the barometer readings in inches of mercury (i.e., 14.696 ÷ 29.921 = 0.49116). Table 8-1 gives the atmospheric pressure for various readings of the barometer.

The pressure of the atmosphere does not remain constant in any one place. It continually varies depending upon the conditions of the weather and the elevation. With respect to elevation, atmospheric pressure will decrease approximately ½ lb. for each 1000 ft. of ascent. Table 8-1 illustrates the effect of altitude and weather (the barometer reading) on atmospheric pressure.

Absolute pressure is pressure measured from true zero or the point of no pressure. It is important to distinguish absolute pressure from *gauge pressure*, whose scale starts at atmospheric pressure. For example, when the hand of a steam gauge is at zero, the absolute pressure existing in the boiler is approximately 14.7 psi. Thus, 5 lb. pressure measured by a steam gauge (i.e., gauge pressure) is equal to 5 lb. plus 14.7 lb., or 19.7 psi of absolute pressure.

PRINCIPLES OF COMPRESSION

The ultimate objective of compression in air conditioning is to cool the air to be conditioned. It is important to note, however, that it is not the air to be cooled that is compressed, but the refrigerant gas used in the coils of the air conditioning unit. The low temperature is produced by the expansion and contraction of the refrigerant gas.

When a gas (i.e., air or a refrigerant) is compressed, both its pressure and temperature are changed in accordance with Boyle's and Charles's laws.

The English scientist Robert Boyle (1627–1691) determined that the absolute pressure of a gas at constant temperature varies inversely as its volume. Somewhat later, the French scientist Jacques Charles (1746–1823) established that the volume of a gas

Table 8-1. Atmospheric Pressure and Barometer Readings for Various Altitudes

Altitude above Sea Level Feet	Atmospheric Pressure Pounds per Square Inch	Barometer Reading Inches of Mercury
0	14.69	29.92
500	14.42	29.38
1000	14.16	28.86
1500	13.91	28.33
2000	13.66	27.82
2500	13.41	27.31
3000	13.16	26.81
3500	12.92	26.32
4000	12.68	25.84
4500	12.45	25.36
5000	12.22	24.89
5500	11.99	24.43
6000	11.77	23.98
6500	11.55	23.53
7000	11.33	23.09
7500	11.12	22.65
8000	10.91	22.22
8500	10.70	21.80
9000	10.50	21.38
9500	10.30	20.98
10000	10.10	20.58
10500	9.90	20.18
11000	9.71	19.75
11500	9.52	19.40
12000	9.34	19.03
12500	9.15	18.65
13000	8.97	18.29
13500	8.80	17.93
14000	8.62	17.57
14500	8.45	17.22
15000	8.28	16.88

is proportional to its absolute temperature when the volume is kept at constant pressure. These findings came to be known as Boyle's and Charles's laws respectively. In Tables 8-2 and 8-3 a series of relations based on these two laws are tabulated for convenient reference.

A more simplified explanation of the interrelationship of these two laws may be gained with the aid of the cylinder illustrated in

Table 8-2. Summary of Boyle's law

1. **Pressure volume formula**

$$PV = P'V' \quad \dots\dots\dots\dots \text{(1)}$$

$$P = \frac{P'V'}{V} \quad \dots\dots\dots\dots \text{(2)}$$

$$V = \frac{P'V'}{P} \quad \dots\dots\dots\dots \text{(3)}$$

2. **Compression constant**

$$PV = \text{constant} \dots\dots\dots \text{(4)}$$

3. **Pressure at any point**

$$P = \frac{\text{constant}}{V} \dots\dots\dots \text{(5)}$$

4. **Volume at any point**

$$V = \frac{\text{constant}}{P} \dots\dots\dots \text{(6)}$$

5. **Ratio of Compression**

$$R = V_i \div V_f \dots\dots\dots \text{(7)}$$

where,

R = ratio or number of compressions
V_i = initial volume
V_f = final volume

$$R = P_i \div P_f =$$
$$V_i \div V_f \dots\dots \text{(8)}$$

6. **Initial pressure of compression**

$$P_i = R \div P_f \dots\dots\dots \text{(9)}$$

where,

P_i = initial pressure absolute
P_f = final pressure absolute

7. **Final pressure of compression**

$$P_f = P_i \div R$$

Table 8-3. Summary of Charles's Law

1. **At constant volume**

$$\frac{P}{T} = \frac{P'}{T'} \dots\dots\dots\dots\dots\dots\dots\dots\dots\dots\dots\dots\dots \text{(1)}$$

where, P = initial pressure absolute
P' = final pressure absolute
T = initial temperature absolute
T' = final temperature absolute

2. **At constant pressure**

$$\frac{V}{T} = \frac{V'}{T'} \dots\dots\dots\dots\dots\dots\dots\dots\dots\dots\dots\dots\dots \text{(2)}$$

where, V = initial volume (usually in cu. ft.)
V' = final volume (usually in cu. ft.)

Fig. 8-1. If the cylinder is filled with air at atmospheric pressure (14.7 psia) represented by volume *(A)*, and piston *(B)* is moved to reduce the volume by, say, ⅓ of *A*, as represented by *B*, then according to Boyle's law, the pressure will be trebled (14.7 × 3 = 44.1 lb. absolute or 44.1 − 14.7 = 29.4 gauge pressure). According to Charles's law, a pressure gauge on a cylinder would at this point indicate a *higher* pressure than 29.4 gauge pressure because of the increase in temperature produced by compressing the air. This is called *adiabatic compression* if no heat is lost or received externally.

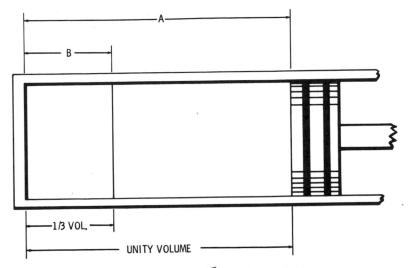

Fig. 8-1. The interrelationship of Boyle's and Charles's laws.

ATMOSPHERIC CONDITIONS AND HUMAN COMFORT

In order to understand the relationship of atmospheric conditions to human comfort, the principal natural laws governing the mixture of air and water vapor that compose the atmosphere, the definitions of the terms involved should be examined first. In examining this relationship, the following items are given special attention:

1. Water vapor.
2. Humidity.
3. Absolute humidity.
4. Relative humidity.
5. Drying effect of air.
6. The dew point.
7. Wet-bulb temperature.
8. Wet-bulb depression.
9. Humidification.
10. Dehumidification.

Water Vapor

In *all* air conditioning calculations it should be understood that the dry air and water vapor composing the atmosphere are separate entities, each with its own characteristics. Water vapor is not dissolved in the air in the sense that it loses its own individuality and merely serves to moisten the air.

Water vapor is actually steam at low temperatures and consequently low pressures; hence its properties are those of steam. According to Dalton's law, in any mechanical mixture of gases, each gas has a partial pressure of its own, which is entirely independent of the partial pressures of the other gases of the mixture.

Humidity

Air is capable of holding, as a mechanical mixture with itself, varying quantities of water vapor, depending upon its temperature. When air absorbs moisture, that is, when it is humidified, the latent heat of evaporation must be supplied either from the air or from another source. And conversely, when the moisture from the air is condensed, the latent heat of condensation (equivalent to the latent heat of evaporation) is recovered.

Air is said to be saturated when it contains all the water vapor it can hold. If partly saturated air is reduced in temperature until the amount of moisture present corresponds to the amount that the air is capable of holding at the given temperature, it will become saturated air.

If the temperature of the air is still further reduced, its ability to hold moisture will be reduced accordingly. As a result, the excess moisture will be condensed, which means that it will be converted from a vapor to a liquid. This is the reverse of the process that occurred as the air absorbed the moisture.

Converting liquid water into water vapor requires a great quantity of heat. The heat necessary for this process is used only in performing the conversion, the temperature of the liquid and the vapor being the same at the end of the process. If the conversion is from liquid to vapor, this involves the latent heat of *evaporation*. The latent heat of *condensation* is involved if the conversion is from vapor to liquid.

Cold air is saturated when it contains very small quantities of water vapor, whereas warm air is not saturated until it contains larger quantities of vapor. For example, air at zero degrees Fahrenheit is saturated when it contains but one-half of one grain (1/7000 lb.) of water vapor per cubic foot. Air at 70°F is saturated when it contains 8 grains of vapor per cubic foot, while at 83°F, 12 grains per cubic foot are required to saturate.

Absolute Humidity

Absolute humidity is defined as the quantity of water vapor actually present in a given volume of air, without regard to its temperature. Thus, if air at *any* temperature contains 4 grains of water vapor per cubic foot, its absolute humidity is expressed as 4 grains per cubic foot.

In the example used above, air at 88°F is saturated when it contains 12 grains of water vapor per cubic foot. Assume that it contains 8 grains per cubic foot. Its absolute humidity, then, is also 8 grains per cubic foot; its relative humidity is 66⅔ percent. Such air will exert a greater moistening effect than air at 92°F containing the same 8 grains of vapor per cubic foot.

Relative Humidity

Since air may contain varying quantities of water vapor at the same temperature, it is necessary to express its degree of saturation as related to its temperature. *Relative humidity*, then, is an expression in percentage of the degree of saturation of air at any given temperature.

A relative humidity of 50 percent, for example, means that the air at its given temperature contains 50 percent of the water vapor required to saturate it at that temperature. The air (at 70°F) is saturated when it contains 8 grains of vapor per cubic foot. Assume that it contains 4 grains per cubic foot. Its absolute humidity is 4 grains per cubic foot, and its relative humidity is 50 percent (4 is 50 percent of 8). Similarly, if the same air is heated to 83°F, its absolute humidity will still be 4 grains per cubic foot, but its relative humidity will be only 33⅓ percent, because 12 grains of vapor per cubic foot are required to saturate air at 83°F. In other words:

$$\frac{4 \times 100}{12} = 33\tfrac{1}{3} \text{ percent}$$

It is obvious that relative humidity depends equally upon both the factors of temperature and moisture content. A variation in either alters the relative humidity.

A clear conception of the relation between temperature and moisture content is essential to an appreciation of air conditioning and the effects of air upon the materials it surrounds, because it is the relative humidity (not the absolute humidity) that determines the effects of air upon such materials. The relative humidity is the governing factor because the drying or moistening effect of air depends upon the relation between the vapor pressure of the moisture in the material exposed to the air and the vapor pressure of the moisture mixed with the air itself. The vapor pressure of the moisture mixed with the air is proportional to the relative humidity of the air. Since this relation is true, calculations are usually based upon relative humidity rather than upon relative vapor pressure, although the latter might be considered the more direct relation.

Air at high relative humidities (regardless of the absolute humidity) exerts a greater moistening effect than air at lower relative humidities. The moistening effect of the air varies approximately with its relative humidity, with regard to the actual weight of water vapor present.

In a textile mill, for instance, where one of the chief functions of air conditioning is to control the moisture in the yarns in course of manufacture, it should now be obvious that temperature control is equally as important as moisture control, since it is upon the relative humidity (water vapor content as related to temperature) that the moistening effect of the air depends.

Drying Effect of Air

The drying effect of air varies approximately inversely with its relative humidity. In other words, the drying effect decreases as the relative humidity increases.

It should be noted that it is relative humidity that determines the drying effect of air, and this effect depends upon both the temperature and the water content of the air—since relative humidity depends upon both these factors.

The quantity of heat that dry air contains is very small because its specific heat is low (0.2415 for ordinary purposes), which means that 1 lb. of air falling 1°F will yield but .2415 of the heat that would be available from 1 lb. of water reduced one degree in temperature.

The presence of water vapor in the air materially increases the total heating capacity of the air because of the latent heat of the vapor itself.

Most hygroscopic materials in the presence of dry air, even at high, dry-bulb temperatures, may actually be cooled rather than heated. This occurs because the dry air immediately begins to evaporate moisture form the material.

The Dew Point

The *dew point* is the temperature of saturation for a given atmospheric pressure. In other words, for a given atmospheric pressure (barometer), it is the temperature at which moisture begins to condense in the form of tiny droplets, or dew.

The saturation temperature for any given quantity of water vapor in the atmosphere is known as the dew point. Any reduction in temperature below the dew point will result in condensation of some of the water vapor present, with a consequent liberation of the latent heat of the vapor, which must be absorbed before any further condensation can take place.

If the vapor pressure of the water vapor in a given space is the same as the vapor pressure of saturated steam at the prevailing dry-bulb temperature, the space contains all the water vapor it can hold at that temperature. The term *saturated water vapor* is applied to water in this state.

329

Wet-Bulb Temperature

Wet-bulb temperature is the temperature at which the air would become saturated if moisture were added to it without the addition or subtraction of heat. It is the temperature of evaporation.

The wet-bulb temperature in conjunction with the dry-bulb temperature is an exact measure of both the humidity of the air and its heat content. In air conditioning the dry-bulb temperature and the wet-bulb temperature must *both* be controlled if the effects of air are to be regulated.

If the bulb of an ordinary thermometer is surrounded with a moistened wick and placed in a current of air and superheated water vapor, it will be found that a reading at some point below the dry-bulb temperature is obtained. The minimum reading thus obtained is the *wet-bulb temperature* of the air.

The reduction in temperature is caused by the sensible heat being withdrawn from the air to vaporize the water surrounding the wet bulb, thus raising the dew point of the air.

The point of equilibrium at which the sensible heat withdrawn balances with the heat of vaporization necessary to bring the dew point up to the same point is the wet-bulb temperature.

In this transformation of energy from sensible heat of vaporization, there is no change in the total amount of energy in the mixture. For this reason, the wet-bulb temperature, once fixed, is an indication of the total heat in any mixture of air and water vapor.

Wet-Bulb Depression

Since outdoor summer air is rarely fully saturated, there is usually a considerable difference between its dry-bulb and its wet-bulb temperatures. This difference is referred to as *wet bulb depression,* and is greatest during the summer.

As previously mentioned, the wet-bulb temperature is that temperature to which air would be cooled by evaporation if the air was brought into contact with water and allowed to absorb sufficient water vapor to become saturated. For example, if the outdoor summer air is drawn through a humidifier and completely saturated, its dry-bulb temperature will be reduced to its wet-bulb temperature, and the air will leave the humidifier at the outdoor wet-bulb temperature. This cooling is accomplished entirely by evaporation, and is due to the latent heat required to convert the liquid water vapor. This conversion occurs the instant the air is brought into contact with the mist in the spray chamber of the humidifier, the heat being taken from the air.

The spray water in a humidifier is used over and over again, only that quantity being added which is actually absorbed by the air. Thus, without any additional expense, a humidifier will perform the function of cooling the air through the wet-bulb depression in the summer.

The extent of the wet-bulb depression in some localities is as much as 25° or 30°. Even in localities adjacent to large bodies of water where the humidity is high and the wet-bulb depression correspondingly low, the latter will quite commonly range from 10° to 15°.

In the vicinity of New York, for instance, the maximum outdoor wet-bulb temperature is about 78°. On such a day the dry-bulb temperature would probably be about 90°, making the wet-bulb depression 12° (90° − 78° = 12°).

In Denver, on the other hand, the maximum outdoor wet-bulb temperature is usually less than 78°. Because the coincident dry-bulb temperature is usually much higher then 90°, it results in a greater wet-bulb depression, which means that more cooling can be accomplished by evaporation.

Humidification

Humidification may be defined as the *addition* of moisture to the air. The conditioning machine that functions to add moisture to the air is called a humidifier.

A humidifier is commonly a low-pressure, low-temperature boiler in which the water is evaporated and the vapor (low-pressure steam) thus formed is caused to mix with air.

In a sense, water functions as a natural humidifier by acting as the medium that conveys heat to the air and as the source of the water vapor required to saturate the heated air. Contrast this with what takes place in a humidifier unit. A machine functions as a

331

humidifier when the temperature of the spray water is above that at which the moisture in the air will condense.

Dehumidification

Dehumidification may be defined as the *removal* of moisture from the air. A machine that functions to remove moisture from the air is called a dehumidifier.

The removal of moisture from the air is accomplished by condensation, which takes place when the temperature of the air is lowered below its dew point. The condensation thus formed falls into the tank of the conditioning machine. In this case the water acts solely as a conveyor of heat from the air (in addition to its cleansing action) and, as such, the finely divided mist is extraordinarily effective (practically 100 percent).

Some conditioning machines can function both as humidifiers or dehumidifiers. This can often be done without alteration to the unit except that the valves in the control line from the dew point thermostat on some designs are adjusted to connect the steam control of the water heater for winter operation, and to connect the three-way mixing valve to the water supply line for summer operation.

Whether the requirement is humidification or dehumidification, the unit always operates under accurate automatic control, maintaining the required indoor conditions winter and summer, regardless of the outdoor weather.

MEASURING DEVICES

A number of instruments are used for measuring the physical properties of air. These include:

1. The thermometer.
2. The barometer.
3. The psychrometer.
4. The pressure gauge.

A *thermometer* (Figs. 8-2 and 8-3) is a device used to measure temperature, and consists of a glass tube terminating in a bulb

Fig. 8-2. A pencil-style stack thermometer.

charged with mercury or colored alcohol. It measures the temperature by the contraction or expansion of the liquid with temperature changes, causing the liquid to rise or recede in the tube. The scale of an ordinary thermometer, either Fahrenheit or Celsius, is simply an arbitrary standard by means of which comparisons can be established.

An ordinary thermometer is used to measure dry-bulb temperature. Dry-bulb temperature is the degree or intensity of heat. In other words, dry-bulb temperature measures the degree of effort that the heat will exert to move from one position to another.

A specially designed thermometer is used to measure wet-bulb temperature. The latter represents the temperature at which the air becomes saturated if moisture is added to it *without* a change of heat. The bulb of an ordinary thermometer is surrounded with

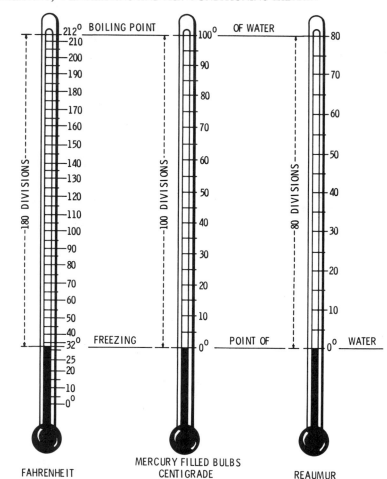

Fig. 8-3. Various thermometer scales.

a moistened wick, placed in a current of air, and superheated with water vapor. Essentially this represents a "wet-bulb" thermometer.

A *barometer* (Fig. 8-4) is an instrument designed to measure atmospheric pressure. Early barometers consisted of a 30-in. long glass tube open at one end and filled with mercury. The open end was submerged in a bowl of mercury, and the mercury in the

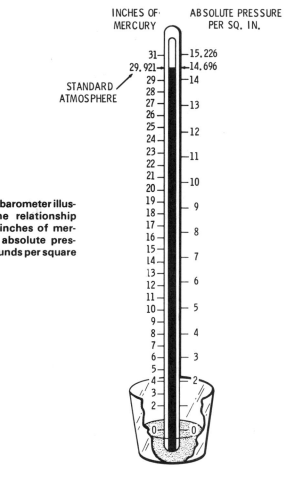

INCHES OF
MERCURY

ABSOLUTE PRESSURE
PER SQ. IN.

STANDARD
ATMOSPHERE

Fig. 8-4. Mercurial barometer illustrating the relationship between inches of mercury and absolute pressure in pounds per square inch.

glass tube would assume a level in accordance with the existing atmospheric pressure. Thus, the height of the mercury column in the tube is a measure of the atmospheric pressure. *Standard* atmospheric pressure at sea level is 29.921 in. of mercury.

A *psychrometer* (or *sling psychrometer)* is an instrument used to measure relative humidity. It consists of a dry-bulb (for air temperature) and a wet-bulb thermometer mounted side by side. The reading on the wet-bulb thermometer is determined by the rate at which the moisture evaporates from its bulb. If the psy-

335

chrometer is working correctly, the reading on the wet-bulb thermometer will be lower than the one on the dry-bulb thermometer. The *difference* between the two readings serves as a basis for determining the relative humidity.

Pressure gauges (Figs. 8-5, 8-6, and 8-7) are used to measure pressure. A *compound gauge* (Fig. 8-6) is used to measure low pressures above atmospheric pressure in psig and below atmospheric pressure in vacuum in inches of mercury. A *high-pressure gauge* (Fig. 8-5) is used to measure pressures ranging from zero to 300 or 400 psig.

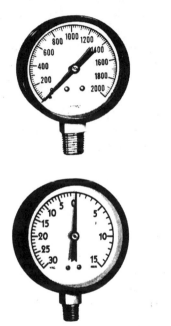

Fig. 8-5. A high-pressure gauge.

Fig. 8-6. A compound gauge that measures both pressure and vacuum.

CLEANING AND FILTERING

The purpose of cleaning and filtering the air is to remove dust and other contaminants that could be harmful to the health or discomforting. Many bacteria that cause diseases are carried on dust particles.

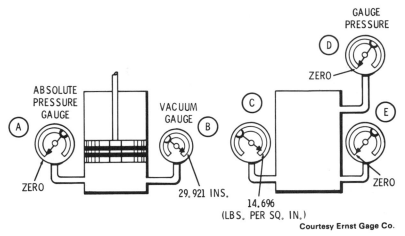

Courtesy Ernst Gage Co.

Fig. 8-7. Gauge illustrating absolute and zero pressure.

Cleaning and filtering may be accomplished with equipment using one of the three following methods:

1. Filtering.
2. Washing.
3. Combined filtering and washing.
4. Electrostatic field.

Filters trap particles by bringing them in contact with specially coated surfaces or by straining them through dry materials of particularly close texture. The filters used in air conditioning equipment may be either dry or wet (viscous) types. Depending upon the type used, air cleaning filters may be replaceable or periodically cleanable. In the latter case, either manual or automatic cleaning is possible.

Air washers form a part of the cooling and humidifying apparatus of the air conditioning system. They operate by passing the air first though fine sprays of water and then past baffle plates upon the wetted surface of which is deposited whatever dust and dirt not caught by the sprays.

Electronic air cleaners employ an electrostatic ionizing field to remove dust particles from the air. The particles are given an electrical charge when passing through the field and are subsequently attracted to metal plates having an opposite polarity.

337

Cleaning and filtering processes are described in greater detail in Chapter 14 (Air Cleaners and Filters).

AIR CONDITIONING SYSTEMS

One of the several ways in which air conditioning systems can be classified is on the basis of *how* and *when* they are to be used. Using this as a basis for classification, the following three broad categories can be determined:

1. Winter air conditioning.
2. Summer air conditioning.
3. Year-round air conditioning.

A *winter air conditioning system* is designed to heat, humidify, and clean the air of the space to be conditioned. It is limited to use in colder weather, particularly during the winter months. No provision is made in this type of air conditioning system to cool the air. Most modern oil- and gas-fired furnaces are designed to function as winter air conditioning systems.

A *summer air conditioning system* is built to cool, humidify or dehumidify, and clean the air. This system calls for the refrigeration of the air with no provision for heating it. Room air conditioners are among the most common types of equipment used for summer air conditioning.

A *year-round air conditioning system* is designed to condition the air on a year-round basis. It heats, cools, humidifies, dehumidifies, and cleans the air according to need throughout the year. This type of air conditioning system is examined in detail in Chapter 10 (Central Air Conditioning Systems), Chapter 11 (Air Conditioning Equipment), and Chapter 12 (Heat Pumps).

CHAPTER 9

Air Conditioning Calculations

This chapter is intended only as an *introduction* to the proce-dures and problems involved in estimating the total cooling load of a structure or a space. Accordingly, it is directed toward the layman, not the engineer. Much more detailed information can be obtained from the publications of the American Society of Heating, Refrigeration and Air-Conditioning Engineers. The relevant materials in the latest edition of the *ASHRAE Handbook of Fundamentals* will be current for at least five years.

The kind of calculations used in determining the size and load requirements of an air conditioning system will depend largely upon the type of system and the purpose for which it is to be used. For example, a summer air conditioning system is princi-pally concerned with supplying cooler air to the interior of a structure. It is also frequently designed to humidify or dehumidify the entering air as conditions require. In any event, calculating the maximum probable cooling load is the primary consideration when designing this type of system. Winter air conditioning, on the other hand, is concerned with supplying warmer air to the interior of a structure during the heating season. In this instance, calculating the *heating* load assumes major importance, although the other aspects of comfort air conditioning (e.g., ventilating, humidifying, air cleaning) are also considered. Finally, year-round air conditioning systems include all the design conditions of both summer and winter air conditioning (i.e., heating, cooling, humidifying, dehumidifying, air cleaning) and the required calculations to determine the probable heating and cooling loads.

Some of the terms used in this chapter have already been introduced in Chapter 8 (Air Conditioning Principles). It might be a good idea to reread that chapter at this point, and also the materials contained in Chapter 2, Volume 1 (Heating Fundamentals). You will find that much of this information overlaps, and it may cause some confusion in the terminology for you. For example, both "heat loss" and "heat leakage" involve the rate at which heat will flow through different materials or combination of materials. These terms are essentially synonymous, the difference between the two being the *direction* in which the heat flows. Most of this terminology will become less confusing with use.

DUCTS AND DUCT SYSTEMS

The ductwork of a warm-air heating system may not be sized for air conditioning. The cooler and heavier air used in air conditioning will usually require ducts having a larger diameter than the existing ductwork. The exact size can be determined by making the necessary calculations.

Duct sizing methods are described in Chapter 7 of Volume 2 (Ducts and Duct Systems). Bear in mind that an efficiently designed duct system involves the following four basic steps:

1. Determining the probable heating, cooling, and/or ventilating load for the structure or space.

2. Planning a suitable distribution system for the air (placement of ducts, cool air outlets, etc.).
3. Calculating the flow rate of the air so that each outlet delivers the required amount of air.
4. Selecting an air conditioning unit capable of meeting load and airflow rate requirements.

TEMPERATURE, MOISTURE, AND AIR MOVEMENT

Most cooling systems are designed to maintain *comfortable* conditions within a structure or a space. The exceptions to this are cooling systems designed for medical laboratories and similar structures in which the protection of products or materials is the primary concern. Whatever the purpose of the cooling system, an understanding of the interrelationship of air temperature, moisture, and movement is essential to the design of an efficient system. These concepts are described in several sections of this chapter (see, for example, "Standards of Comfort" and "The Comfort Chart"). This particular section is concerned with a review and explanation of the terminology and definitions that contribute to a clearer understanding of those concepts.

Air is composed of water vapor and dry air. These two components are combined in such a way that neither loses its distinct characteristics. A number of different terms are used to describe the qualities or properties of air, but the two terms essential in heating and cooling calculations are "humidity" and "temperature."

Humidity is a general term used to refer to the water vapor (moisture) content of air. When one uses this term, it is usually in reference to the *sensation* (or lack of sensation) of moisture in the air. For purposes of heating and cooling applications, the more narrowly defined terms of absolute humidity, relative humidity, and specific humidity are used.

Absolute humidity is the actual mass of water vapor in one cubic foot of air (i.e., the weight of water vapor per unit volume of air) and is expressed in grains or pounds per cubic foot (1

341

lb. = 7000 grains), or grams per cubic centimeter. Absolute humidity is equivalent to the density of the air.

Relative humidity is the ratio of absolute humidity to the maximum possible density of water vapor in the air at the same temperature. In other words, it is a *percentage* or *ratio* of water vapor in the mixture of dry air and water vapor at a certain temperature relative to the maximum quantity that the volume of air could possibly carry at that temperature. The relative humidity at any given temperature can be obtained by first using a sling psychrometer to determine the amount of moisture (i.e., water vapor) actually present in the air and then dividing this figure by the amount of moisture that the air can hold at that temperature, and multiplying the result by 100 in order to obtain the percentage factor.

Specific humidity is the *weight* of water vapor per pounds of dry air. Do not confuse specific humidity with relative humidity. The latter term indicates the percentage of water vapor, the former the weight.

Sometimes both temperature and humidity are used in conjunction with one another as a calculation factor, and the temperature-humidity index is an example of this. By definition, the *temperature-humidity index* (formerly called the *Discomfort Index*) is a numerical indicator of human *discomfort* resulting from temperature and moisture. It is calculated by adding the indoor dry-bulb and the indoor wet-bulb thermometer readings, multiplying the sum by 0.4 and adding 15. The results you obtain are the same as those used for the effective temperature index (see "Standards of Comfort"). This can be worked out from the data provided in Table 9-1.

Temperature is a general term used to express the sensation (or lack of sensation) of heat in the air. Among the more specific terms used in the heating and cooling calculations to describe the air temperature are dry-bulb temperature and wet-bulb temperature.

Dry-bulb temperature is the actual temperature of air as measured by an ordinary thermometer. *Wet-bulb temperature* is the temperature at which the air would become saturated if moisture were added to it without the addition or subtraction of heat. In actual practice, the wet-bulb temperature reflects humidity con-

Table 9-1. Recommended Scale of Interior Effective Temperatures for Various Outside Dry-Bulb Conditions

Degrees Outside	Degrees Inside			
Dry-Bulb	Dry-Bulb	Wet-Bulb	Dew Point	Effective Temperature
100	82.5	69.0	62.3	76.0
95	81.0	67.7	60.8	74.8
90	79.5	66.5	59.5	73.6
85	78.1	65.3	58.0	72.5
80	76.7	64.0	56.6	71.3
75	75.3	63.0	55.6	70.2
70	74.0	62.0	54.5	69.0

ditions in the area. A high wet-bulb reading, for example, means that the humidity is also high.

The *daily temperature range* is the difference between the maximum and minimum dry-bulb temperatures during a 24-hour period on a typical day for a heating or cooling system. It is used in determining the factors used in making Btu tabulations. The Btu tabulation cooling form illustrated in Fig. 9-1 shows their use. In Fig. 9-1 you will note that the tables labeled "wall factors" and "ceiling factors" each have a column reserved for four different degrees of daily temperature range (i.e., 15°F, 20°F, 25°F, and 30°F). Reading across from left to right, the different daily temperature ranges intersect with a number of other columns representing differences in the dry-bulb temperatures. The selection used will depend upon the type (or absence) of insulation.

The distinction between dry air and the moisture content of air and between dry-bulb and wet-bulb temperatures is extended to the two types of heat conveyed by the entering air and the air already contained in the space: sensible and latent heat.

Sensible heat is the amount of heat in air that can be measured by an ordinary thermometer (i.e., a dry-bulb thermometer). The daily weather report gives us sensible heat temperatures, but it does not represent the total heat we experience. It constitutes a *portion* of the heat resulting from air infiltration and ventilation, and internal heat sources such as people, electric lights, and electric motors. Sensible heat also results from heat leakage (or heat loss in the case of heating calculations) and solar radiation.

Latent heat is the amount of heat contained in the water vapor (moisture) of the air. It constitutes a portion of the heat resulting from infiltration and ventilation and any internal sources capable of adding water vapor to the air (e.g., cooking vapors, steam, people). The amount of latent heat in the air can be determined

Amana Refrigeration Inc., Amana, Iowa

BTU TABULATION
(BASED ON 24 HOUR PER

Job Name _____   Date _____

Location _____   Computed By _____

ROOM		LIVING ROOM		DINING ROOM		KITCHEN		BED ROOM 1	
Room Size (LxWxH)									
Linear Ft. Exposed Wall									
Floor-Ceiling Area									
	FACTOR	AREA OR QUAN.	BTUH	AREA OR QUAN.	BTUH	AREA OR QUAN.	BTUH	AREA OR QUAN.	BTUH
Windows DIRECTION									
Walls Shade									
Sunlit									
Ceiling									
Cooking						1200			
People	380	3						1	
Room Sensible Heat									
Room CFM									

WINDOW & DOOR FACTORS

Temp. Diff. F	15	20	25	30
North	12	18	25	31
NE & NW	23	29	36	42
East & West	32	38	44	51
SE & SW	35	41	48	55
South	26	32	39	45

Above Factors Assume Shades Or Venetian Blinds. If No Shades Double Factors.

For Outside Shading Or Awnings Multiply Factor x .60.

For Double Glazed Or Storm Windows Multiply Factor x .80.

NOTES:
PEOPLE LOAD — One Person in each bedroom and three persons in living room.

ROOF OVERHANG — South walls **only**. 36" overhang provides complete shade to wall and windows. Use shade values for windows and walls.

DOORS — Treat all outside doors as windows.

WALL FACTORS   Light Color

Temp. Diff. F		15	20	25	30
15 Daily Temp. Range	No Insul.	3.6	5.4	7.2	9.0
	1½" Insul.	2.1	3.0	4.0	5.1
	3" Insul.	1.5	2.1	2.9	3.6
20 Daily Temp. Range	No Insul.	2.7	4.5	6.5	8.2
	1½" Insul.	1.5	2.6	3.6	4.7
	3" Insul.	1.1	1.8	2.6	3.3
25 Daily Temp. Range	No Insul.	1.4	3.6	5.4	7.4
	1½" Insul.	0.9	2.1	3.0	4.1
	3" Insul.	0.6	1.5	2.1	2.9

For Masonry Walls Multiply Factors x 1.2. For ½ Value in Table.

CEILING FACTORS

	Temp. Diff. F	15
15 Daily Temp. Range	2" Insul.	5.7
	4" Insul.	3.6
	6" Insul.	3.2
20 Daily Temp. Range	2" Insul.	5.0
	4" Insul.	3.2
	6" Insul.	2.9
25 Daily Temp. Range	2" Insul.	4.4
	4" Insul.	2.9
	6" Insul.	2.6

Factors assume dark color roof with attic having vents.
If light color roof — take 75% of values in tables.
If no vents in attic space double values.

Fig. 9-1. Btu tabulation

by using a psychrometric chart (see Appendix B, Psychrometric Charts). The amount of *excess* latent heat so determined will indicate the amount of moisture that must be removed from the air in order to obtain comfortable conditions.

Both sensible heat gain and latent heat gain are expressed in

COOLING (Residential)
DAY OPERATION OF EQUIPMENT)

Outside Design Temp _____

Inside Design Temp _____

Temp Diff. _____

_____ Outside Humidity Factor _____ Daily Temp Range _____

	BED ROOM 2						BATH		BTUH TOTALS
	AREA OR QUAN.	BTUH	AREA OR QUAN.	BTUH	AREA OR QUAN.	BTUH	AREA OR QUAN.	BTUH	
	1								

Dark Color

15	20	25	30
5.1	6.9	8.9	10.7
2.7	3.6	4.7	5.1
2.0	2.6	3.3	4.1
4.2	6.0	7.8	9.8
2.1	3.2	4.2	5.1
1.5	2.3	3.0	3.6
3.3	5.1	6.9	8.9
1.7	2.7	3.6	4.7
1.2	2.0	2.6	3.3

North or Shaded Wall Figure

	20	25	30
	6.9	8.7	9.6
	4.4	5.1	5.7
	4.0	4.6	5.1
	6.2	7.5	8.6
	3.7	4.7	5.4
	3.5	4.2	4.9
	5.5	6.9	7.6
	3.6	4.4	5.0
	3.2	4.0	4.5

HOUSE TOTAL SENS. BTUH _____

DUCT GAIN_____% _____

VENTILATION LOAD BTUH _____

TOTAL SENSIBLE BTUH _____

GRAND TOT. = TOT. SENS. x HUM. FACTOR _____

% DUCT GAIN

	1 Story	1½ Story	2 Story
Ducts Insulated	8	6	4
Ducts Not Insulated	18	15	12

NOTE: All ducts in attic spaces must be insulated with 3" minimum thickness insulation with vapor barrier. Ducts in crawl spaces and damp basements must be covered with 2" minimum thickness insulation with vapor barrier.
Do not figure duct gain for concrete slab under floor ducts.

HUMIDITY FACTORS

% Outside Hum.	Below 40	40 to 45	Above 45
Factor	1.25	1.30	1.35

VENTILATION LOAD

House Volume Cu. Ft. (LxWxH) x .40 BTUH/CU. Ft.

Courtesy Amana Refrigeration, Inc.

cooling forum.

Btu's. When the total of the two are added together, their sum represents the *total* heat gain in Btu's that must be removed from the air each hour by the air conditioner.

Sensible heat gain (or load) is represented by a change in the *dry-bulb* temperature readings, whereas latent heat gain is represented by a change in the *web-bulb* temperature.

STANDARDS OF COMFORT

The influence of air temperature, moisture, and movement on physical comfort have been very thoroughly investigated. Once again, the most authoritative sources of information on this subject are the results of research conducted by the American Society of Heating, Refrigeration and Air-Conditioning Engineers. The most current edition of the *ASHRAE Guide* should be consulted for details because some revisions have been made.

The sensations of warmth or cold experienced by the human body depend not only upon the dry-bulb temperature but also upon the moisture content of the air. Cooling applications that remove only the sensible heat fall short of establishing comfortable conditions if the latent heat gain is particularly high. The air will be cooler under these conditions, but it will feel damp and uncomfortable. In order to meet minimal standards of comfort, both sensible and latent heat must be reduced to an acceptable level.

The average comfort conditions in summer and winter are considerably different, although the two zones overlap to some degree. This difference is caused largely by differences in clothing, and the natural inclination of the body to acclimate itself to somewhat higher temperatures in the summer.

The *effective temperature* is an arbitration index of the degree of warmth or cold as apparent to the human body, and takes into account the temperature, moisture content, and motion of the surrounding air.

Effective temperatures are not strictly a degree of heat; at least not in the same sense that dry-bulb temperatures are. For instance, the effective temperature could be lowered by increasing the rate of air flow even though wet- and dry-bulb temperatures remain the same. Consequently, effective temperature is more correctly defined as the *body sensation* of warmth or cold resulting from the combined effect of temperature, humidity, and air movement. For space cooling and heating, however, the air movement factor is considered a constant at approximately 20 ft. per minute, and under this condition effective temperature is determined by the wet-bulb and dry-bulb thermometer readings only.

THE COMFORT CHART

The *comfort chart* (Fig. 9-2) is an empirically determined effective temperature index that has been published by the *ASHRAE* since 1950.

The purpose of the comfort chart is to indicate the percentage of people feeling comfortable at various effective temperatures in the winter and summer. This serves only as an approximate standard of comfort, because individual reactions to warmth and cold are much too variable, but it *is* the most precise and scientific form of measurement available.

From the chart, one can obtain an approximate idea of the various effective temperatures at which a majority of people will feel comfortable (i.e., the summer and winter comfort zones).

Most air conditioning systems are designed with a recommended indoor design relative humidity of about 50 percent or slightly lower. Budget jobs will range as high as 60 percent relative humidity. The indoor dry-bulb temperature will range from 75°F or slightly below to about 80°F, depending upon the degree of occupancy and whether it is a budget job or not. In any event, the indoor design conditions *should* fall within the comfort zone.

More information about the use of the comfort chart is included in Appendix B (Psychrometric Charts).

COOLING LOAD ESTIMATE FORM

Air conditioning equipment manufacturers often provide their local representatives with forms for making cooling load esti-

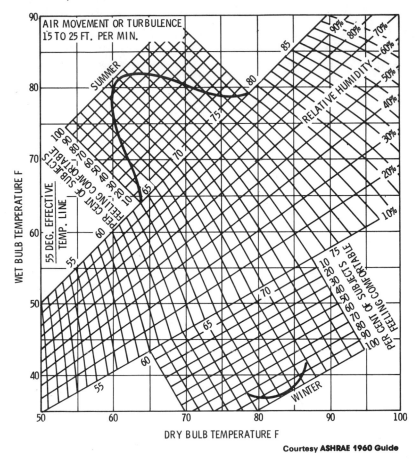

Courtesy ASHRAE 1960 Guide

Fig. 9-2. Comfort chart for still air.

mates. Along with these forms they also provide tables, slide charts, tabulation sheets, and other aids for computing the cooling load.

Typical tabulation sheets and cooling load estimate forms are shown in Fig. 9-1 and 9-3. These forms contain lists of factors that represent *approximate* values for a variety of different items (e.g., walls, ceilings, people). For example, the form illustrated in Fig. 9-1 requires that you select the appropriate design dry-bulb temperature and its column of factors. After calculating the area

348

Customer ....
Buyer ....
Estimate Number ....
Equipment Selected ....; Model ....; Size ....
Direction House Faces ....; Gross Floor Area ....sq ft; Gross Inside Volume .... cu ft

Address ....
Installation by ....
Estimate by .... Date ....
Direction House Faces ....; Gross Floor Area ....sq ft; Gross Inside Volume .... cu ft

Design Conditions:

Degrees North Latitude	Dry-Bulb Temperature (F)	Wet-Bulb Temperature (F)
....	....	....

ITEM	AREA (sq ft)	FACTOR (Circle the factors applicable.)									BTU/HR (Area x Factor)
1. (a) WINDOWS, Gain from Sun (Figure all windows for each exposure, but use only the exposure with the largest load.)		For glass block, reduce factors by 50%; for storm windows or double-glass, reduce factors by 15%.									
		No Shading		Inside Shades		Outside Awnings		Load for Each Exposure (Area x Factor)			
Northeast	....	60		25		20		....			
East	....	100		40		25		.... Use			
Southeast	....	75		30		20		.... only			....
South	....	75		35		20		.... the			
Southwest	....	110		45		30		.... largest			
West	....	150		65		45		.... load.			
Northwest	....	120		50		35		....			

For calculating gain from sun through windows under overhanging roofs, see example given in Instructions.

		DESIGN DRY-BULB TEMPERATURE (F)									
		90	92	95	97	100	102	105	110	115	
(b) WINDOWS, Heat Gain (Total of all windows)											
Single-glass	....	13	15	19	22	25	27	30	36	42	...|.....
Double-glass or glass block	....	7	8	9	10	11	12	13	16	19	...|.....
2. WALLS											
No insulation (brick veneer, frame, stucco, etc.)	....	4	4	5	6	6	7	8	9	10	...|.....
1 in. insulation or 25/32 in. insulation sheathing	....	3	3	4	4	5	5	6	7	9	...|.....
2 in. or more insulation	....	2	2	2	2	3	3	3	4	4	...|.....
3. PARTITIONS (Between conditioned and unconditioned space)	....	2	2	3	3	4	4	5	6	7	...|.....
4. ROOFS (a) Pitched or flat with vented air space, and:											
No insulation	....	18	18	19	20	21	21	22	24	25	...|.....
No insulation, with attic fan	....	9	11	12	14	16	17	19	22	25	...|.....
2 in. insulation	....	5	5	5	6	6	6	7	7	7	...|.....
4 in. insulation	....	3	3	4	4	4	4	4	5	5	...|.....
(b) Flat with no air space, and:											
No insulation	....	28	29	30	31	33	34	35	38	40	...|.....
1 in. or 25/32 in. insulation	....	14	14	15	16	16	17	18	19	20	...|.....
1½ in. insulation	....	8	9	9	9	10	10	11	11	12	...|.....
3 in. insulation	....	6	6	6	6	7	7	7	8	8	...|.....
5. CEILING (Under unconditioned rooms only)	....	3	3	4	4	5	5	6	7	8	...|.....
6. FLOORS (Omit if over basement, enclosed crawl space, or slab.)											
Over unconditioned room	....	2	2	2	3	3	4	4	5	6	...|.....
Over open crawl space	....	3	3	4	5	5	6	7	8	9	...|.....
7. OUTSIDE AIR Total sq ft of floor area	....	2	2	2	2	3	3	4	4	5	...|.....
8. PEOPLE (Use minimum of 5 people)		.... (number of people) x 200									...|.....
9. SUB-TOTAL											
10. LATENT HEAT ALLOWANCE		30 per cent of Item 9									
11. TOTAL		Sum of Items 9 and 10									

Fig. 9-3. Residential cooling load estimate form.

(in sq. ft.) of the structural section, you must multiply this figure by its factor to determine the Btu's per hour of heat gain, and enter the result in the column on the extreme right.

Because a manufacturer's cooling estimate form will probably not be available, it may be necessary to design your own. If this should be the case, then provisions should be made on the cooling estimate form for distinguishing between sensible and latent heat sources.

Your cooling load estimate form should contain the following basic categories of heat gain:

A. Sensible heat gain.
   1. Heat leakage.
   2. Solar radiation.
   3. Internal heat sources.
      a. Infiltration.
      b. Ventilation.
      c. Electric lights.
      d. Electric motors.
      e. People.
      f. Appliances.
B. Latent heat gain.
   1. Infiltration.
   2. Ventilation.
   3. People.
C. Ventilation heat gain (from outside sources).
   1. Sensible heat.
   2. Latent heat.

The sum of these categories will represent the total heat gain expressed in Btu's per hour. A balance must be established between this hourly heat gain within the conditioned space and the hourly capacity of the air conditioning unit to remove this heat gain in order to maintain the inside design temperature.

Fig. 9-4 is a floor plan (not drawn to scale) of a house that will serve as a basis for most of the cooling calculations used in this chapter.

The house is located 30° North, and complete exposure to the sun occurs at 1:00 P.M. The structure contains 740 sq. ft. of floor space divided into four rooms and a bath. There are ten 3′ × 4′

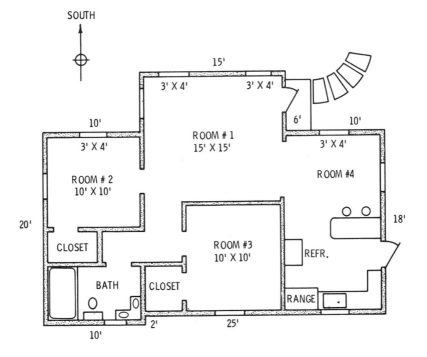

Fig. 9-4. Floor plan of a one-story residence constructed on concrete slab.

windows, with four of them facing south. The indoor design temperature is 80°F.

INDOOR-OUTDOOR DESIGN CONDITIONS

The indoor and outdoor design conditions must be established *before* any cooling load calculations can be made. This will be the first step in your procedure. As you will see, the *difference* between the indoor and outdoor design temperatures will eventually serve as a basis for selecting the correct size air conditioner.

Designing a system to meet the *maximum* outdoor summer temperature is generally not necessary because these temperatures either rarely occur or occur for a comparatively short duration of time. It is the general practice to design the system for slightly less severe conditions and thereby save on equipment and installation costs.

VENTILATION REQUIREMENTS

Each conditioned space requires a specific amount of outside fresh air to be circulated through it in order to remove objectionable odors (e.g., cooking odors, tobacco smoke, body odors) and to maintain comfort standards.

Ventilation standards indicating recommended air changes for a variety of space usages and occupancy have been established by the ASHRAE. Local codes are often based on ASHRAE research and data. These ventilation standards are easy and convenient to use, but are objected to by some authorities because the sources of contamination seldom bear any relationship to cubic area.

The air conditioning system must be capable of supplying enough outside fresh air to maintain air purity and comfort standards. The ventilation requirements for a given structure or space are based either on the desired number of air changes or the number of occupants. The volume rate of ventilation air is expressed in cubic feet per minute (CFM).

According to the ASHRAE, the outside fresh air requirement for both residences and apartments ranges from 10 CFM (minimum) to 20 CFM (recommended). Fresh air requirements for other types of structures are listed in Table 9-2.

The *amount* of outside fresh air required for each air change per hour equals the amount of inside air that must be removed from the structure or space during the same time span. The following formula can be used to determine the amount of air supply per minute:

$$\text{CFM} = \frac{\text{area in cubic feet}}{\text{minutes of air change}}$$

The floor plan of the residence shown in Fig. 9-4 contains 740 sq. ft. The height of each room is 7 ft., which gives an area of 5180 cu. ft.

Table 9-2. Fresh Air Requirements

Type of Building or Room	Minimum Air Changes per Hour	Cubic Feet of Air per Minute per Occupant
Attic spaces (for cooling)	12–15	
Boiler room	15–20	
Churches, auditoriums	8	20–30
College classrooms		25–30
Dining rooms (hotel)	5	
Engine rooms	4–6	
Factory buildings (ordinary manufacturing)	2–4	
Factory buildings (extreme fumes or moisture)	10–15	
Foundries	15–20	
Galvanizing plants	20–30	
Garages (repair)	20–30	
Garages (storage)	4–6	
Homes (night cooling)	9–17	
Hospitals (general)		40–50
Hospitals (children's)		35–40
Hospitals (contagious diseases)		80–90
Kitchens (hotel)	10–20	
Kitchens (restaurant)	10–20	
Libraries (public)	4	
Laundries	10–15	
Mills (paper)	15–20	
Mills (textile—general buildings)	4	
Mills (textile dyehouses)	15–20	
Offices (public)	3	
Offices (private)	4	
Pickling plants	10–15	
Pump rooms	5	
Restaurants	8–12	
Schools (grade)		15–25
Schools (high)		30–35
Shops (machine)	5	
Shops (paint)	15–20	
Shops (railroad)	5	
Shops (woodworking)	5	
Substations (electric)	5–10	
Theaters		10–15
Turbine rooms (electric)	5–10	
Warehouses	2	
Waiting rooms (public)	4	

Table 1 (Chapter 7) lists the number of average air changes per minute required for good ventilation for a number of different applications. You will note that each listed air change ranges from a minimum to a maximum number of changes. The rate of air change you select will depend upon the following:

353

1. Geographical location.
2. Occupancy.
3. Ceiling height.

Warmer climates and larger numbers of occupants require a greater rate of air change. Conversely, an 8-ft. ceilng will require more ventilation than a 15-ft. ceiling.

Using a 3-minute rate of air change for the residence illustrated in Fig. 9-4, the required number of air changes necessary to give proper ventilation can be determined as follows:

$$CFM = \frac{5180 \text{ cu. ft.}}{3} = 1726.7$$

Thus, 1726.7 CFM is needed to change the air every 3 minutes (or 20 air changes every hour).

Always check the local health department codes or local building codes for required ventilation standards first. If none exist, use the recommended air changes listed in this table, or data available from ASHRAE research (Table 9-2).

HEAT LEAKAGE

Heat leakage refers to the amount of heat flow through structural sections and is stated in Btu's per hour per degree Fahrenheit temperature per square foot of exposed surface. A significant portion of the total heat gain of a space is due to heat from the outside of a structure or from a nonconditioned space passing (i.e., leaking) through walls, ceilings, floors, and roofs to the interior of the structure.

The coefficient of heat transmission (U-factor) is the specific value used in determining the amount of heat leakage. It has already been described in Chapter 4 of Volume 1 (Heating Calculations) in the section "Heat Loss." Actually the only difference between heat leakage and heat loss is the direction of heat flow. Both are concerned with the same thermal properties of construction materials and the rate at which heat flows through them. Insofar as heat leakage is concerned, the direction of heat flow is from the outside to the inside of the structure. In the case of heat loss, the reverse is true.

The formula used for calculating heat leakage is identical to the one used for heat loss, and may be stated as follows:

$$Q = UA(t_o - t_i).$$

where,

Q = Amount of heat transmission in Btuh.

U = Overall coefficient of heat transfer (U-factor) between the adjacent space and the conditioned space (stated in Btu's per hour per square foot per degree Fahrenheit).

t_o = Air dry-rule temperature in degrees Fahrenheit of adjacent space or outdoors.

t_i = Air dry-bulb temperature in degrees Fahrenheit of conditioned space of the structure.

SOLAR RADIATION

A portion of the heat gain in the interior of a structure can be attributed to solar radiation coming in through the windows. To disregard this factor when calculating the total loads of the structure can result in serious error. It is particularly important to consider the *type* of shading because this will determine the amount of heat gained from solar radiation.

One method of determining heat gain from solar radiation *through window glass* is illustrated by the following equation:

Total Instantaneous Heat Gain Through Window Glass = Area of Window Glass × Factor for Shading × Heat Gain by Solar Radiation + Heat Gain by Convection and Radiation

The ASHRAE has conducted extensive research into all aspects of solar radiation, and makes the results of this research available through its publications. Tables 9-3 and 9-4 were adapted by permission from the *ASHRAE 1960 Guide* to illustrate the equation for determining total instantaneous heat gain through window glass given above.

Calculating the amount of heat gain due to solar radiation through a glass window can be even more clearly illustrated by using the above equation to solve a problem. Let us assume that you have a residence containing four 3′ × 4′ windows *facing south*. The structure is located 30° North, and complete exposure to the sun occurs at 1:00 P.M. The indoor temperature is 80°F. The windows are shaded by dark brown canvas awnings that are open at the sides. As you will note by looking carefully at Tables 9-3 and 9-4, each of these points in the description of the structure is important in the solution. The total instantaneous heat gain through these windows can be determined as follows:

1. Window glass area = 3′ × 4′ = 12 sq. ft.
2. Factor for shading = 0.25.
3. Heat gain by solar radiation = 45 Btuh per sq. ft.
4. Heat gain by convection and radiation = 17 Btuh per sq. ft.

Total instantaneous heat gain through the four windows equals

$$12 \times (0.25 \times 45 + 17)$$
$$12 \times 28.25$$
$$339 \text{ Btuh (for one window)} \times 4 \text{ windows}$$
$$1356 \text{ Btuh}$$

In Table 9-4 note that the shading factors are given not only for such items as shading screens and awnings located on the outside of the structure but also for window shades, venetian blinds, and draperies located on the inside. It is *always* recommended that windows of air conditioned spaces be shaded in some manner.

INTERNAL HEAT GAIN

A number of heat sources within a conditioned space contribute to heat gain independently from outside sources. The most important internal heat gain sources are:

1. People.
2. Electric lighting.
3. Appliances.
4. Electric motors.
5. Steam.

The occupants of conditioned spaces give off both sensible and latent heat. The *amount* of heat gain will depend upon a number of variables, including: (1) the duration of occupancy, (2) the number of people, and (3) their principal activity. Table 9-5 lists estimated heat gains from a variety of activities performed by individuals.

HEAT GAIN FROM INFILTRATION AND VENTILATION

A certain amount of warm outdoor air will enter the interior of a structure by means of infiltration and ventilation. Both phenomena are described in considerable detail in Chapter 6 (Ventilation Principles), and in the section "Ventilation Standards" in this chapter. These materials should be read before proceeding any further.

Insofar as cooling load calculations are concerned, infiltration (i.e., natural ventilation) is the leakage of warmer outdoor air into the interior of a structure usually as the result of wind pressure. This occurs primarily through cracks around windows and doors. As a result, the *crack method* is considered the most accurate means of calculating heat leakage by air infiltration (see the appropriate section of Chapter 4 of Volume 1, Heating Calculations). A rule-of-thumb method for calculating heat leakage around doors located in exterior walls is to allow *twice* the window heat leakage.

Some authorities feel that infiltration will fulfill the ventilation requirements of small structures (e.g., houses, offices, and small shops or offices) and that no special provisions need be made for a mechanical ventilating system. Unfortunately, there is little or no infiltration on days during which the outdoor air is perfectly still.

The amount of ventilation is generally determined by the

Table 9-3. Heat Gain Due to Solar Radiation (Single Sheet of Unshaded Common Window Glass)

Latitude	Sun Time A.M. →	Sun Time	N	NE	E	SE	S	SW	W	NW	Horiz.
					Instantaneous Heat Gain in Btu per hour (sq. ft.)						
30° North	6 A.M.	6 P.M.	25	98	108	52	5	5	5	5	17
	7	5	23	155	190	110	10	10	10	10	71
	8	4	16	148	205	136	14	13	13	13	137
	9	3	16	106	180	136	21	15	15	15	195
	10	2	17	54	128	116	34	17	16	16	241
	11	1	18	20	59	78	45	19	18	18	267
	12		18	19	19	35	49	35	19	19	276
40° North	5 A.M.	7 P.M.	3	7	6	2	0	0	0	0	1
	6	6	26	116	131	67	7	6	6	6	25
	7	5	16	149	195	124	11	10	10	10	77
	8	4	14	129	205	156	18	12	12	12	137
	9	3	15	79	180	162	42	14	14	14	188
	10	2	16	31	127	148	69	16	16	16	229
	11	1	17	18	58	113	90	23	17	17	252
	12		17	17	19	64	98	64	19	17	259
50° North	5 A.M.	7 P.M.	20	54	54	20	3	3	3	3	6
	6	6	25	128	148	81	8	7	7	7	34
	7	5	12	139	197	136	12	10	10	10	80
	8	4	13	107	202	171	32	12	12	12	129
	9	3	14	54	176	183	72	14	14	14	173
	10	2	15	18	124	174	110	16	15	15	206
	11	1	16	16	57	143	136	42	16	16	227
	12		16	16	18	96	144	96	18	16	234
		P.M. →	N	NE	E	SE	S	SW	W	NW	Horiz.

Courtesy ASHRAE 1960 Guide

Table 9-4. Shade Factors for Various Types of Shading

Type of Shading	Finish on Side Exposed to Sun	Shade Factor
Canvas awning sides open	Dark or medium	0.25
Canvas awning top and sides tight against building	Dark or medium	0.35
Inside roller shade, fully drawn	White, cream	0.41
Inside roller shade, fully drawn	Medium	0.62
Inside roller shade, fully drawn	Dark	0.81
Inside roller shade, half drawn	White, cream	0.71
Inside roller shade, half drawn	Medium	0.81
Inside roller shade, half drawn	Dark	0.91
Inside venetian blind, slats set at 45°	White, cream	0.56
Inside venetian blind, slats set at 45°	Diffuse reflecting aluminum metal	0.45
Inside venetian blind, slats at 45°	Medium	0.65
Inside venetian blind, slats set at 45°	Dark	0.75
Outside venetian blind, slats set at 45°	White, cream	0.15
Outside venetian blind, slats set at 45° extended as awning fully covering window	White, cream	0.15
Outside venetian blind, slats set at 45° extended as awning covering 2/3 of window	White, cream	0.43

	Dark	Green tint
Outside shading screen, solar altitude 10°	0.52	0.46
Outside shading screen, solar altitude 20°	0.40	0.35
Outside shading screen, solar altitude 30°	0.25	0.24
Outside shading screen, solar altitude, above 40°	0.15	0.22

Courtesy ASHRAE 1960 Guide

Table 9-5. Rate of Heat Gain from Occupants of Conditioned Spaces

Degree of Activity	Typical Application	Total Heat Adults, Male Btu/Hr	Total Heat Adjusted Btu/Hr	Sensible Heat Btu/Hr	Latent Heat Btu/Hr
Seated at rest	Theater—Matinee	390	330	180	150
	Theater—Evening	390	350	195	155
Seated, very light work	Offices, hotels, apartments	450	400	195	205
Moderately active office work	Offices, hotels, apartments	475	450	200	250
Standing, light work; or walking slowly	Department store, retail store, dime store	550	450	200	250
Walking: seated Standing; walking slowly	Drugstore, Bank	550	500	200	300
Sedentary work	Restaurant	490	550	220	330
Light bench work	Factory	800	750	220	530
Moderate dancing	Dance hall	900	850	245	605
Walking 3 mph; moderately heavy work	Factory	1000	1000	300	700
Bowling Heavy work	Bowling alley Factory	1500	1450	465	985

Courtesy *ASHRAE 1960 Guide*

number of air changes (inside air replaced by air from the out-doors) required by a structure. This is most commonly based upon the number of occupants and building use. In structures having air conditioning, *most* of the outdoor air used for ventilation will pass through the air conditioning unit. A small portion of the air will bypass the air conditioning coils and add to the sensible and latent heat levels of the interior spaces of the structure. The temperature and humidity of the air that does pass over the cooling coils of the unit are reduced to room conditions or below.

CALCULATING INFILTRATION AND VENTILATION HEAT GAIN

It has been shown that outdoor air enters a structure by means of both infiltration and ventilation. Because air is composed of a mixture of dry air and moisture particles, the heat gain produced by the entering air will be expressed in terms of both its sensible heat gain and its latent heat gain in Btu's per hour.

The sensible and latent heat gain resulting from entering air represents only a small portion of the total heat gain involved in determining the design cooling load of a structure. For many applications, however, the calculation of this portion of the heat gain is crucial to a well designed system. The following formulas, adapted from ASHRAE materials, are used for making these calculations:

$$(1) \quad Q_S = cfm \times 1.08(t_o - t_i)$$
$$(2) \quad Q_L = cfm \times .68(w_o - w_i)$$
$$(3) \quad Q_T = Q_S + Q_L$$

where,

Q_S = Sensible load.

Q_L = Latent load.

Q_T = Total load.

CFM = Rate of entry of outdoor air (cubic feet per minute).

t_o = Dry-bulb temperature of outside (entering) air.

t_i = Dry-bulb temperature of inside air.

w_o = Outdoor wet-bulb temperature.

w_i = Indoor wet-bulb temperature.

In order to use these formulas, it is first essential to determine the outdoor and indoor design conditions (i.e., the dry-bulb and wet-bulb temperatures) and the maximum rate of entering air (in cubic feet per minute).

Rule-of-Thumb Methods

Manufacturers of air conditioning equipment and mail order houses (e.g., Sears, Montgomery Ward) that sell air conditioning equipment through their catalogs provide rule-of-thumb methods for calculating the size of the air conditioner required by a structure. The Btu calculation formulas are based on recommended coefficient factors for different types of construction and conditions. The responsibility for calculating the cooling load (and ultimately selecting a suitable air conditioner) lies with the purchaser of the equipment.

The problem of using *any* rule-of-thumb method is that the results are not precise. In other words, the results represent an approximate estimate; not the results one would expect from an engineer's calculations. There is always the danger of oversizing or undersizing the air conditioner. Under normal conditions, however, this method of calculating the size of a central air condtioner is reasonably accurate.

The Btu tabulation and cooling estimate forms illustrated in Fig. 9-1 and 9-3 are typical examples of the forms provided by manufacturers of air conditioning equipment. Note that these forms not only provide coefficient factors for various types of construction but also recommend standard loads for different activities. For example, the cooling load estimate form shown in Fig. 9-3 provides for a latent heat allowance of 30 percent of the total heat gain from all other sources. Another example of this practice is the commonly used standard load of 1500 Btu for kitchen activities. People are usually given a 200 Btu allowance per person, and residences are calculated on the basis of 2 people per bedroom. Thus, a three bedroom house would have a 1200 Btu allowance for people ($3 \times 2 = 6 \times 200 = 1200$ Btu).

The Btu calculation formulas used with the various rule-of-thumb methods also provide for temperature and humidity adjustments where conditions differ from the recommended levels. This is sometimes accomplished by providing humidity factors and a range of dry-bulb temperature differences (see Fig. 9-1) or by providing humidity and temperature adjustment allowances. In the latter case, the coefficient factors are usually based on a specific wet-bulb temperature and dry-bulb temperature difference. If, for example, the former were 75°F and *your* wet-bulb reading were 80°F, the instructions might require that you add 10 percent of the total heat gain of the structure for the humidity adjustment. Similar allowances are provided for temperature adjustments.

One of the easier and more popular rule-of-thumb methods employed for calculating the size of *large* central air conditioners is to use one ton of refrigeration for each 500–700 sq. ft. of floor area, or each 5000–7000 cu. ft. of space. A *ton of refrigeration* is equivalent to 12,000 Btu per hour. This figure is based on the fact that 1 lb. of melting ice will absorb 144 Btu of heat over a 24-hour period. Therefore, a ton of ice will absorb 288,000 Btu during the same period of time (i.e., 144 Btu × 2000 lb.). 288,000 ÷ 24 hours = 1 ton of refrigeration.

THE ARI METHOD

The Air-Conditioning and Refrigeration Institute (ARI) has created a standard form with instructions for calculating the size of the air conditioner required by a structure or space. The package contains a form for listing all the information about the structure necessary for calculating heat gain, instructions for using the form, data tables, and calculating sheets.

These materials offer a very convenient and easy to follow method of estimating the size of the air conditioner. It is not the most precise method, but it is much more accurate than the so-called rule-of-thumb methods described above. For more information, write:

Air-Conditioning and Refrigeration Institute
1346 Connecticut Avenue, N.W.
Washington, DC 20036

CONTRACTOR'S ESTIMATE

If you are not satisfied with your own cooling load estimate, you should invite several air conditioning contractors to give their bids. This is particularly true if you are considering a central air conditioning system. For a central air conditioning system, each contractor should include the estimated Btu per hour required to cool the structure. Each of the bids should be fairly close in estimated Btu and cost. Your choice will be based largely upon availability of replacement parts, the reputation of the local dealer for quick and reliable service, and the estimated Btu output. You will naturally want an air conditioning system that will efficiently remove the requried amount of heat. If this proves to be more expensive than another type, you would be wise to choose the more expensive one because your operating costs will be cheaper over the long run. Any contractor's bid that shows a wide variation from the others either in cost or estimated Btu required to cool the structure should be regarded with some suspicion. It may simply be an attempt to win the contract. Most reliable bids will be fairly close.

CHAPTER 10

Central Air
Conditioning

The term "central air conditioning" refers not so much to the method of cooling used in a structure as it does to the type of installation and its location. A central air conditioning system is one that is generally *centrally* located in a structure in order to simultaneously serve a number of rooms and spaces. Furthermore, a typical central air conditioning system is assembled in the field rather than at the factory.

This chapter examines first the principal cooling methods used in central air conditioning systems and then provides a description of several of the more common central air conditioning installations.

COOLING METHODS

Central air conditioning can be accomplished by means of a variety of different cooling methods. Those described in this chapter are:

1. Evaporative cooling.
2. Cold-water-coil cooling.
3. Gas compression refrigeration.
4. Gas absorption refrigeration.
5. Thermoelectric refrigeration.
6. Cooling with steam.

In a majority of air conditioning installations, and almost exclusively in the smaller horsepower range found in residences and small commercial buildings, the vapor or gas compression method of cooling is used (see "Gas Compression Refrigeration").

A comparatively recent entry into the field of residential air conditioning is thermoelectric refrigeration. This method of cooling is based on the thermocouple principle; the cool air is produced by the cold junctions of a number of thermocuoples wired in series (see "Thermoelectric Refrigeration").

Absorption refrigeration and cooling with steam are cooling methods generally found in large commercial and industrial applications (see "Gas Absorption Refrigeration" and "Cooling with Steam").

EVAPORATIVE COOLING

An evaporative cooling system cools the indoor air by lowering its dry-bulb temperature. In effect, it cools by evaporation, and it accomplishes this function by means of an evaporative cooler. Figs. 10-1, 10-2, and 10-3 show the basic components of three evaporative coolers.

As shown in Fig. 10-4, an evaporative cooler consists of a blower and blower motor, water pump, water distribution tubes, water pads, and a cabinet with louvered sides. In operation, the blower draws air through the louvers of the cabinet where it comes into contact with the moisture in the pads. The air passes

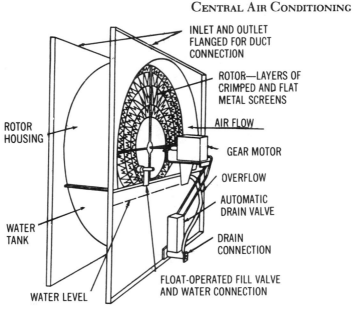

INLET AND OUTLET
FLANGED FOR DUCT
CONNECTION

ROTOR—LAYERS OF
CRIMPED AND FLAT
METAL SCREENS

AIR FLOW

GEAR MOTOR

OVERFLOW

AUTOMATIC
DRAIN VALVE

DRAIN
CONNECTION

FLOAT-OPERATED FILL VALVE
AND WATER CONNECTION

ROTOR
HOUSING

WATER
TANK

WATER LEVEL

Courtesy *1965 ASHRAE Guide*

Fig. 10-1. Typical rotary evaporative cooler.

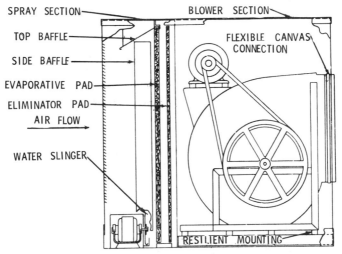

SPRAY SECTION

BLOWER SECTION

TOP BAFFLE

SIDE BAFFLE

EVAPORATIVE PAD

ELIMINATOR PAD

AIR FLOW

WATER SLINGER

FLEXIBLE CANVAS
CONNECTION

RESILIENT MOUNTING

Courtesy *1965 ASHRAE Guide*

Fig. 10-2. Typical spray evaporative cooler.

367

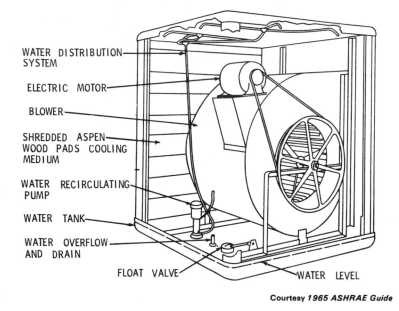

WATER DISTRIBUTION SYSTEM

ELECTRIC MOTOR

BLOWER

SHREDDED ASPEN WOOD PADS COOLING MEDIUM

WATER RECIRCULATING PUMP

WATER TANK

WATER OVERFLOW AND DRAIN

FLOAT VALVE

WATER LEVEL

Courtesy *1965 ASHRAE Guide*

Fig. 10-3. Typical drip evaporative cooler.

through these moist water pads and into the interior of the structure.

The water in the pads absorbs heat from the air as it passes through them. This causes a portion of the water to evaporate and lowers the dry-bulb temperature of the air as it enters the room or space. It is this lower dry-bulb temperature that produces the cooling effect.

In an evaporative cooling system, the water is recirculated and used over and over again. Only enough water is added to replace the amount lost by evaporation. The pump supplies water to the pads through the distribution tubes.

The air is never recirculated because it contains too much moisture once it has passed through the evaporative cooler. New air must always be drawn from the outdoors.

An evaporative cooling system is generally not very effective in a humid climate because the outdoor air is not dry enough. Evaporative coolers have been used for years with excellent

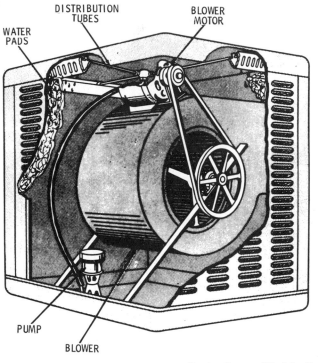

WATER PADS

DISTRIBUTION TUBES

BLOWER MOTOR

PUMP

BLOWER

Fig. 10-4. Evaporative cooler.

results in New Mexico, Arizona, Nevada, and similar areas with dry climates.

COLD-WATER-COIL COOLING

Indoor air temperatures can also be reduced by passing warm room air over a cold surface, such as a water-cooled coil, and then recirculating it back into the room. When a water-cooled coil is used for this purpose, the system is referred to as *cold-water-coil cooling* (Fig. 10-5).

A water-cooled coil is effective only when there is a sufficient supply of cold water. The temperature of the water should range

369

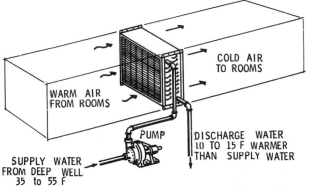

COLD AIR
TO ROOMS

WARM AIR
FROM ROOMS

PUMP

DISCHARGE WATER
10 TO 15 F WARMER
THAN SUPPLY WATER

SUPPLY WATER
FROM DEEP WELL
35 to 55 F

Courtesy Honeywell Tradeline Controls

Fig. 10-5. Cold-water-coil cooling.

from 35° to 55°F, and the most common source is a deep well. A pump supplies the cold water to the coil where it picks up heat from the air passing through it. This warmer water is then discharged to a storm sewer, dry well, or some other outdoor receiver. The discharge water is approximately 10° to 15°F warmer than the supply water.

GAS COMPRESSION REFRIGERATION

A gas compression refrigeration air conditioning system operates on the direct-expansion cooling principle. Basically the system consists of a compressor, condenser coil, receiver, expansion device, and evaporator coil. A refrigerant flowing through the system is affected by temperature and pressure acting simultaneously in such a way that heat is transferred from one place to another. In other words, heat is *removed* from the room air for cooling and added to it for heating.

Freon and ammonia represent two of the more popular refrigerants used in gas compression refrigeration systems. Ammonia compression systems are commonly found in large industrial plants, whereas gas compression systems using Freon as a refrigerant dominate residential air conditioning.

The operating principle of a typical gas compression air conditioning system is explained in the next section.

MECHANICAL REFRIGERATION CYCLE

The *mechanical refrigeration cycle* is illustrated schematically in Fig. 10-6. The *liquid* refrigerant is contained initially in the receiver, which is usually located in the lower section of the condenser, although it can be a separate tank. The compressor, acting as a pump, forces the liquid refrigerant under high pressure through the liquid line to the expansion device.

The function of the expansion device is to regulate the flow of refrigerant into the evaporator coil. This expansion device may be in the form of an expansion valve or a capillary tube.

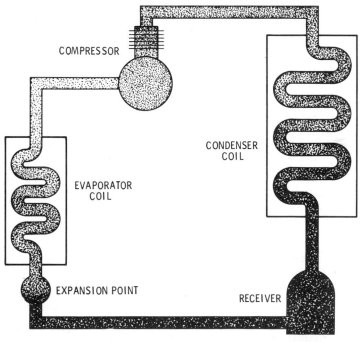

Courtesy Honeywell Tradeline Controls

Fig. 10-6. Mechanical refrigeration cycle.

As the high-pressure liquid refrigerant is forced through the expansion device, it expands into a larger volume in the evaporator, thus reducing its pressure and consequently its boiling temperature. Under this low pressure the liquid refrigerant boils until it becomes a vapor. During this change of state, the refrigerant absorbs heat from the warm air flowing across the outside of the evaporator.

After the refrigerant has boiled or vaporized, thus removing its quota of heat, it is of no more value in the evaporator coil and must be removed to make way for more liquid refrigerant. Instead of being exhausted to the outdoor air, the low-pressure heat-laden refrigerant vapor is pumped out of the evaporator through the suction line to the compressor. The compressor then compresses the refrigerant vapor, increasing its temperature and pressure and forces it along to the condenser.

At the condenser, the hot refrigerant vapor is cooled by lower temperature air passing over the condenser coils, thus absorbing some of the refrigerant heat. As a result, the air temperature increases and the refrigerant temperature decreases until the refrigerant is cooled to its saturation condition. At this condition, the vapor will condense to a liquid. The liquid, still under high pressure, flows to the expansion device, thus completing the cycle.

Note that cold is *never* created during the mechanical refrigeration cycle. Instead, heat is merely transferred from one place to another. When the refrigerant passes through the evaporator, it absorbs heat from the room air, thereby cooling it. When the higher temperature refrigerant passes through the condenser, it gives up heat to the air entering the room, thereby warming it.

GAS ABSORPTION REFRIGERATION

The *gas absorption refrigeration* method of cooling uses heat as its energy instead of electricity. This heat can be in the form of steam from a gas-fired or oil-fired atmospheric steam generator, or it can come from a gas or oil burner applied directly to the refrigeration generator. Normally, water is used as the refrigerant and lithium bromide as an absorbent. The absorption unit operates under a vacuum that gives the water a boiling temperature low enough for comfort cooling.

The absorption refrigeration system shown in Fig. 10-7 is charged with lithium bromide and water, the lithium bromide being the absorbent and the water the refrigerant. This solution is contained within the refrigeration generator.

As steam heat is applied to the generator, a part of the refrigerant (water) is evaporated or boiled out of the solution. As this water vapor is driven off, absorbent solution is raised by vapor lift action to the separating chamber (5) above the generator.

Refrigerant (water) and absorbent separate in the vapor separating chamber (5), the refrigerant vapor rises to the condenser (6), and the separated absorbent solution flows down through a tube (8) to the liquid heat exchanger and thence to the absorber.

The refrigerant (water vapor) passes from the separating chamber to the condenser through a tube (6), where it is condensed to a liquid by the cooling action of water flowing through the condenser tubes. The cooling water that flows through the condenser is brought from some external source, such as a cooling tower, city main, or well.

The refrigerant vapor thus condensed to water within the condenser then flows through a tube (7) into the cooling coil. This tube contains a restriction that offers a resistance and therefore a pressure barrier to separate the slightly higher absolute pressure in the condenser from the lower pressure within the cooling coil. The refrigerant (water) entering the cooling coil vaporizes due to the lower absolute pressure (high vacuum) that exists within it. The high vacuum within the evaporator lowers the boiling temperature of water sufficiently to produce the refrigeration effect.

The evaporator or cooling coil is constructed with finned horizontal tubes, and the air being cooled flows horizontally over the coil surface. Evaporation of the refrigerant takes place within the cooling coil, the heat of evaporation for the refrigerant is extracted from the air stream, and cooling and dehumidifying is accomplished.

In the absorber, the solution absorbs the refrigerant vapors which were formed in the evaporator directly adjacent. To explain the presence of the the absorbent at this point, it is neces-

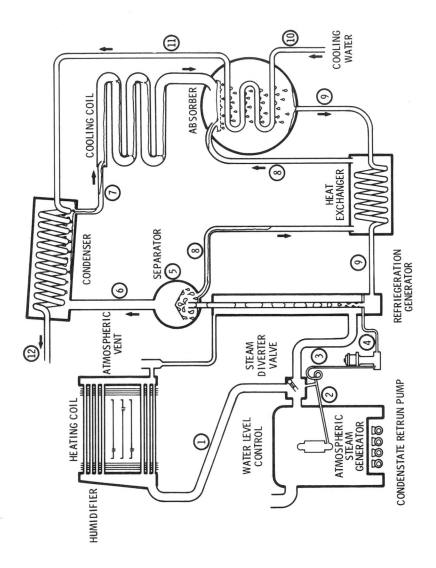

Fig. 10-7. Gas absorption refrigeration unit.

374

sary to divert attention back to the generator. The absorbent was separated from the refrigerant by boiling action. The absorbent then drains from the separator (5) down to the liquid heat exchanger and then to the absorber through the tube (8) designed for this purpose. The flow of solution in this circuit can actually exist by gravity action only because the absorber is slightly below the level of the separating chamber.

It must be understood at this point that lithium bromide in either dry or in solution form has a very strong affinity for water vapor. It is because of this principle that the refrigerant vapor is absorbed back into solution again. Because the *rate* of absorption is increased at lower temperatures, a water-cooling coil is provided within the absorber shell.

The resultant mixture of refrigerant and absorbent drains back through the heat exchanger through another tube (9) to the refrigeration generator where it is again separated into its two component parts to repeat the cycle.

The liquid heat exchanger serves to increase operating efficiency. The absorbent solution leaves the refrigeration generator at a relatively high temperature. Because its affinity for water vapor is increased as its temperature is reduced, precooling is desirable before it enters the absorber. Conversely, the combined solution of refrigerant and absorbent leaving the absorber and flowing toward the generator is relatively cool.

Because heat is applied in the generator to drive off water vapor, it is desirable to preheat this liquid before it enters the generator. With counterflow action in the liquid heat exchanger, both precooling and preheating are accomplished within the solution circuit.

As stated previously, a high vacuum exists throughout all circuits. However, a slightly higher absolute pressure exists in the generator and condenser than in the cooling coil and absorber. This difference in pressure is maintained by a difference in height of the solution columns (or restrictor) in the various connecting tubes.

The effect of heat applied to the refrigeration generator raises the absorbent solution to the vapor separator located at the top of the generator. The absorbent solution is able to flow from the generator to the absorber by gravity, aided by the slight pressure differential between the two chambers.

In the absorber, water vapor is taken into the solution, which then flows back to the bottom of the generator. Because the pressure in the absorber is slightly below that in the generator, solution flow from a low-pressure area to one of relatively higher pressure is accomplished by the higher elevation of the absorber.

The water vapor (refrigerant) released from the generator rises to the condenser, where it is condensed to a liquid. Elevation of the condenser permits gravity flow of the refrigerant to the evaporator, aided by the slight pressure differential between the two chambers. Thus, by taking advantage of differences in fluid temperatures, density, and height of columns, continuous movement in the same direction throughout all circuits is accomplished without moving parts.

The cooling-water circuit can be traced in Fig. 10-7 by noting that it enters the absorber coil at point 10. Flow is then directed through tube 11 to the condenser and leaves the unit at point 12.

Fig. 10-8 shows a schematic sectional diagram of a typical absorption year-round air conditioner. Inlet air enters a plenum chamber that contains filter elements. The air, after having been cleaned, passes through the cooling coil (evaporator) and heating coil, and is then returned to the rooms or spaces being cooled.

During the cooling cycle, the room air is cooled and dehumidified. Heat is extracted from the air and moisture is condensed on the cooling coil. Thus, both functions of cooling and dehumidification are performed simultaneously. During the heating cycle, the air is warmed by the steam heating coil, and moisture is added by the humidifier.

Steam is provided for both heating and cooling cycles by the steam generator located in the base of the conditioner. Steam at atmospheric pressure flows from the generator into a two-position steam diverter valve, which automatically directs the flow of steam to either the heating coil or the absorption refrigeration unit. The steam diverter valve is positioned by an electric motor governed by a remote heating and refrigeration switch.

After the air leaves the heating coil, it passes through the humidifier, which functions during the heating cycle only. The humidifier, in this installation, consists of a number of horizontal trays,

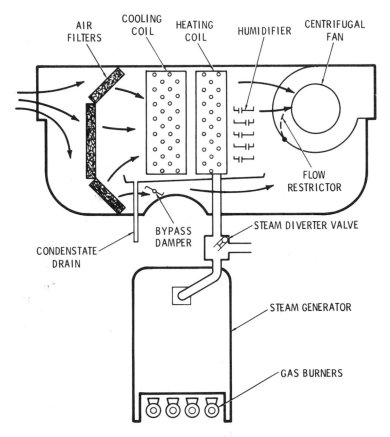

Fig. 10-8. Typical absorption year-round air conditioner.

each equipped with an overflow tube which feeds water to the next lower tray. This arrangement provides a large evaporative surface for positive humidification. Water is supplied at a predetermined and controlled rate of flow to the humidifier trays.

The air is drawn through the unit by a centrifugal fan that delivers the heated or cooled air through a duct-distributing system to the various rooms or spaces being conditioned. Because less air is normally required for winter heating than for summer cooling, a flow-restricting device is mounted in the fan scroll and functions on the heating cycle to automatically reduce the flow of air by adding resistance. This device usually consists of a pivoted blade, and its location is indicated by the dotted line in Fig. 10-8.

When maximum air delivery is desired, the air-restricting device is positioned tightly against the inside column of the fan scroll. When the restrictor is pivoted toward the fan wheel, thereby reducing normal wheel clearance, a resistance is thus imposed which alters fan performance to reduce air delivery. The airflow-restricting device is moved automatically between predetermined summer and winter positions by the motor that governs its operation.

Correct air distribution practices dictate a need in some cases for the handling of a greater quantity of air on the cooling cycle than required by the refrigeration unit for full-rated cooling capacity. Should this excess air be needed, the amount in excess of rated quantity should be handled through the bypass, thus assuring correct cooling coil performance. The bypass damper is located in the air circuit just beyond the cooling coil. During the heating season, the bypass damper will automatically be repositioned by the controlling motor to the closed position.

THERMOELECTRIC REFRIGERATION

Thermoelectric refrigeration is a cooling method developed comparatively recently for use in residential air conditioning, although the operating principle is not new. Thermoelectric refrigeration has been used for years as a cooling method for small refrigerators. It is also the cooling used on nuclear submarines.

Thermoelectric refrigeration is based on the thermocouple principle. In a thermoelectric refrigeration cooling system, electricity in the form of direct current power is applied to the thermocouple and heat is produced. The heat is transferred from one junction to the other producing a hot and cold junction. The direction of current flow determines which junction is hot and which is cold. If the power supply connections are reversed, the positions of the hot and cold junctions are also reversed.

The amount of cooling or heating that can be produced with a single thermocouple is small. For that reason, a number of thermocouples are wired in series to produce the quantity of heating and cooling required by the installation.

The schematic of a typical thermoelectric cooling system shown in Fig. 10-9 illustrates how direct current power is applied to a number of thermocouples wired in series. Heat is transferred from one side to the other depending on the direction of current flow.

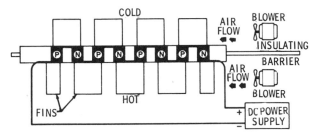

Fig. 10-9. Typical thermoelectric refrigeration cooling system.

COOLING WITH STEAM

Cooling with steam is based on the well-known fact that the boiling point of any liquid depends on the pressure to which it is subjected. By lowering the pressure, the boiling point is also lowered. When a liquid at a certain temperature and corresponding boiling point is sprayed into a closed vessel in which the pressure is lower, the entering liquid is above its boiling point at the new reduced pressure, and rapid evaporation takes place.

The basic components of a typical steam cooling system are shown in Fig. 10-10. In operation, the water to be cooled, returning from the air conditioning or other heat-exchanging apparatus at a temperature of 50°F, is sprayed into the evaporator. Because of the large surface of the spray, the boiling (or flashing) is very rapid and the unevaporated water falls to the bottom of the evaporator chilled to 40°F. At this temperature it is withdrawn from the evaporator by the chilled water pump and pumped to the heat-exchanging apparatus to again absorb heat, thus completing the cycle.

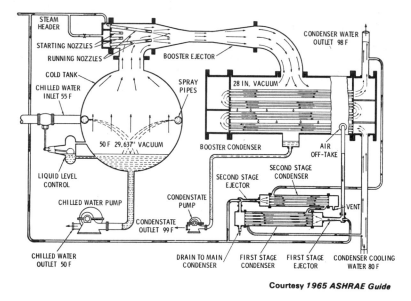

Fig. 10-10. Basic components of a steam cooling system.

In the booster ejector, steam flowing at high velocity through nozzles located in the ejector head is expanded in the venturi shaped diffuser. The kinetic energy of the steam is in part utilized in imparting velocity to the water vapor liberated in the evaporator and in compressing this vapor over a compression range from the evaporator pressure to the condenser pressure. In this compression, the temperature of the water vapor is raised so that it can be readily condensed by condensing water at temperatures normally available.

The initial evacuation and constant purging of air and other noncondensable gases is handled by a secondary group of ejectors and condensers. These are relatively small in size, but of the greatest importance, and normally consist of two steam-operated ejectors in series, each with its own condenser in which condensing water condenses the propelling water and entrained vapor. Condensers used in a steam cooling system may be either of the surface type, with water passing through condenser tubes over which the mixture of operating steam and water vapor flows, or

380

of the jet or barometric type, with condensing water sprayed directly into the steam mixture.

The amount of condensing water and condensing surface employed is such that the vapor temperature in the condenser is normally 5°F above the condensing water discharge temperature. This fixes the terminal pressure condition to which the steam mixture is compressed. The initial pressure condition (in the evaporator) is determined by the chilled water temperature desired. Variation in either of these two conditions affects the compression range and therefore the amount of operating steam required.

COOLING APPLICATIONS

There are many different ways to apply central air conditioning to a structure. The method used depends on a number of factors, but the type of heating system is probably the dominant one. The most common types of cooling applications are:

1. Water chillers.
2. Split-system cooling.
3. Year-round air conditioning.
4. Central cooling packages.
5. Cooling coils.

Water Chillers

A water chiller is used to add cooling and dehumidification to steam or hot water heating systems. A refrigeration-type water chiller consists of a compressor, condenser, thermal expansion valve, and evaporator coil. The water is cooled in the evaporator coil and pumped through the system.

A typical system in which a water chiller is used is shown in Fig. 10-11. The boiler and water chiller are installed as separate units, each with its own circulator (pump), or with one circulator in the return line. Hot water from the boiler is circulated through the convectors for heating, and cold water is circulated through the same piping from the water chiller for cooling purposes.

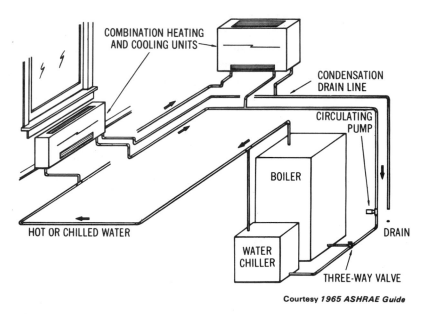

Fig. 10-11. Water chiller and boiler installed as separate units.

Each convector unit contains a blower to force the air across the convector coil (Fig. 10-12). Water condensed from the coil during the cooling operation is trapped in a drip pan and discarded through a drain connected to the convector. Some convectors also contain a filter for air cleaning. The room convectors in a water chiller cooling system are usually designed for individual control.

The same piping carries both the chilled and hot water to the room convectors, but it must be insulated to minimize condensation during the cooling operation.

Water chillers are available as separate units or as a part of a complete package containing the boiler. Separate water chiller units are used when cooling must be applied to an existing steam or hot water heating system.

Split-System Cooling

Another method of applying central air conditioning to a steam or hot-water heating system is to add forced-air cooling. This type of cooling application is sometimes referred to as a *split-system* installation; that is to say, a system split or divided between one type of heating (conventional steam or hot water) and another type of cooling (forced air). This results in some confusion because the term "split system" is also used to refer to the separation of components in a year-round central air conditioning system using forced warm-air heating and cooling (see below: "Location of Equipment").

Three typical methods of applying forced-air cooling to a steam or hot-water heating system are shown in Fig. 10-13.

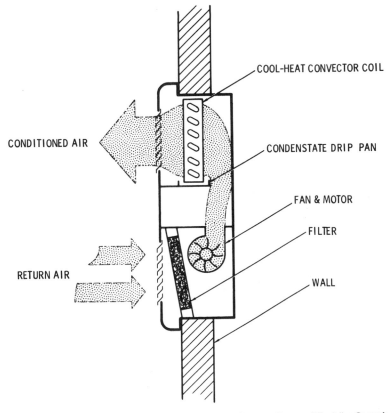

Fig. 10-12. Air forced by convector across convector coil.

383

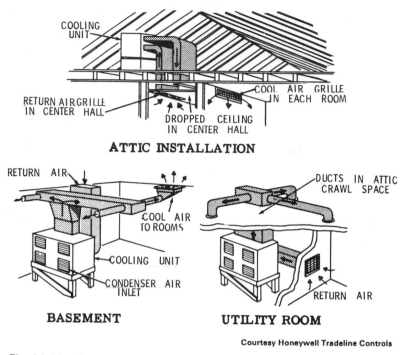

ATTIC INSTALLATION

BASEMENT   UTILITY ROOM

Courtesy Honeywell Tradeline Controls

Fig. 10-13. Methods of applying forced-air cooling to steam or hot-water heating systems.

Year-Round Air Conditioning

In a year-round air conditioning system, the heating and cooling units are combined in a single cabinet (Fig. 10-14). This combined package heats, cools, humidifies, dehumidifies, and filters the air in the structure as required. The unit may have an air-cooled, water-cooled, or evaporative type condenser. The arrangement of the ductwork will depend in part on the type of condenser used in the unit.

Central Cooling Packages

A central cooling package is a unit designed for central air conditioning applications. It consists of a cooling coil and the refrigeration equipment and will provide the necessary cooling

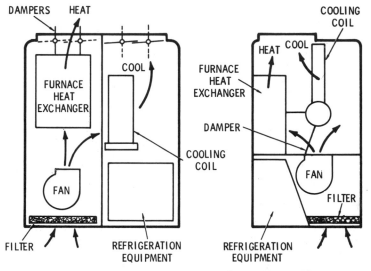

Fig. 10-14. Heating and cooling units contained in a single cabinet.

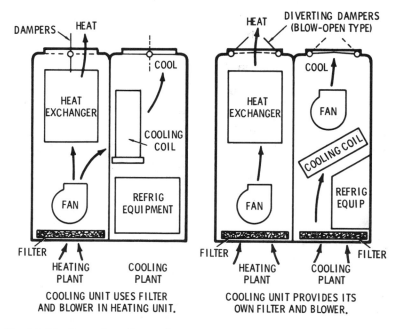

Fig. 10-15. Central cooling package.

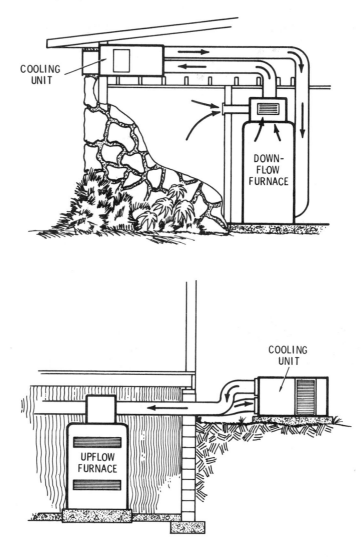

Fig. 10-16. Applications of central cooling

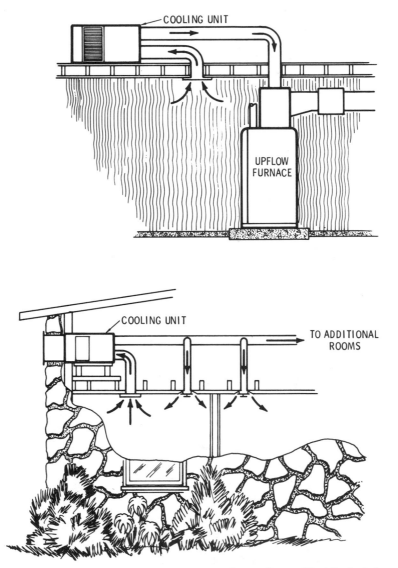

package having an air-cooled condenser.

and dehumidification as conditions require. These units are available with their own fans and filters, or they may be installed to use the filter and blower of the existing heating equipment (Fig. 10-15).

The method used to install a central cooling package will depend on whether the unit has an air-cooled or water-cooled condenser. If an air-cooled condenser is used, provisions must be made to carry outdoor air to and away from the condenser. Typical installations in which an air-cooled condenser is used are shown in Fig. 10-16. Fig. 10-17 illustrates some typical methods of

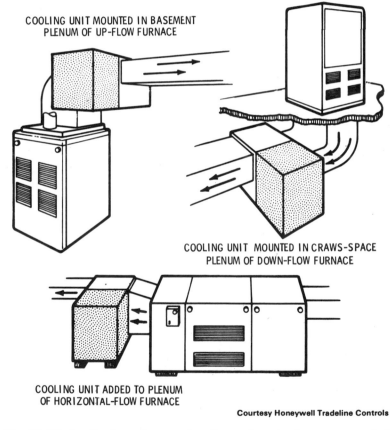

COOLING UNIT MOUNTED IN BASEMENT
PLENUM OF UP-FLOW FURNACE

COOLING UNIT MOUNTED IN CRAWS-SPACE
PLENUM OF DOWN-FLOW FURNACE

COOLING UNIT ADDED TO PLENUM
OF HORIZONTAL-FLOW FURNACE

Courtesy Honeywell Tradeline Controls

Fig. 10-17. Applications of a central cooling package having a water-cooled condenser.

applying a central cooling package in which a water-cooled condenser is used.

Cooling Coils

Cooling can be applied to a warm-air heating system by installing an evaporator coil or a cold-water coil in the ductwork.

The evaporator coil is the low-side section of a mechanical refrigeration system. As shown in Fig. 10-18, the evaporator coil is installed in the ductwork above the furnace. It is connected by refrigerant piping to the condenser coil and compressor installed outdoors.

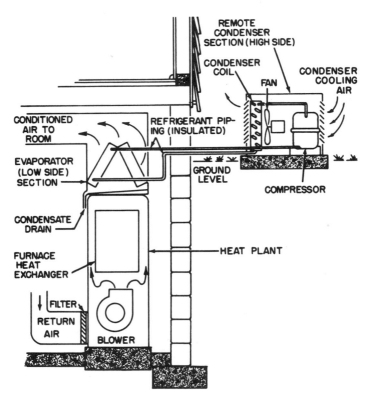

Courtesy Honeywell Tradeline Controls

Fig. 10-18. Evaporator coil installed in ductwork of a warm-air heating system.

A thermostatic expansion valve and condensation drip pan (with drain) are included with the evaporator coil. Sometimes a fan is added to the coil to supplement the furnace blower.

A cold-water coil may be used instead of an evaporator coil in the ductwork. Cold water is supplied to the coil by a water chiller, which can be located in the basement, a utility room, or outdoors. If the water chiller is installed outdoors, a gas engine can be used to drive the compressor (Fig. 10-19).

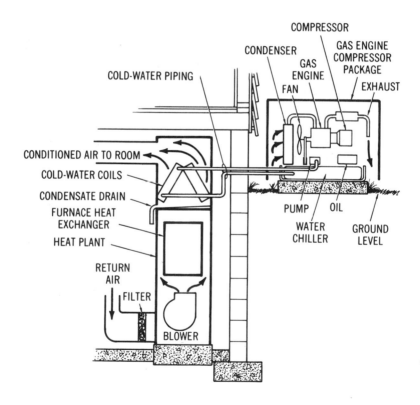

Fig. 10-19. Cold-water coil used with outdoor gas engine compressor.

LOCATION OF EQUIPMENT

The location of the various equipment used in a central air conditioning system will depend on the type of heating system, the type of equipment, and the design of the structure. The equipment in a central air conditioning system can be arranged in one of the following ways:

1. Single-package system.
2. Remote, or split, system.

In a single-package system, the compressor, evaporator, condenser, blower, and heating equipment are contained in one compact unit in the same cabinet. A typical example of a single-package system is the year-round air conditioning unit shown in Fig. 10-14. A cooling unit mated to a heating unit as shown in Fig. 10-15 may also be regarded as a single-package system.

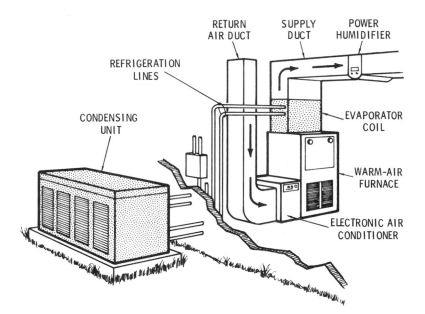

Courtesy Mueller Climatrol Corp.

Fig. 10-20. Typical remote split-type air conditioning system.

A typical remote, or split, system is shown in Fig. 10-20. In this installation, the compressor and air-cooled condenser unit is located outdoors. The evaporator coil, fan, and heating plant are located indoors in the conditioned space.

CHAPTER 11

Air Conditioning Equipment

The gas compression method of mechanical refrigeration used in conjunction with an air-cooled condenser is probably the most common cooling system used in residences and small commercial buildings. For that reason, this chapter is devoted almost exclusively to a description of mechanical refrigeration equipment.

MECHANICAL REFRIGERATION EQUIPMENT

The equipment of a gas compression mechanical refrigeration system can be divided into mechanical and electrical components. The electrical components are described below (see "Electrical Components") and in Chapter 6 of Volume 2 (Other Automatic Controls). The principal mechanical components include:

1. Compressor.
2. Condenser.
3. Receiver.
4. Evaporator.
5. Liquid refrigerant controls.

COMPRESSORS

A *compressor* is a device used in a mechanical refrigeration system to receive and compress low-pressure refrigerant vapor into a smaller volume at higher pressure. Thus, the primary function of a compressor is to establish a pressure difference in the system to create a flow of the refrigerant from one part of the system to the other.

Compressors are manufactured in many different sizes for a variety of different applications. They are classified by their method of operation into the three following types:

1. Centrifugal compressors.
2. Rotary compressors.
3. Reciprocating compressors.

Another method of classifying the compressors used in residential cooling systems is on the basis of how accessible they are for field service and repair. The following three types are recognized:

1. Open-type compressors.
2. Hermetic compressors.
3. Semihermetic compressors.

As shown in Fig. 11-1, an *open-type compressor* is usually driven by a separately mounted electric motor. Both the compressor and motor are easily accessible for field service and repair.

The *hermetic compressor* shown in Figs. 11-2 and 11-3 differ from the open type in being *completely* sealed, usually by welding. No provision is made for service access.

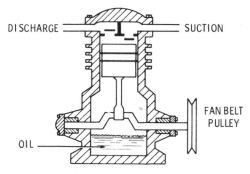

DISCHARGE — SUCTION

FAN BELT PULLEY

OIL

Fig. 11-1. Diagram on an open-type compressor.

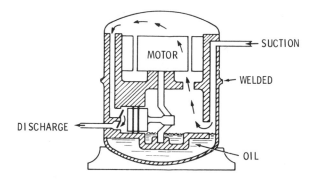

MOTOR — SUCTION

— WELDED

DISCHARGE ◄—

OIL

Fig. 11-2. Diagram of a hermetic compressor.

The *semihermetic compressor* is similar in construction to the hermetic type, except that field service and repairs are possible on the former through bolted access plates.

Both the compressor and electric motor are sealed in the same casing in hermetic and semihermetic compressors. As a result, the motors are cooled by a refrigerant that flows through and around the motors. Quick-trip overload relays provide additional protection against overheating should the refrigerant flow be cut off.

Semihermetic and hermetic compressor motors of a given size are designed and constructed to operate on a heavier current without overheating. Hermetically sealing the compressor and

395

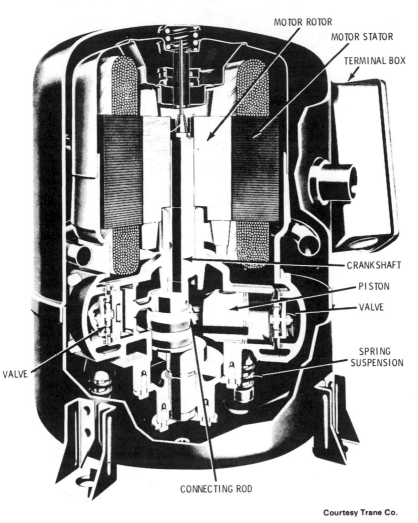

MOTOR ROTOR

MOTOR STATOR

TERMINAL BOX

CRANKSHAFT

PISTON

VALVE

SPRING SUSPENSION

VALVE

CONNECTING ROD

Courtesy Trane Co.

Fig. 11-3. Sectional view of a typical vertical hermetic compressor.

electric motor in the same casing also results in a greater output. The principal disadvantage of a hermetic compressor is that it must be replaced with a new unit when it malfunctions. Because this usually occurs in the cooling season, the homeowner may be without air conditioning for a day or so when it is most needed.

A *centrifugal compressor* is a nonpositive-displacement compressor that relies in part on centrifugal effect for pressure rise. Compression of the refrigerant is accomplished by means of centrifugal force. As a result, this type of compressor is generally used in installations having large refrigerant volumes and low-pressure differentials.

A *rotary compressor* is a hermetically sealed, direct-drive compressor that compresses the gas by movement of the roll in relation to the pump chamber. Rotary compressors are manufactured in large quantities for use in residential cooling systems (Fig. 11-4).

A *reciprocating compressor* is a positive-displacement compressor with a piston (or pistons) moving in a straight line but alternately in opposite directions. Both open and hermetic reciprocating compressors are manufactured for use in refrigeration systems. A typical reciprocating compressor is shown in Fig. 11-5.

TROUBLESHOOTING COMPRESSORS

When a compressor is suspected of being defective, a czomplete analysis should be made of the system before the compressor is replaced. In some cases, the symptoms encountered in servicing an air conditioner may lead the serviceman to suspect the compressor when actually the trouble is in another section of the system. For example, noise and knocking are often attributed to a faulty compressor when the trouble may be a loose compressor flywheel, incorrect belt alignment, or air in the system or a large quantity of oil being pumped through the compressor because of liquid refrigerant in the crankcase.

The troubleshooting chart that follows lists the most common operating problems associated with air conditioning compressors. For each observable symptom, a possible cause and remedy are suggested.

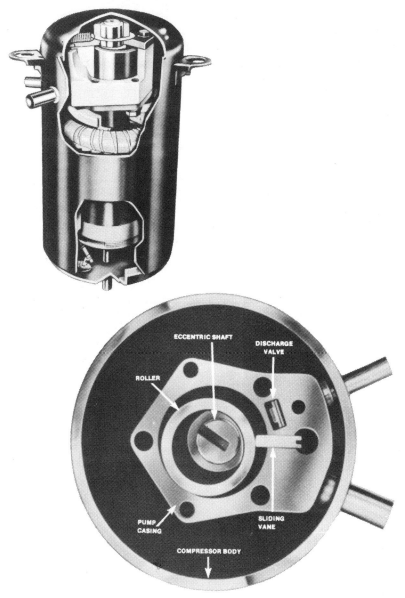

Fig. 11-4. Rotary compressor.

398

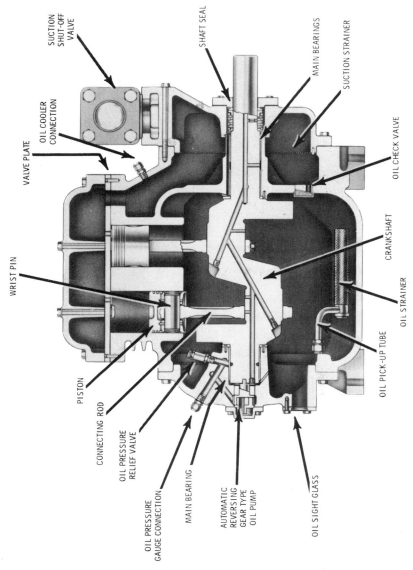

Fig. 11-5. Direct-drive reciprocating compressor showing arrangement of component parts.

Symptom and Possible Cause   *Possible Remedy*

Compressor Does Not Start; No Hum

1. Open power switch.	1. Close switch.
2. Fuse blown.	2. Replace fuse.
3. Broken electrical connection.	3. Check circuit and repair.
4. Overload stuck.	4. Wait for reset; check current.
5. Frozen compressor or motor bearings.	5. Replace the compressor.
6. High head pressure; cut out open due to high pressure.	6. Push high pressure button and check for air circulation in condenser.
7. Central contacts in open position.	7. Repair and check control.
8. Open circuit in compressor stator.	8. Replace the compressor.
9. Thermostat set too high.	9. Reset to proper level.
10. Solenoid valve closed.	10. Examine holdng coil; if burned out, replace.

Compressor Starts but Motor Will Not Get Off of Starting Windings; High Amperage and Rattle in the Compressor.

1. Compressor improperly wired.	1. Check wiring against wiring diagram; rewire if necessary.
2. Low line voltage.	2. Check line voltage and correct (decrease load on line or increase wire size).
3. Relay defective.	3. Replace relay.
4. Run capacitor defective.	4. Replace run capacitor.
5. Compressor motor starting and running windings are shorted.	5. Replace compressor.

AIR CONDITIONING EQUIPMENT

Symptom and Possible Cause:	Possible Remedy:
6. High discharge pressure.	6. Correct excessive high pressure.
7. Starting capacitor weak.	7. Check capacitor; replace if necessary.
8. Tight compressor.	8. Check oil level and correct; or replace compressor.

Compressor Will Not Start; Hums and Trips on Overload

1. Compressor improperly wired.	1. Check wiring against wiring diagram; rewire if necessary.
2. Low line voltage.	2. Check line voltage and correct.
3. Starting capacitor defective.	3. Replace capacitor.
4. Relay contacts not closing.	4. Check contact points; replace if defective.
5. Grounded compressor motor or motor with open winding.	5. Replace compressor.
6. High discharge pressure.	6. Check excessive high pressure. Check air.
7. Tight compressor.	7. Check oil level and correct, or replace compressor.

Compressor Starts and Runs But Short Cycles

1. Low line voltage.	1. Check line voltage; correct.
2. Additional current passing through overload protector.	2. Check wiring diagram; fan motors may be connected to the wrong side of the protector.

401

Symptom and Possible Cause: *Possible Remedy:*

3. Suction pressure high.

3. Check compressor for possibility of misapplication.

4. High discharge pressure.

4. Correct excessive high pressure.

5. Run capacitor defective.

5. Check capacitor and replace.

6. Compressor too hot; inadequate motor cooling.

6. Check refrigerant charge; add if necessary.

7. Compressor motor windings are shorted.

7. Replace compressor.

8. Overload protector defective.

8. Check current, give reset time; if it does not come back, replace compressor.

9. Compressor tight.

9. Check oil level and correct, or replace compressor.

10. Discharge valve defective.

10. Replace compressor.

Compressor Short Cycling

1. Thermostat differential set too closely.

1. Widen differential.

2. Dirty air filter.

2. Replace.

3. Refrigerant charge too low.

3. Recharge system with correct charge.

4. Dirty strainer or dryer in liquid line.

4. Replace.

5. Restricted capillary tube or expansion valve.

5. Replace.

6. Dirty condenser.

6. Clean condenser.

7. Too much refrigerant.

7. Discharge some refrigerant.

8. Air in system.

8. Purge system.

9. Compressor valve leaks.

9. Replace compressor.

Symptom and Possible Cause: *Possible Remedy:*

10. Overload protector cutting out.

10. Check current; give reset time; if it does not come back, replace compressor.

Compressor Runs Continuously

1. Shortage of refrigerant.

1. Test at refrigerant test cock; if short of gas add proper amount. Test for leaks.

2. Compressor too small for load.

2. Increase capacity by increasing speed or using larger compressor.

3. Discharge valve leaks badly.

3. Test valve; if leaking, remove head of compressor and repair or service.

Compressor Noisy

1. Vibration because unit not bolted down properly.

1. Examine bolts and correct.

2. Too much oil in circulation, causing hydraulic knock.

2. Check oil level, check for oil refrigerant test cock; correct.

3. Slugging due to flooding back of refrigerant.

3. Expansion valve open too wide. Close.

4. Wear of parts such as piston, piston pins.

4. Locate cause. Repair or replace compressor.

High Suction Pressure

1. Overfeeding of expansion valve.

1. Regulate expansion check bulb attachment.

2. Compressor too small for evaporator or load.

2. Check capacity. Try to increase speed or replace with larger-size compressor.

Symptom and Possible Cause: *Possible Remedy:*

3. Leaky suction valves.   3. Remove head, examine valve discs, or rings; replace if worn.

Low Suction Pressure

1. Restricted liquid line and expansion valve or suction screens.
2. Compressor too big for evaporator.

3. Insufficient gas in system.

4. Too much oil circulating in system.
5. Improper adjustment of expansion valves.

1. Pump down, remove, examine and clean screens.
2. Check capacity against load, reduce speed if necessary.
3. Check for gas shortage at test cock.
4. Remove oil.

5. Adjust valve to give more flow. If opening valve does not correct, increase size to give greater capacity.

Each compressor should be equipped with internal devices to provide protection against the following operating problems:

1. Motor overload.
2. Locked rotor.
3. Extreme voltage supply.
4. Excessive winding temperature.
5. Excessive pressure.
6. Loss of refrigerant charge.
7. Compressor cycling.

If these devices are operating properly, the compressor will provide efficient and troublefree service.

COMPRESSOR REPLACEMENT

Before replacing a hermetic compressor, be sure to check other possible causes of system malfunction (see "Troubleshooting Compressors"). Do not replace the compressor unless you are absolutely certain it is the source of the trouble.

Many manufacturers will provide instructions for replacing the compressor along with their installation, servicing, and operating literature. Carefully read these instructions before attempting to disconnect the compressor.

Disconnect the power supply, remove the fuses, and check the liquid refrigerant for oil discoloration or an acrid odor. These are indications that a compressor burnout has contaminated the system. If the system is not properly cleaned up, the replacement compressor will also burn out.

The system can be checked for contamination by discharging a small amount of refrigerant and oil through the high-side port onto a clean white cloth and checking it for discoloration and odor. Perform the same test on the low-side gauge port. If the system shows signs of contamination, discharge the remainder of the refrigerant through the liquid line gauge port (on a factory-charged system) or the high-side gauge port (on a field-charged system). Inspect the refrigerant lines to determine the exact extent of contamination.

Examine the refrigerant lines connected to the evaporator for contamination. A rapid compressor burnout will usually leave the evaporator coil unaffected. If the burnout has been particularly slow and the refrigerant and oil have been circulated through the system, the evaporator will also be contaminated. A contaminated evaporator can be cleaned by flushing it with R11.

ELECTRIC MOTORS

The electric motors used to power mechanical refrigeration equipment are commonly of the following two types:

1. Single-phase induction motors.
2. Three-phase induction motors.

The single-phase induction motors are usually classified by the method used to start them. Among the more common ones are:

1. Split-phase motors.
2. Capacitor-start motors.
3. Permanent-split capacitor motors.
4. Capacitor-start, capacitor-run motors.

Capacitor-start and capacitor-start, capacitor-run motors are described in Chapter 6 of Volume 2 (Other Automatic Controls).
Either capacitor-start, capacitor-run or three-phase induction motors can be used to power compressors. The latter are normally used when three-phase current is available.

TROUBLESHOOTING ELECTRICAL MOTORS

The troubleshooting chart that follows lists the most common operating problems associated with electric motors. For each symptom, a possible cause and remedy are suggested.

Symptom and Possible Cause	*Possible Remedy*
Motor Blows Fuses, Trips Overload	
1. Fuses and/or overload too small.	1. Install larger sizes if necessary within safe limit for motor.
2. Poor switch contacts.	2. Check and replace contacts. Replace entire switch if necessary.
3. Low voltage.	3. Check voltages with meter; if more than 10 percent low, notify power company to correct condition.

4. Leaky discharge valve.	4. Replace.
5. Overloaded motor.	5. Check Bhp load against back pressure and compressor speed; if motor too small, increase size.

Motor Hot

1. Low voltage.	1. Check voltage with meter; if more than 10 percent low, notify power company to correct condition.
2. Bearings need oil.	2. Oil bearings to reduce friction.
3. Overloaded motor.	3. Check Bhp load against back pressure and compressor speed; if motor too small, increase size.

GAS ENGINES

A four-cylinder water-cooled gas engine using natural gas as the fuel can be used to power the compressor (and sometimes the condenser fan). It is normally mounted in a weatherproof cabinet outside the residence where venting is not a problem. Because an internal combustion engine must be vented, electric motors are by far the most popular type used for powering mechanical refrigeration equipment.

ELECTRICAL COMPONENTS

The principal electrical components of a mechanical refrigeration system include the following:

1. Electric compressor motor.
2. Compressor contactor or relay.

3. Compressor starter.
4. Overload protector.
5. Capacitor.
6. Potential relay.
7. Pressure switch.
8. Evaporator fan motor.
9. Condenser fan motor.
10. Evaporator fan relay.

The wiring diagram in Fig. 11-6 illustrates the relationship of these various components. Switches, relays, and capacitors have already been described in considerable detail in Chapter 6 of Volume 2 (Other Automatic Controls).

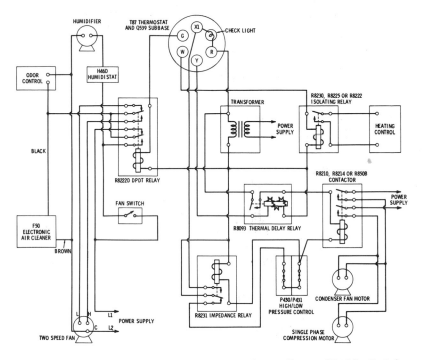

Courtesy Honeywell Tradeline Controls

Fig. 11-6. Impedance relay, thermal delay relay, isolating relay, heat control, electronic air cleaner, and humidity controls added to a basic cooling control system.

The room thermostat is sometimes included when the electrical components of a refrigeration system are listed. The reader is referred to Chapter 4 of Volume 2 (Thermostats and Humidistats) for detailed information about thermostats.

TROUBLESHOOTING
ELECTRICAL COMPONENTS

The troubleshooting charts that follow list the most common problems associated with the operation of the electrical components in air conditioning equipment. For each symptom, a possible cause and remedy are suggested.

Symptom and Possible Cause	*Possible Remedy*
Starting Capacitor Is Open, Shorted, or Burned Out	
1. Relay contacts not operating properly.	1. Clean contacts or replace.
2. Improper capacitor.	2. Check for proper MFD rating and voltage.
3. Low voltage.	3. Check and correct.
4. Improper relay.	4. Check and replace.
5. Short cycling.	5. Replace starting capacitor.
Running Capacitor Is Open, Shorted, or Burned Out	
1. Improper capacitor.	1. Check for proper MFD rating and voltage.
2. Excessive high line voltage.	2. Correct line voltage to not more than 10 percent of rated motor voltage.

**Relay Is Shorted or
Burned Out**

1. Line voltage is too low or too high.
2. Incorrect running capacitor.
3. Relay is loose.
4. Short cycling.

1. Check and correct.
2. Replace with correct MFD capacitor.
3. Tighten relay.
4. Replace relay.

CONDENSER

A *condenser* is a device used to liquify gas by cooling. In operation, hot discharge gas (refrigerant vapor) from the compressor enters the condenser coil at the top and, as it is condensed, drains out of the condenser to a receiver located at a lower level.

The condenser coil is located along with the compressor and controlling devices in the *condensing unit.* In a remote or split-system air conditioning installation, the condensing unit is located outdoors (Fig. 11-7). Condensers are available in a variety of designs, including plain tube, finned tube, plate type, and as series-pass and parallel-pass units. A number of different condensers are illustrated in Figs. 11-8, 11-9, 11-10, and 11-11.

Condensers may be classified with respect to the cooling method used into the following three types:

1. Air-cooled condensers.
2. Water-cooled condensers.
3. Combined air- and water-cooled condensers.

An *air-cooled condenser* consists of a coil of ample surface across which air is blown by a fan or induced by natural draft (Fig. 11-12). This type of condenser is universally used in small-capacity refrigerating units.

A *water-cooled condenser* is similar to a steam surface condenser in that cooling is accomplished by water alone that circulates through tubes or coils enclosed in a shell. The refrigerant circulates through the annular space between the tubes or coils.

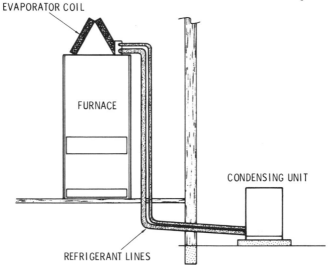

EVAPORATOR COIL

FURNACE

CONDENSING UNIT

REFRIGERANT LINES

Fig. 11-7. Location of the condensing unit in a remote or split-system air conditioning installation.

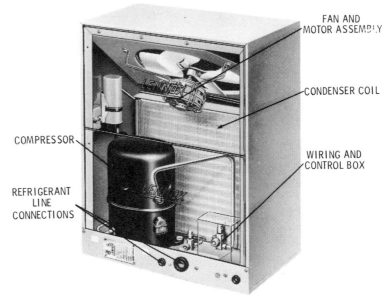

FAN AND
MOTOR ASSEMBLY

CONDENSER COIL

COMPRESSOR

WIRING AND
CONTROL BOX

REFRIGERANT
LINE
CONNECTIONS

Courtesy Lennox Air Conditioning and Heating

Fig. 11-8. Lennox HSW4 series air conditioning condenser unit.

411

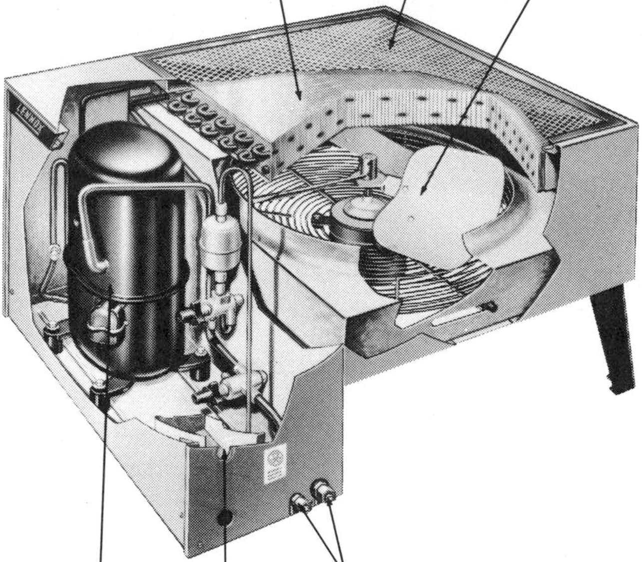
CONDENSER COIL
CONDENSER COIL GRILLE
FAN AND MOTOR ASSEMBLY
COMPRESSOR
WIRING AND CONTROL BOX
REFRIGERANT LINE CONNECTIONS

Courtesy Lennox Air Conditioning and Heating

Fig. 11-9. Lennox HS8 series air conditioning condenser unit.

Because of its construction, a water-cooled condenser is also sometimes called a *double-pipe condenser* (Fig. 11-13).

Maximum temperature differences can be obtained by connecting the condenser for counterflow. The type of arrangement usually gives the best operating results.

Shell and tube construction is recommended for a small condensing unit. In medium-size units, shell and coil construction works very well (Figs. 11-14 and 11-15).

A *combined air- and water-cooled condenser,* more commonly known as an *evaporative condenser,* consists of a coil cooled by water sprayed from above and cold air entering from below (Fig. 11-16).

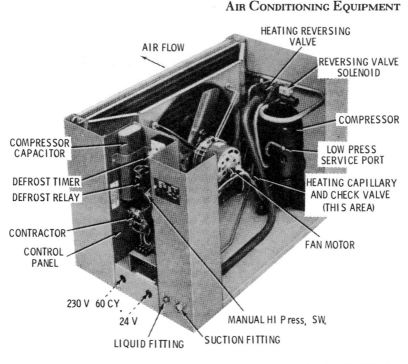

AIR FLOW

HEATING REVERSING VALVE

REVERSING VALVE SOLENOID

COMPRESSOR

COMPRESSOR CAPACITOR

LOW PRESS SERVICE PORT

DEFROST TIMER
DEFROST RELAY

HEATING CAPILLARY AND CHECK VALVE (THIS AREA)

CONTRACTOR
CONTROL PANEL

FAN MOTOR

230 V 60 CY

24 V

LIQUID FITTING

SUCTION FITTING

MANUAL HI Press, SW.

Courtesy Bard Mfg. Co.

Fig. 11-10. Bard 36HPQ1 condensing unit.

As water evaporates from the coil, it brings about a cooling effect, which condenses the refrigerant within the coil. The hot refrigerant gas within the coil is thus changed to the liquid state by the combined action of the sprayed water and the large volume of moving air supplied by the fan. The water that does not evaporate is recirculated by means of a pump.

Because an evaporative condenser is not wasteful of water, large compressor installations are possible in areas where water is scarce. Tests have shown that the amount of water required will not exceed .03 gpm per ton of refrigeration. Evaporative condensers also eliminate waste water disposal problems and provide the most economical means of cooling refrigerant gases.

Air-cooled condensers, like evaporators, should be kept free from dirt, lint, and other foreign materials because they tend to

413

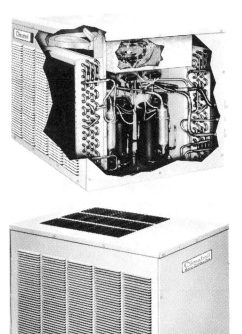

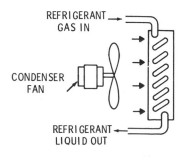

Courtesy Mueller Climatrol Corp.

Fig. 11-11. Climatrol 938-1 dual compressor condensing unit.

REFRIGERANT GAS IN

CONDENSER FAN

REFRIGERANT LIQUID OUT

Fig. 11-12. Air-cooled condenser.

Courtesy Honeywell Tradeline Controls

414

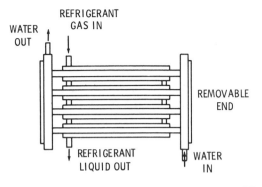

Fig. 11-13. Double-pipe condenser.

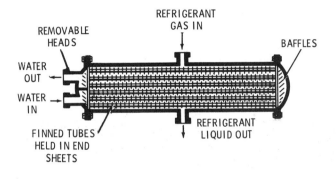

Fig. 11-14. Shell-and-tube condenser.

reduce the airflow around tubes and fins if they are allowed to accumulate.

TROUBLESHOOTING CONDENSERS

Common operating problems associated with air conditioning condensers are listed in the troubleshooting charts that follow. For each symptom, a possible cause and remedy are suggested.

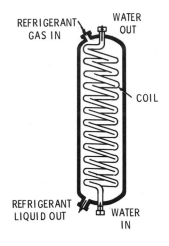

Fig. 11-15. Shell-and-coil condenser.

Courtesy Honeywell Tradeline Controls

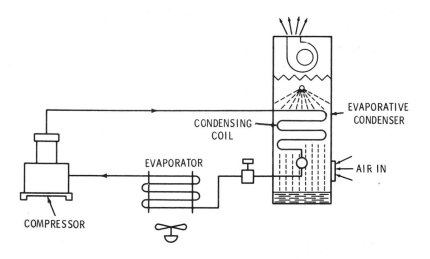

Courtesy Honeywell Tradeline Controls

Fig. 11-16. Evaporative condenser.

416

Symptom and Possible Cause   *Possible Remedy*

Compressor Pressure Too High

1. Air in system.
2. Dirty condenser.
3. Refrigerant too high.
4. Unit location is too hot.
5. Condenser air is off.

1. Purge system.
2. Clean condenser.
3. Discharge some refrigerant.
4. Change unit location.
5. Check condenser motor connections for burnout.

Condenser Pressure Too Low

1. Refrigerant charge too low.

2. Compressor discharge or suction valves defective.
3. Entering temperature to evaporator is low.

1. Check for leak, repair, and recharge with correct amount of refrigerant.
2. Replace compressor.

3. Raise temperature.

RECEIVER

As the name suggests, the *receiver* is the reservoir for any excess liquid refrigerant not being used in the system. The liquid receiver must be large enough to hold the *total* amount of refrigerant used in the system. Receivers are commonly constructed of drawn steel shells welded together to form a single unit.

EVAPORATOR

An *evaporator* is a device used in either mechanical or absorption type refrigeration systems to transfer or absorb heat from the air surrounding the evaporator to the refrigerant. In so doing, the

417

liquid refrigerant is evaporated or boiled off as it passes through the evaporator.

Evaporators are made of copper tubing with or without closely spaced aluminum fins designed to increase the heat transfer surface. Because of its function (i.e., removing heat from room air) and construction, an evaporator is also referred to as an *evaporator coil, cooling coil, blower coil,* or *direct-expansion coil.*

Fig. 11-17 illustrates a typical evaporator coil used in the bonnet or plenum of a warm-air furnace. The coil capacity must be matched to the condensing unit for efficient cooling. This is particularly important to remember when converting an existing heating system to year-round air conditioning.

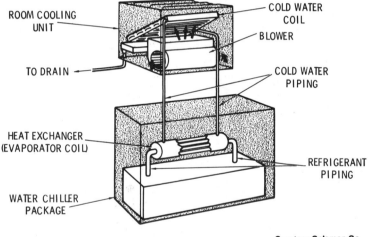

Courtesy Coleman Co.

Fig. 11-17. Evaporator coils.

An evaporator coil may also be placed in the heat exchanger of a water chiller as shown in Fig. 11-18. The water cooled in the water chiller is piped to the cold-water coil over which the room air is circulated. The coil may be duct mounted (in a warm-air heating system) or located in a room cooling unit (in a forced hot-water heating system).

Fig. 11-18. Evaporator coil and water chiller.

TROUBLESHOOTING EVAPORATORS

An evaporator must be kept clean and free of dirt and dust so that the flow of air through the tubes remains unrestricted. If the evaporator is damaged or leaking to such an extent that it cannot be successfully repaired, it should be replaced by a new assembly. If repairs or replacement are necessary, the complete coil assembly must be removed from the system.

Before removing a damaged or leaking evaporator, the refrigerant lines must first be disconnected and the evaporator retaining bolts (if used) loosened and removed.

The new evaporator is bolted or otherwise secured in place and connected to the refrigerant lines. After all connections are made, the entire system is evacuated, recharged with refrigerant, and tested for leaks.

419

REFRIGERANTS

A *refrigerant* is any substance that produces a refrigerating effect by absorbing heat as it expands or vaporizes. A desirable refrigerant should possess chemical, physical, and thermodynamic properties that permit efficient and safe operation in refrigerating systems. Among the properties possessed by a good refrigerant are:

1. Low boiling point.
2. Nontoxic and nonirritating.
3. Nonexplosive.
4. Nonflammable.
5. Mixes well with oil.
6. Operation on a positive pressure.
7. High latent heat value.
8. Not affected by moisture.

Two of the more commonly used refrigerants are Freon 12 and Freon 22. These are clear, almost colorless liquids at temperatures below their boiling points.

Freon 12 has a boiling point of –21.8°F at atmospheric pressure and is characterized by moderate pressure differentials between suction and discharge. A moderate volume of Freon 12 is required per ton of refrigeration.

Freon 22 has a boiling point of –41.4°F at atmospheric pressure. In contrast to Freon 12, it has a considerably higher pressure differential between suction and discharge. As a result, it requires a smaller volume of refrigerant per ton of refrigeration.

LIQUID REFRIGERANT CONTROL DEVICES

Each refrigeration system may be described in terms of a low side and a high side of operating pressure. The *low side* is that part of the refrigeration system that normally operates under low pressure (as opposed to the high side). It is identified as that part of a refrigeration system lying between the expansion valve and the intake valve in the compressor, and includes the evaporating or cooling surface, the intake line, and the compressor crankcase. In other words, that part of the refrigeration equipment under intake pressure. The term "low side" is sometimes used to designate the evaporator coils.

The *high side* is that part of the refrigeration system operating under high pressure. The term "high side" is sometimes used to designate the condensing unit.

Some form of expansion device is necessary to control the flow of liquid refrigerant between the low and high sides of a refrigeration system. The various expansion devices designed to provide automatic control of refrigerant flow are:

1. Automatic expansion valves.
2. Thermostatic expansion valves.
3. Float valves.
4. Capillary tubes.

AUTOMATIC EXPANSION VALVES

An *automatic expansion valve* is a pressure-actuated diaphragm valve used to maintain a constant pressure in the evaporator of a direct-expansion mechanical refrigeration system (Fig. 11-19). It accomplishes this function by regulating the flow of refrigerant from the liquid line into the evaporator. In this way, the evaporator is always supplied with the proper amount of refrigerant to meet conditions. An automatic expansion valve does not respond well to load fluctuations. For this reason, it is not recommended for use in air conditioning (see "Thermostatic Expansion Valves" below).

THERMOSTATIC EXPANSION VALVES

The *thermostatic expansion valve* is designed to automatically control the flow of liquid refrigerant entering the evaporator coil. The valve mechanism must operate freely and without restriction in order to allow the proper amount of refrigerant to enter the evaporator (Figs. 11-20 and 11-21).

Failure of any part of the thermostatic expansion valve will

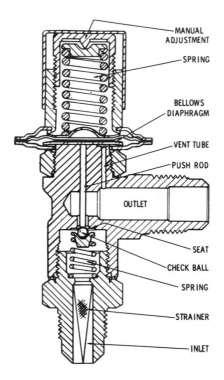

Fig. 11-19. Automatic expansion valve.

affect the refrigerating capacity of the unit and the cooling capacity of the system. Faulty operation of the expansion valve may be caused by mechanically frozen internal parts clogging the strainer or valve orifice, or by failure of the regulating sensor bulb, which operates the needle valve.

If the expansion valve is frozen partly open, the capacity of the unit will be affected because the flow of refrigerant is restricted. If the expansion valve is frozen in fully open position, the liquid refrigerant may flow through the entire system causing very cold inlet line, high inlet pressure, pounding of the compressor, and a cold compressor head.

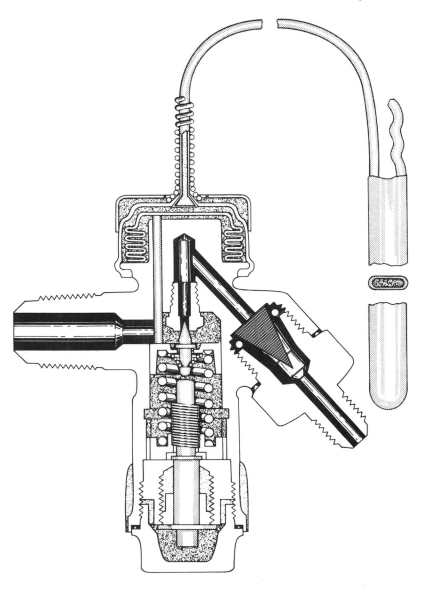

Fig. 11-20. Typical thermostatic expansion valve.

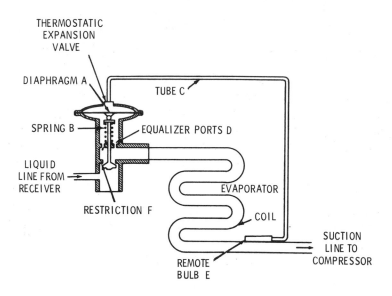

Fig. 11-21. Thermostatic expansion valve operating principle.

The pressure in the sensor bulb or feeler of the expansion valve must be somewhat higher than the pressure in the evaporator, which means that the inlet line where the bulb is clamped must be at a higher temperature than within the evaporator. Accordingly, the vapor in the inlet line at this point must be in a somewhat superheated state. This superheat should be at the minimum which will allow the valve to regulate the flow of refrigerant.

For higher-temperature work, such as comfort cooling work, the amount of superheating will vary between 5° and 10°. For lower temperature work, such as product cooling work, the superheating will vary between 4° and 6° or in some cases even lower.

Excessive superheat indicates a lack of sufficient refrigerant flowing through the expansion valve, a condition that reduces the capacity of the evaporator.

The sensitivity and response of the thermostatic expansion valve is largely dependent upon the proper installation of the sensor (feeler) bulb. The sensor bulb should always be firmly attached to the inlet line.

When properly installed and adjusted, the thermostatic expansion valve will maintain all the evaporator surface effective and in contact with the boiling refrigerant regardless of the change in load on the evaporator, provided of course that the evaporator valve has sufficient capacity for peak loads.

Under normal operating conditions, the sensor (feeler) bulb will cause the thermostatic expansion valve to close during the shutdown period. There are, however, certain conditions that will affect this and may cause overflooding of the evaporator. For this reason, the inlet lines should be trapped.

If the evaporator in an air conditioning system is connected to outside air or where it may receive air at a temperature lower than that surrounding the sensor (feeler) bulb, the difference in temperature between the evaporator and the bulb will cause the valve to open and overflood the low side. Under such conditions, it is essential that the liquid line be equipped with a solenoid valve to positively shut off the refrigerant during the shut-down period.

FLOAT VALVES

A *float valve* is one actuated by a float immersed in a liquid container (Fig. 11-22). Both low-side and high-side float valves are used to control the flow of liquid refrigerant in a refrigeration system.

A *low-side* float valve is one that is operated by a low-pressure liquid. It opens at a low level and closes when the liquid is at a high level. In other words, when there is no liquid in the evaporator, the float and lever arm are positioned so that the valve is left open. When liquid refrigerant under pressure from the compressor again enters the float chamber, the float rises until a predetermined level is reached and the valve is closed.

A *high-side* float valve is one that is operated by a high-side

425

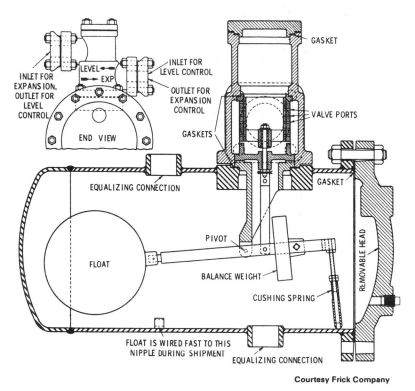

Fig. 11-22. Float valve construction details.

pressure. The valve opens on an increase of the liquid level in the float chamber and admits liquid to the low side.

CAPILLARY TUBES

A *capillary tube* is a tube of small internal diameter used in refrigeration air conditioning systems as an expansion device between the high and low pressure sides. It can also be used to transmit pressure from the sensor bulb of some temperature controls to the operating element (Fig. 11-23).

The use of a capillary tube as a liquid refrigerant expansion device is largely limited to completely assembled factory refrigeration units because the bore diameters and the length of the tube are critical to its efficiency. The pressure reduction that occurs between the condenser and evaporator results from the pressure drop or friction loss in the long, small-diameter passage provided by the capillary tube. No pressure-reducing valve is necessary between the high-side and low-side pressure zones when a capillary system is used.

REFRIGERANT PIPING

The refrigerant travels between the various components of a mechanical refrigeration unit or system in small-diameter copper tubing sometimes referred to as the *refrigerant lines.*

The *suction line* is the piping between the evaporator and compressor inlet. Its function is to carry the refrigerant vapor to the compressor. It is important that the suction line be correctly sized for a practical pressure drop at full load. Under minimum load conditions, the suction line should be able to return oil from the evaporator to the compressor. Two other desirable features that should be incorporated in the design of a suction line are:

1. The prevention of oil drainage into a nonoperating evaporator from an operating one.
2. The prevention of liquid drainage into a shut-down compressor.

The *liquid line* is the piping that carries the liquid refrigerant from the condenser or receiver to a pressure-reducing device.

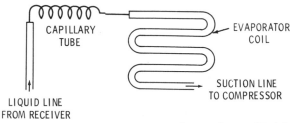

Fig. 11-23. Capillary tube and connections to mechanical refrigeration system.

The refrigerant piping (lines) should be carefully checked to make certain they will function properly. All connections should be examined for leaks, and all bends checked to make certain the tubing has not been squeezed together. A squeezed or pinched line will restrict the flow of refrigerant.

The *slope* of the suction line is also important in remote or split-system installations. When the evaporator coil is higher than the condensing unit, the suction line should be sloped with a continuous fall of at least ¼ in. per foot toward the condensing unit.

If the evaporator coil is higher than the condensing unit and the excess line is coiled, the excess tubing must be coiled horizontally in such a way that the flow of refrigerant is from the top to the bottom of the coil and toward the condensing unit in a continuous fall (Fig. 11-24).

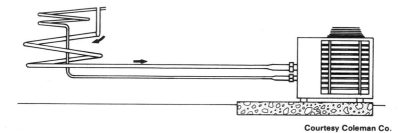

Courtesy Coleman Co.

Fig. 11-24. Flow of refrigerant toward the condensing unit.

Check the liquid and suction lines to make certain that they do not contact one another. Heat will transfer to the suction line if there is bare contact between the two.

Check refrigerant line connections for proper seat at the evaporator coil. These are non-reusable connecting valves that must be 100 percent seated for effective operation. If these valves have not been 100 percent seated, then the metal diaphragm will obstruct the line and restrict the flow of refrigerant. If this condition is suspected, use a wrench on the stationary fitting of the valve as shown in Fig. 11-25 while tightening the nut with another wrench.

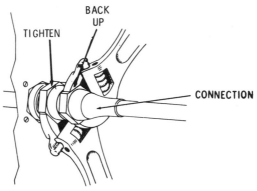

Fig. 11-25. Using wrenches to tighten connection.

TROUBLESHOOTING REFRIGERANT PIPING

The troubleshooting chart that follows lists the most common operating problems associated with the refrigerant piping. For each observable symptom, a possible cause and remedy are suggested.

Symptom and Possible Cause   *Possible Remedy*

Frosted or Sweaty Suction Line

1. Capillary tube or expansion valve passes excess refrigerant.
2. Expansion valve is stuck.

3. Evaporator fan not running.
4. Overcharge of refrigerant.
5. Ambient temperature too low.

1. Check the size and bore of capillary tube. Readjust the expansion valve.
2. Clean valve; replace if necessary.
3. Repair or replace.

4. Correct.

5. Block the condenser to increase the suction pressure or stop the unit.

Symptom and Possible Cause: *Possible Remedy:*

Hot Liquid Line

1. Low refrigerant charge.
2. Expansion valve stuck or open too wide.

1. Fix leak and recharge.
2. Clean valve and replace if necessary.

Frosted or Sweating Liquid Line

1. Restriction in dryer.

1. Replace dryer.

Frost on Expansion Valve or on Capillary Tube

1. Ice plugging capillary tube or expansion valve.

1. Apply hot wet cloth to capillary tube or expansion valve; a suction pressure increase indicates moisture present. Replace dryer.

FILTERS AND DRYERS

Filters and dryers are devices that provide very important functions in a mechanical refrigeration system. A *filter* is a device used to remove particles from the liquid refrigerant and from the oil by straining the fluid. For this reason, it is also sometimes referred to as a *strainer*. If a filter or strainer was not used, these particles trapped in the fluid could block small passages in the thermostatic valve or capillary tube, thereby seriously affecting the operation of the cooling system, or they could eventually damage mechanical parts.

A *dryer* is a device designed to remove moisture from the refrigerant in a mechanical refrigeration system. It is also referred to as a *dehydrator* or a *drier* (a spelling variant used by some authorities).

A typical combination filter-dryer is shown in Fig. 11-26. The

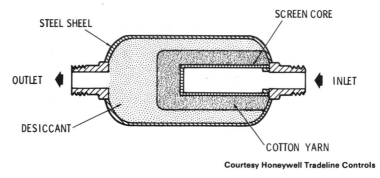

Fig. 11-26. Combination filter-dryer.

desiccant surrounding the filter core is usually a silica gel and functions as a drying agent. This unit is usually installed in the liquid line (either at the liquid receiver outlet or at the expansion valve outlet).

Filters and dryers are not included with small-capacity cooling units (residential types) filled and hermetically sealed at the factory. Filters and dryers are usually installed in systems where the refrigerant circuit is designed for field service.

PRESSURE-LIMITING CONTROLS

Certain pressure-limiting controls are used in cooling systems to protect them from extremes in refrigeration suction and discharge line pressures. Whenever the pressures in the system deviate from the normal operating range, the pressure control breaks the circuit to the compressor until the pressure returns to normal. High-side and low-side pressure switches are described in Chapter 6 of Volume 2 (Other Automatic Controls).

REVERSING VALVES

A *reversing valve* is a device used in a heat pump to change or reverse the direction of refrigerant flow. In this way, the heat pump can provide either cool air or heat to a structure, depending upon whether the cooling or heating cycle is in operation. Further information on reversing valves is contained in Chapter 12 (Heat Pumps).

WATER-REGULATING VALVES

Temperature-actuated water-regulating valves are used on water-cooled condensers to maintain condensing pressures within the desired range. This is accomplished by increasing or decreasing the rate of water flow as required by conditions in the system. Most water-regulating valves may be classed as either *direct acting* or *pilot operated.*

In the case of a direct-acting valve, the deflection of the bellows caused by an increase in refrigerant pressure overcomes the force of the springs and pushes the disc away from the seat allowing water to flow. When the unit shuts down, the refrigerant pressure becomes less than the spring pressure and the water valve closes off (Fig. 11-27).

In the pilot-operated valve, the main plunger to which the disc is attached is actuated by water pressure. Opening and closing the pilot port causes the differential pressure across the hollow plunger to vary. The amount of water that will flow through any given size and type of orifice will depend on the pressure differential across the orifice.

Water-regulating valves are rated at a certain quantity flow under a given pressure differential and the amount of valve opening. The amount of opening is controlled by refrigerant pressure, but if the pressure differential is insufficient, no amount of opening will provide the necessary water for the condenser. Ignorance of these facts has resulted in the condemnation of many valves, although they may have been in perfect operating condition.

AUTOMATIC CONTROLS

Refrigeration and air conditioning systems consist of refrigeration equipment and an electrical control circuit. These components are interconnected to produce and control the required

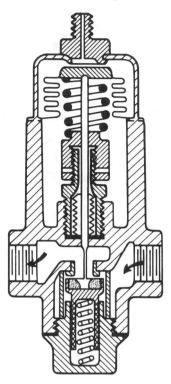

Fig. 11-27. Diagram of a direct-acting water-regulating valve.

cooling. Wiring diagrams of some typical automatic cooling control circuits used in refrigeration and air conditioning systems are illustrated in Figs. 11-28 and 11-29.

A cooling control circuit can be divided into the following principal components:

1. Basic controller or thermostat.
2. Limit control.
3. Primary control.
4. Power supply.

The relationships and functions of these various control system components are illustrated in Fig. 11-6. Detailed descriptions of these components are given in a number of different chapters, especially Chapters 4 and 6 of Volume 2. Refer to the Index for additional information.

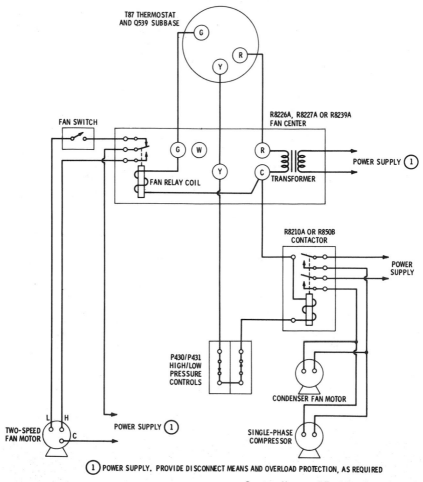

Courtesy Honeywell Tradeline Controls

Fig. 11-28. Basic cooling control system with system power supply by fan center transformer.

SYSTEM TROUBLESHOOTING

The troubleshooting chart that follows lists the most common operating problems associated with the system as a whole. For each problem and symptom, a possible cause and remedy are suggested.

434

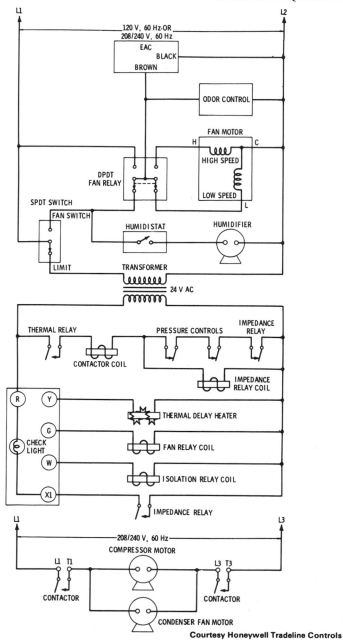

Fig. 11-29. Ladder diagram of basic cooling system control circuit.

Courtesy Honeywell Tradeline Controls

435

Symptom and Possible Cause   *Possible Remedy*

Unit Operates Low or Continuously

1. Shortage of refrigerant.	1. Fix leak and recharge.
2. Control contacts frozen or stuck closed.	2. Clean points or replace.
3. Insufficient air or dirty condenser.	3. Check and correct.
4. Air conditioner space poorly insulated or excess load in structure.	4. Replace with larger unit.
5. Compressor valves are defective	5. Replace compressor.
6. Restriction in refrigerant system.	6. Check and correct.
7. Filter dirty.	7. Clean or replace.
8. Air is bypassing the coil or service load.	8. Check return air; keep windows and doors closed.

Space Temperature Too High; Not Enough Cooling

1. Refrigerant charge low.	1. Check for leaks and recharge.
2. Control set too high.	2. Reset control.
3. Cap tube, expansion valve, or dryer plugged.	3. Repair or replace.
4. Iced or dirty coils.	4. Defrost or clean.
5. Unit too small.	5. Replace with larger unit.
6. Insufficient air circulation.	6. Correct air circulation.
7. Cap tube or expansion valve does not allow enough refrigerant.	7. Reset or replace.
8. Cooling coils too small.	8. Replace.
9. Restrictions or small gas lines.	9. Correct restrictions; increase line size.

Symptom and Possible Cause:	*Possible Remedy:*
10. High and low pressures approaching each other; compressor valves are defective.	10. Replace compressor.
11. Low line voltage.	11. Check line voltage and correct.
12. Dirty air filter.	12. Replace.
13. Dirty condenser.	13. Clean condenser.
14. Air circulator size too small.	14. Replace with larger air circulator.
15. Ductwork too small.	15. Increase size of ductwork.

Noisy Unit

1. Tubing rattle.	1. Fix so it is free from contact.
2. Fan blade causing vibration.	2. Check for bend; replace if necessary.
3. Refrigerant overcharged or oil is too high.	3. Check for correct refrigerant charge and maintain oil level. If necessary, replace expansion valves or capillary tube.
4. Loose parts or mountings.	4. Fix and tighten.
5. Motor bearings worn.	5. Replace motor.
6. Lack of oil in the compressor.	6. Add required oil.

No Air Delivery out of Register

1. System not set for summer cooling.	1. Read operating instructions and make required adjustment.
2. Fan motor not operating.	2. Repair or replace.
3. Open power switch.	3. Close switch.

Symptom and Possible Cause: *Possible Remedy:*

4. Fuse blown.	4. Replace with same-size fuse.
5. Broken connection.	5. Check circuit and repair.
6. Register closed.	6. Open register.
7. Evaporator fan motor leads are not connected to line voltage.	7. Connect the leads.

GENERAL SERVICING AND MAINTENANCE

It must be thoroughly understood that the greatest precautions must be taken to exclude air and moisture from a refrigeration system and that it should not be opened to the atmosphere without first removing the refrigerant from that part of the system to be serviced or repaired.

The information on general service operation contained in the sections that follow is based on the use of Freon12 since this is the most common refrigerant encountered. The same instructions apply to methylchloride with the added precaution that methylchloride is inflammable in concentrations of 8 to 18 percent with air, and suitable care must be observed when using an open flame to prevent fire.

PUMPING DOWN

Whenever a refrigeration system is to be opened to the atmosphere for service operations or repairs, it is necessary to remove the refrigerant from that part of the system to be opened. By *pumping down* the system prior to servicing, the refrigerant can be saved.

The manufacturer of the equipment will usually include detailed instructions concerning the pump-down procedure. Read and follow these instructions carefully.

Pumping down the system usually involves confining the refrigerant to the receiver by closing off the liquid line stop valve and (with a gauge attached to the intake stop valve) operating the compressor. By doing this, all the gas is drawn back to the compressor and condensed in the condenser, but is prevented from going further by the liquid line stop valve.

The compressor should be run until the suction pressure is reduced to holding steady at approximately 2 to 5 lb. pressure. Do *not* draw a vacuum on the system. A vacuum will cause moisture to be drawn into the system when it is opened.

PURGING

Purging refers to the release of air or noncondensable gases from a system, usually through a cock placed on or near the top of the receiver. This term is also applied to the sweeping of air out of a newly installed part or connection by releasing refrigerant gas into the part and allowing it to escape from the open end, thus pushing the air ahead of it.

Follow the instructions for purging contained in the manufacturer's installation and operating literature.

EVACUATING THE SYSTEM

Sometimes it becomes necessary to remove the *entire* refrigerant charge from the system. This operation is referred to as *evacuating the system* and is accomplished as follows:

1. Close the discharge service valve stem by turning it clockwise and remove the gauge connection plug.
2. Open the inlet service valve by turning the stem counterclockwise and attach a compound gauge to the inlet service valve.
3. Close the valve ½ turn clockwise and start the unit discharging through the open gauge connection in the discharging service valve.

If a vacuum is pumped too rapidly, the compressor will have a tendency to pump oil out of the compressor crankcase. Attach a copper tube to the gauge port and bend it so that any oil pumped out may be drained into a container. During this operation, the compressor may knock.

If knocking occurs, stop the compressor for about a minute and then restart. Continue the process until the gauge indicates a 20-in. vacuum or better. At this point, leaks in the system may be detected by putting sufficient oil in the container to cover the end of the tubing and continuing the pumping operation. When the system is entirely evacuated, no more bubbles should appear in the oil container. After the system is fully evacuated, replace the discharge gauge connection plug or attach a pressure gauge as desired.

CHARGING

Charging is the addition of refrigerant to a system from an external drum. There are a number of ways to add a refrigerant charge to a system, but the safest method usually is to introduce the refrigerant through the liquid line. The procedure may be outlined as follows:

1. Connect the refrigerant cylinder to the liquid line port at the condenser or compressor.
2. Purge the connecting line and tighten the last connection.
3. Disconnect the compressor so that it will not run during the charging operation.
4. Turn on the blower to the condenser.
5. Warm the refrigerant cylinder by placing it in a bucket of warm water (Fig. 11-30). Do *not* immerse any refrigerant connections.
6. Remove the cylinder from the water when you are satisfied it has been thoroughly warmed.
7. Wipe the cylinder dry and invert it.
8. Open the cylinder valve and the charging port valve.

The refrigerant should flow very readily into the high-pressure side of the system if Steps 1-8 were carefully followed.

After the system has been charged, close the refrigerant cylinder valve, allow 2 or 3 minutes to pass, and disconnect the

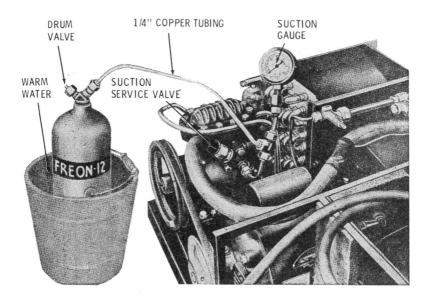

DRUM VALVE

1/4" COPPER TUBING

SUCTION GAUGE

WARM WATER

SUCTION SERVICE VALVE

FREON-12

Fig. 11-30. Warming the refrigerant cylinder.

cylinder. Reconnect the compressor and operate the unit, using gauges to determine if the charge is sufficient.

Always exercise caution when making connections or disconnections on the liquid line. This line is under high pressure, and the refrigerant is in a liquid form. Guard against refrigerant spraying into the face and eyes. Any minor leakage that may occur around the refrigeration hose in disconnection will be cold and at high pressure. To minimize pressures, liquid line disconnections should be made *after* the unit has been shut down for at least 5 minutes.

If the refrigerant charge is introduced in the low-pressure side of the system, the charging cylinder should always be kept in a vertical position. This precaution prevents the refrigerant liquid from flowing into the crankcase of the compressor.

Never heat a refrigerant cylinder with a torch or any other type of flame. Warming a cylinder in this manner generates excessive pressures which can result in an explosion. *Always* use warm water to heat a refrigerant cylinder.

SILVER BRAZING REPAIRS

For the repair of tubing condensers, evaporators, and parts made of light metal, *silver brazing* is the ideal process. It was formerly known as silver soldering, a term still frequently used. The term silver soldering is used to avoid confusing the use of silver brazing alloys with the soft solders. Some silver brazing alloys contain a certain amount of silver alloyed with copper and zinc. Others contain silver, copper, and phosphors. These alloys are available in forms having melting temperatures ranging from 1175° to 1500°F.

The use of silver brazing alloys enables the serviceman to obtain strong joints without danger of burning or overheating the base metals. Apart from the skill acquired through practice, the two most important requirements of a good silver brazing job are clean surfaces and enough heat to make the silver flow freely, but not so much that the silver burns to form scale. The best source of heat is an oxyacetylene or compressed air and illuminating gas torch.

The various operations to be performed in silver brazing are as follows:

1. Preparation.
2. Preparing swaged joints.
3. Preparing different-size tubing for connection.
4. Applying the flux.
5. Applying the brazing alloy.

Silver brazing has very little tensile strength of its own. The total strength of the joint is derived from the union of the two surfaces as a result of the action of the alloys used. Accordingly, surfaces must fit together tightly.

The tubing should be expanded or swaged to a depth at least equal to its diameter for tubes up to a ½ in. and not less than ½-in. deep for tubes of larger diameter. Special swaging blocks and drifts which accurately size and shape the inside of the tube and the outside of the expanded section are recommended for this operation where there is sufficient volume to warrant the investment. For occasional jobs, the tube can be held on a flare block and a swaging drift driven into it to form the bell end.

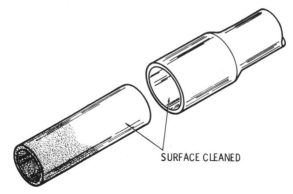

Fig. 11-31. Surface preparation. The surface must be clean.

Where two tubes of different sizes, such as ⅜ in. and ½ in., are to be joined, the smaller tube can be expanded by the previously described method until the outside diameter of the smaller tube is sized to fit snugly into the inside of the larger tube.

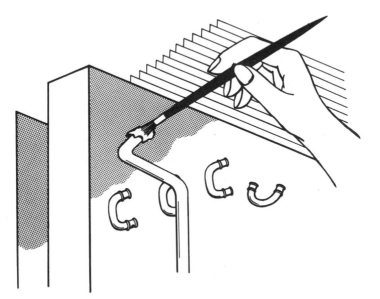

Fig. 11-32. Applying flux with brush.

443

When the surfaces of the tube ends to be joined have been thoroughly cleaned with steel wool or sandpaper, they should be fitted together and clamped, or firmly held together so that no movement will occur while they are being brazed (Fig. 11-31).

Apply enough flux with a brush to cover the surfaces to be joined, but not so much that it will run down the tubing. Make certain, however, that the flux is inside the joint all the way around (Fig. 11-32).

After the flux has been applied, the heat should be concentrated on one side of the joint and the silver brazing applied (Figs. 11-33 and 11-34). The temperature of the parts to be joined should be high enough to melt the silver by touching it to the heated surfaces near the flame. When the silver melts, apply it to the heated surfaces near the flame, but not under the flame.

Fig. 11-33. Applying heat.

Fig. 11-34. Applying brazing alloy.

Move the flame around the heated surface following it with silver until silver has been applied to the entire joint. Do not use too much silver and try to keep it from running down inside the tubing. Apply only enough heat to cause the silver brazing alloy to flow freely in order to avoid the formation of scale or the burning of the surfaces.

Most of the heat should be applied to the heavier parts of the joint where it will be conducted through the metal to the location where the silver alloy is to be applied.

Flames should never be applied directly to the point where brazing is being done. For thorough inspection of the joint, all flux must be carefully removed. Pin holes may exist under the film of melted brazing flux, and are not readily noticed until the flux has been removed.

445

Cleaning the joint can be done either by washing it with water while the joint is still hot, or by thoroughly brushing and scraping it with a wire brush, or emery cloth after the joint has cooled.

Where silver brazing is being done near an enameled or painted surface, or such materials as wood, insulating material, and other combustible surfaces, the surface should be protected with sheet asbestos during the brazing operation.

Valves, controls, or other apparatus to which a tube is being joined by silver brazing must be protected from damage by heat. Either remove the internal parts of the valve or protect the entire assembly with a wet cloth.

If a joint has previously been soldered with soft solder, all traces of the soft solder must be removed because the tin in soft solder amalgamates with copper at the temperature necessary for a silver brazing operation.

CHAPTER 12

Heat Pumps

A *heat pump* is a refrigeration device used to transfer heat from one space or substance to another. The heat pump is designed to take heat from a medium-temperature source, such as outdoor air, and convert it to higher-temperature heat for distribution within a structure. By means of a specifically designed reversing valve, the heat pump can also extract heat from the indoor air and expel it outdoors.

Because a heat pump system uses the reverse-cycle principle of operation, it is sometimes referred to a *reverse-cycle conditioning* or *reverse-cycle refrigeration*. The latter term is not correct because there are fundamental differences between the operating principles of a heat pump and a true refrigeration unit. The confusion probably stems from the fact that during the cooling cycle, the operation of a heat pump is identical to that of the mechanical refrigeration cycle in a packaged cooling air conditioning unit. The indoor coil functions as an evaporator, cooling the indoor air. The outdoor coil is a condenser, in which the hot refrigerant gas rejects heat to the outside air.

The introduction of heat pumps in which all components (fans, pumps, refrigeration units, and the necessary automatic controls) were contained within the same housing did not occur until after the end of World War II. Until the late 1950's, heat pump technology was such that they could only be used for heating purposes in climates that experienced mild winters. More recent technological developments have made them effective heat-generating sources in climates subject to periods of severe cold.

A heat pump uses the same basic components as the mechanical refrigeration system. For that reason, air conditioning components common to other systems are covered very briefly in this chapter. More detailed descriptions can be found in Chapter 11 (Air Conditioning Equipment).

HEATING CYCLE

The heating cycle of a heat pump begins with the circulation of a refrigerant through the outdoor coils (Fig. 12-1). Initially, the refrigerant is in a low-pressure, low-temperature liquid state, but it soon absorbs enough heat from the outdoor air to raise its temperature to the boiling point. Upon reaching the boiling point, the refrigerant changes into a hot vapor or gas. This gas is then compressed by the compressor and circulated under higher pressure and temperature through the indoor coils where it comes into contact with the cooler room air that circulates around the coils. The cooler air causes the gas to cool, condense, and return to the liquid state. The condensation of the refrigerant vapor releases heat to the interior of the structure. After the refrigerant has returned to a liquid state, it passes through a special pressure-reducing device (an expansion valve) and then back through the outdoor coils where the heating cycle begins all over again. The temperature of the room air that originally cooled the higher-temperature refrigerant vapor is itself increased by the process of heat transfer and recirculated throughout the room to provide the necessary heat.

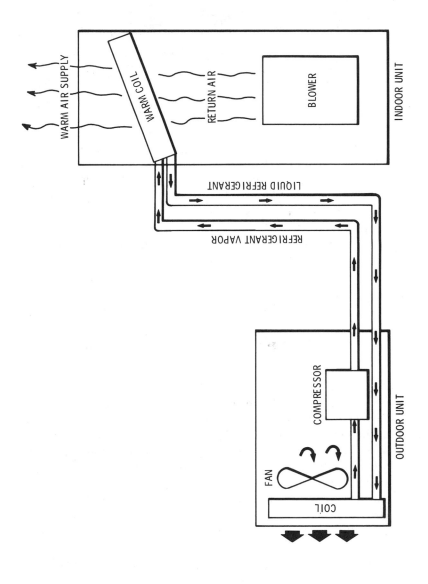

Fig. 12-1. Operation of heat pump heat cycle.

449

COOLING CYCLE

In the cooling cycle, the four-way reversing valve causes the flow of the refrigerant to be reversed. As a result, the compressor pumps the refrigerant in the opposite direction so that the coils that heat the building or space in cold weather, cool it in warm weather. In other words, the heat is extracted from the interior, cycled through the heat pump, and then expelled outside the building or space during the condensation of the refrigerant (that is, its change from a gaseous to a liquid state (Fig. 12-2).

HEAT SINK

The heat given off by the process of condensation is received by the *heat sink*. This is true for both the heating and cooling cycles. In the former, the air of the rooms or spaces functions as the heat sink. In the cooling cycle, the outside air, a water supply (e.g., a well or a sewage pipe), or the ground commonly serve as heat sinks outside the structure.

DEFROST CYCLE

Because the outdoor air is relatively cool when the heat pump is on the heating cycle, and the outdoor coil is acting as an evaporator, frost forms on the surface of the coil under certain weather conditions of temperature and relative humidity. Because this layer of frost on the coils interferes with the efficient operation of the heat pump, it must be removed. This is accomplished by putting the heat pump through a defrost cycle.

In the *defrost cycle*, the action of the heat pump is reversed at certain intervals and returned to the cooling cycle. This is done to temporarily heat the outdoor coil and melt the frost accumulation. The temperature rise of the outdoor coil is hastened because the operation of the outdoor fan stops when the system switches over to the cooling cycle.

The system will remain in the cooling cycle until the coil

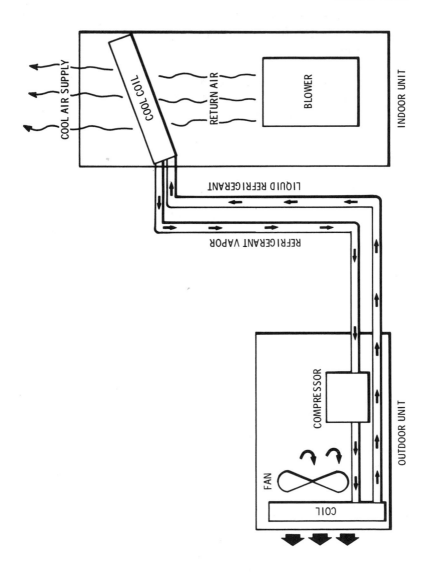

COOL AIR SUPPLY

COOL COIL

RETURN AIR

BLOWER

INDOOR UNIT

LIQUID REFRIGERANT

REFRIGERANT VAPOR

COMPRESSOR

OUTDOOR UNIT

FAN

COIL

Fig. 12-2. Operation of heat pump heat cycle.

451

temperature has risen to 57°F. The time of the defrost cycle will vary, depending upon how much frost has collected on the coil. During this period, the indoor motor continues to operate and blow cool air. This cold condition can be eliminated by installing an electric heating element (see "Auxiliary Electric Heating Elements" in this chapter). The heating element is wired in conjunction with the second stage of a two-stage thermostat, and will come on automatically when the heat pump is in the defrost cycle (terminals 9 and 7 on the defrost relay, Fig. 12-3).

The defrost cycle control system consists of a thermostat, timer, and relay. The defrost thermostat is located at the bottom of the outdoor coil where it can respond to temperature changes in the coil. It makes contact (closes) when the temperature of the outdoor coil drops to 32°F. This action of the thermostat causes the timer motor (located in the unit electrical box) to start. After the accumulative running periods reach either 30 minutes or 90 minutes (depending upon the type of cam installed in the timer), the timer energizes the defrost relays which reverses the reversing valve and stops the *outdoor* fan motor. The unit remains in the defrost cycle (cooling cycle) until the temperature of the outdoor coil reaches 57°F. At that temperature the coil is free of

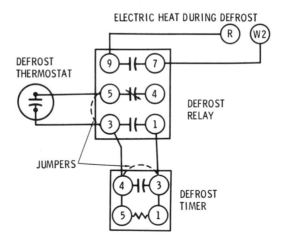

Courtesy Bard Mfg. Co.

Fig. 12-3. Defrost system wiring diagram.

frost and the frost thermostat opens to stop the timer and return the unit to the heating cycle. The timer will not run again until the outdoor coil temperature drops to 32°F. The timer runs *only* when the thermostat contacts are closed.

A defrost timer is shipped with a 30-minute cam installed (Fig. 12-4). With this cam, the unit will defrost once every 30 minutes (of accumulated running time) when the outdoor coil temperature is below 32°F. If there is little or no frost on the coil, the defrost cycle will be correspondingly short (approximately 45 seconds to 1 minute). A 90-minute cam is recommended for mild climates where defrosting is seldom necessary.

TYPES OF HEAT PUMPS

A typical *heat pump system* consists of the heat pump and auxiliary equipment, the external heat source, and the energy required to drive the compressor motor. Heat pump systems can be distinguished from one another by the source they depend on to provide the heat of evaporation (i.e., the external heat source).

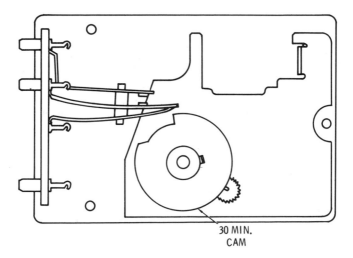

30 MIN.
CAM

Courtesy Heat Controller, Inc.

Fig. 12-4. Thirty-minute cam.

Accordingly, they may be divided into the following five categories:

1. Air-to-air system.
2. Water-to-air system.
3. Solar energy-to-air system.
4. Earth-to-air system.
5. Waste heat-to-air system.

A *water-to-air system* is the most economical, stable, and efficient of the five systems listed above. Well water is generally used although other water sources have also proven acceptable. In a water-to-air system, the source of the water generally acts as both the heat-delivering medium and the heat sink, depending on whether the heating or cooling cycle is being used.

A major techincal problem with heat pumps that rely on outdoor air as a heat source (i.e., an *air-to-air system)* has been that the temperature of the air is commonly lowest when heat requirements are highest; that is, during the cold winter months. When outdoor temperatures drop below 0°F, the heat pump is largely ineffective. For this reason, some sort of supplementary radiant heating system is usually employed until outdoor air temperatures rise to a level suitable for effective use in the heat pump.

Ground-to-air heat pump systems have demonstrated the least reliability because the ground surrounding the evaporator coil eventually insulates the coil from efficiently extracting heat. This insulating effect is created by a layer of frost in the ground. The frost is produced by the cooling effect of the evaporator coil. When the temperature of the ground surrounding the coil approaches that of the coil itself, heat transfer is reduced below an efficient level or ceases altogether.

Solar heat and waste heat as heat sources for a heat pump are still largely in the experimental stages. Heat sources may also be combined to improve overall performance during the heating season. One heat source serves to supplement the other during those periods in the heating season when one of them is not particularly effective. Such combinations as air and water or air and ground are common. The only drawback to using combination heat sources is that the initial cost of the system in which they are used is often higher than in one in which a single heat source is used.

HEAT PUMP INSTALLATIONS

Heat pump installations can be divided into the following two types:

1. Remote heat pumps.
2. Packaged heat pumps.

A *remote heat pump* consists of an outdoor condensing section and an indoor coil, blower, and filter section. The two sections are connected by refrigerant tubing. This type of installation is also referred to as a *split-system heat pump* by some manufacturers (Figs. 12-5, 12-6, and 12-7).

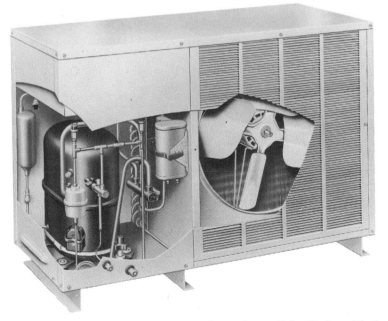

Courtesy Lennox Air Conditioning and Heating

Fig. 12-5. Outdoor heat pump unit.

455

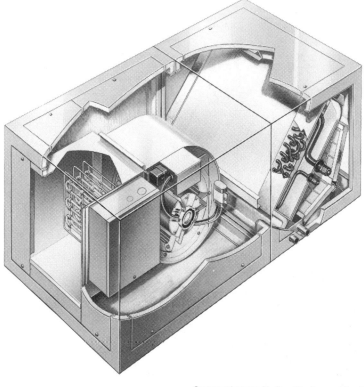

Fig. 12-6. Indoor heat pump unit.

Sometimes the outdoor condensing section is installed on the roof and the indoor section is suspended from the ceiling. This is a very common type of installation in commercial buildings.

In residential installations, the outdoor section is usually placed on a concrete slab next to the house and the indoor section is located either in the attic (installed horizontally) or in a closet space (installed vertically) on the same level as the outdoor unit.

A *packaged heat pump* (Fig. 12-8) contains the condenser, evaporator coils, compressor, blower and motor, automatic controls, and filter all in one unit. These heat pumps are also referred to as *single-packaged units* or *self-contained heat pumps* by various manufacturers.

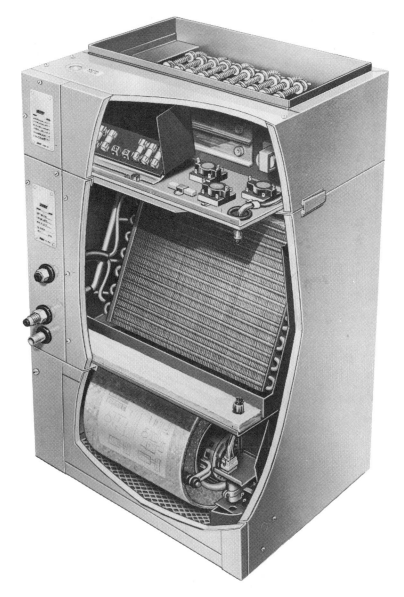

Fig. 12-7. Indoor unit for split-system.

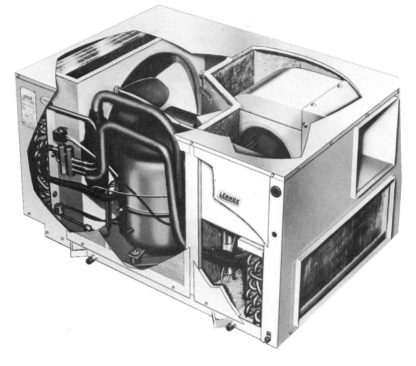

Fig. 12-8. Packaged heat pump unit.

HEAT PUMP COMPONENTS

A typical heat pump will consist of the following basic components:

1. Compressor section.
2. Air handler section.
3. Refrigerant lines.
4. Automatic controls.

The *compressor section* of a heat pump contains the compressor (or compressors), outdoor coil, fan(s) and motor(s), control system box, and four-way reversing valve. In addition, the compressor section will also contain a refrigerant-distributing device and the defrosting controls (automatic timer and terminating thermostat). The compressor section of a remote, or split-system, heat pump is located outdoors.

The *air handler section* consists of the indoor coil, blower and blower motor, check valve, thermal expansion valve, refrigerant distributing device, and the air filter(s).

The refrigerant lines are divided into a *liquid line* containing the refrigerant in a liquid state, and a *vapor* or *suction line* containing the refrigerant vapor. During the cooling cycle, cool refrigerant vapors are moving through the line to the compressor section. During the heating cycle, hot vapors are moving in the opposite direction.

The automatic controls can be divided into the electrical and mechanical components. The electrical components are identical to those found in other types of air conditioning equipment, and are described in Chapter 6 of Volume 2 (Other Automatic Controls) and Chapter 11 (Air Conditioning Equipment). The four-way reversing valve is a mechanical control device specifically designed for use in heat pumps. This valve is energized by the room thermostat and is used to reverse the heat pump cycle. It is described in this chapter (see "Four-Way Reversing Valve").

HEAT PUMP COMPRESSOR

Air conditioning compressors are described in Chapter 11 (Air Conditioning Equipment). The information contained in this chapter pertains for the most part to the operating characteristics of compressors used in heat pumps.

The principal function of the compressor is to receive the refrigerant vapor at low pressure and compress it into a smaller volume at higher pressure and temperature. It then pumps the refrigerant vapor to one of the coils for either the heating cycle or the cooling cycle operation.

In a remote unit installation, the compressor is located in the outdoor section along with the fan and outdoor coils. In a packaged or self-contained heat pump, all the components, including the compressor, are located in the same unit housing. If the compressor is installed indoors, provisions should be made for sufficient air intake and exhaust. Always consult the manufacturer's local field representative for recommendations when the compressor is to be installed indoors.

FOUR-WAY REVERSING VALVE

The redirection of the refrigerant flow in a heat pump system is accomplished with a four-way reversing valve (Fig. 2-9). This valve contains a solenoid coil, which, when energized, moves a pilot valve. The pilot valve then actuates the main valve, which slides to its other position and reverses the flow of refrigerant in the system. The operation of a four-way reversing valve is shown schematically in Figs. 12-10 and 12-11.

HEAT PUMP COILS

A heat pump contains two coils. One coil is located in the compressor section and is sometimes referred to as the "outdoor coil." The other coil, or "indoor coil," is located in the air handler section of the unit. Either coil may function as an evaporator (cooling coil) or condenser (heating coil), depending on which cycle is in operation. In other words, the coils discharge or absorb heat from the air depending on whether the heat pump is operating on the cooling or heating cycle. The function of the coils in a heat pump is to *transfer* heat, *not* to create cold.

During the cooling cycle, the indoor coil functions as an evaporator, cooling the indoor air. The outdoor coil is the condenser during this cycle, discharging heat to the outdoor air.

During the heating cycle, the indoor coil functions as the condenser and heats the indoor air. Warm air is absorbed by the outdoor air, which functions as the evaporator.

AIR FILTERS

Clean air filters are particularly important on the heating cycle of a heat pump. A dirty filter reduces the airflow through the

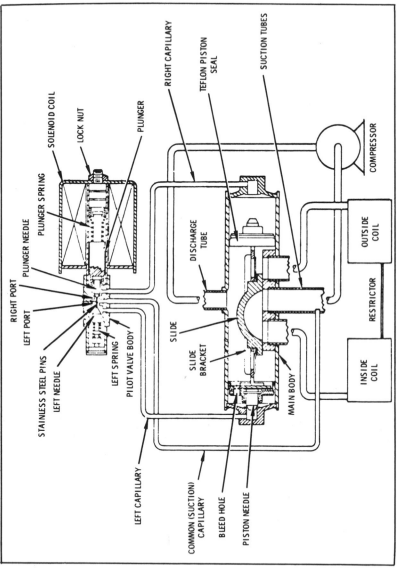

Fig. 12-9. Schematic of a four-way reversing valve.

461

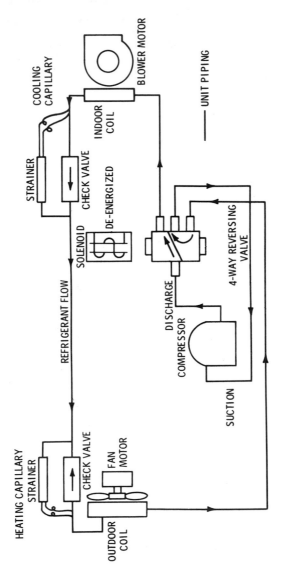

Fig. 12-10. Heating cycle piping diagram.

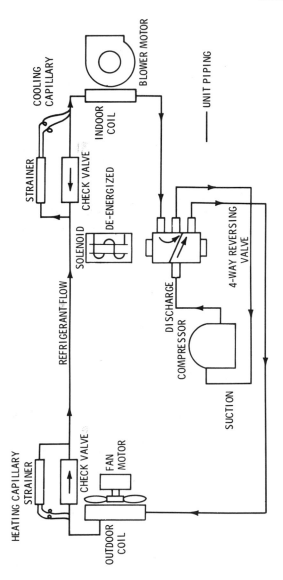

Courtesy Bard Mfg. Co.

Fig. 12-11. Cooling cycle piping diagram.

463

indoor coil. This results in raising the compressor head pressure on the heating cycle. As the airflow reduces to a low level, the compressor head pressure may reach a point high enough to trip the high-pressure cutout of the heat pump. This would then throw the system into straight resistance operation. Some heat pump installations have thermostats equipped with a device on the thermostat subbase deigned to warn that the heat pump has tripped on its high-pressure cutout. After the filters have been cleaned or replaced, the unit may be reset at the subbase by switching to the *off-reset* position and then back to the original switch position.

Check the filters regularly to make sure that they are clean and the airflow is unobstructed. If the filters are dirty, they should be removed and cleaned or replaced. Washable filters can be vacuumed or washed with a mild dishwasher detergent. The filter should then be rinsed thoroughly with clean water and allowed to dry.

AUXILIARY ELECTRIC HEATERS

An auxiliary electric heater (or heaters) can be installed in the blower-coil section of the heat pump or in the ductwork to provide supplementary heat. These units are controlled by the heat pump thermostat and operate automatically only when needed. They are a necessary addition to a heat pump when the unit is used in areas with particularly severe winters.

Slide-in electric heater assemblies similar to the one shown in Fig. 12-12 can be field installed in the heat pump. Each heater assembly is complete with fusible links, limit control, and relay installed and wired. A field-installed outdoor thermostat can be used to energize and regulate the second, third, and fourth stages of supplementary heat (Fig. 12-13). The thermostat must be installed before installing the heater assemblies.

CONDENSATION DRAIN LINE

The *condensation drain line* is used to ensure proper drainage of condensation from the indoor coil. The drain line consists of

¾-in. pipe that extends to the outdoors from a threaded condensation drain connection on the side of the indoor coil unit.

As shown in Fig. 12-14, a trap is installed in the drain line near the unit. The purpose of this trap is to ensure proper drainage of the condensation by preventing air from being drawn back into the unit through the drain line.

During the cooling season the condensation drain line should be checked for free and running condition. If the water does not flow freely, the drain line should be checked for obstructions or leaks and repaired accordingly.

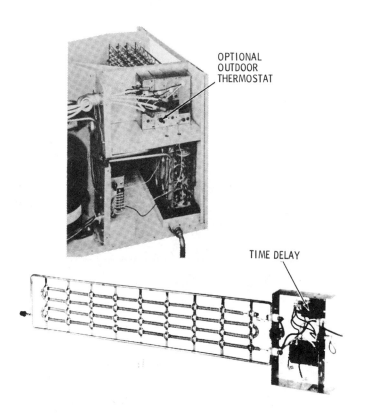

OPTIONAL
OUTDOOR
THERMOSTAT

TIME DELAY

Fig. 12-12. Slide-in electric heater assemblies.

Courtesy Amana Refrigeration, Inc.

Fig. 12-13. Outdoor temperature control used to control auxiliary electric heaters.

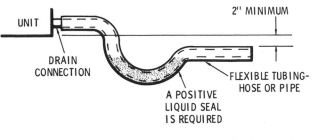

Courtesy Heat Controller, Inc.

Fig. 12-14. Trap installed on drain line.

AUTOMATIC CONTROLS

The various electrical components of an automatic control circuit used to operate a heat pump are shown in Fig. 12-15. Many of these components have already been described in Chapter 6 of Volume 2 (Other Automatic Controls). The Index should also be consulted for additional sources of information.

Typical wiring connections for a heat pump heating and cooling control system are illustrated in Figs. 12-16 and 12-17.

This particular control system utilizes a heat pump panel in which many of the control components are housed. The control system includes a defrost control consisting of a timer, temperature switch, and defrost relay. The combination of a timer and temperature switch provides defrost control during the heating cycle. The defrost system works automatically. If there is any excessive accumulation of frost on the coils, the timer and temperature switch cause the unit to enter the defrost cycle. When the coil is free of frost, the defrost thermostat opens to stop the timer and returns the unit to the heating cycle.

REFRIGERANT LINES

Most heat pump units used in residential heating and cooling systems are shipped with precharged lines; that is to say, the unit is charged with the exact operating charge of refrigerant. Pre-

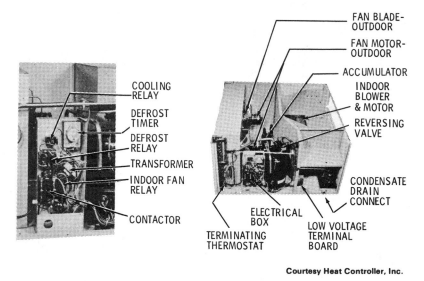

COOLING RELAY

DEFROST TIMER

DEFROST RELAY

TRANSFORMER

INDOOR FAN RELAY

CONTACTOR

FAN BLADE-OUTDOOR

FAN MOTOR-OUTDOOR

ACCUMULATOR

INDOOR BLOWER & MOTOR

REVERSING VALVE

CONDENSATE DRAIN CONNECT

ELECTRICAL BOX

TERMINATING THERMOSTAT

LOW VOLTAGE TERMINAL BOARD

Fig. 12-15. Electrical components of a typical heat pump control system.

charged tubing in standard lengths of 15 ft., 20 ft., and 40 ft. are available from manufacturers when it is necessary to extend the length of the refrigerant piping in a remote (split-system) heat pump installation.

The couplings on precharged lines will contain internal diaphragms to restrict the movement of the refrigerant during shipment. If this should be the case, the coupling will also contain an internal hardened cutter. By tightening the coupling, the cutter will pierce the diaphragm and allow the refrigerant to flow. As the coupling is tightened, a stainless-steel ring completes the seal and makes the joint leakproof. Figs. 12-18, 12-19, 12-20, and 12-21 illustrate how the connection is made.

When the air handler or indoor section of a remote heat pump installation is installed at an elevation higher than the outdoor compressor section, a vertical loop should be provided in the suction line next to the air handler (Fig. 12-22). The top of the loop should be *at least* as high as the top of the indoor coil. The purpose of the loop is to trap the refrigerant in the coil and prevent gravity drainage of the liquid to the compressor when the unit is completely shut down.

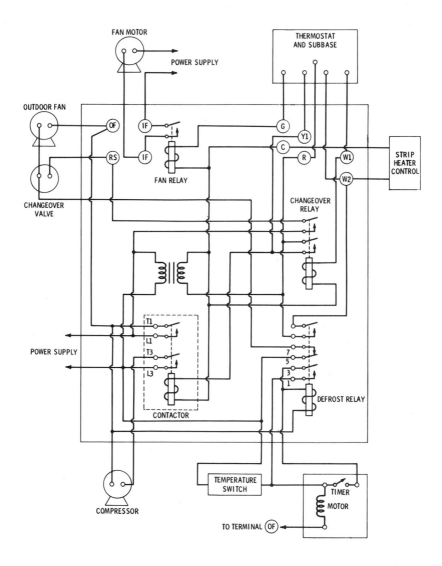

Fig. 12-16. Typical heat pump control system.

469

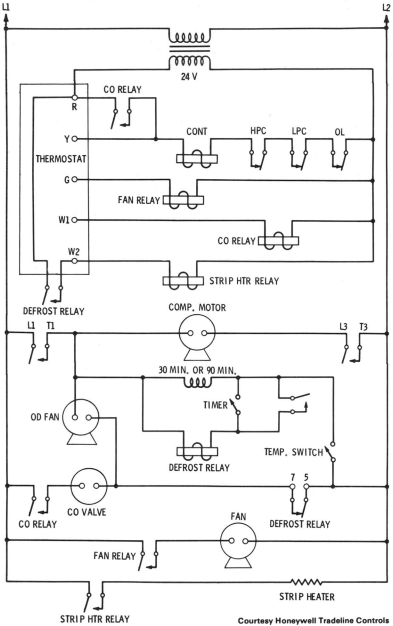

Fig. 12-17. Ladder diagram of heat pump control circuit.

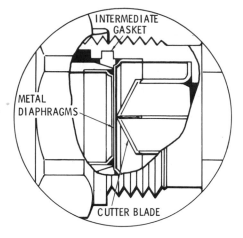

Fig. 12-18. Initial seal and air exclusion. Lines 10 percent connected.

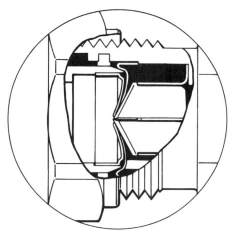

Fig. 12-19. Diaphragms punctured and pressure equalized. Lines 20 percent connected.

In a remote heat pump installation, 75 ft. is the maximum recommended length of refrigerant tubing connecting the indoor and outdoor sections. Lengths greater than 75 ft. have resulted in a reduced capacity for the unit.

Fig. 12-20. Diaphragms cut and folding. Lines 50 percent connected.

A major problem encountered with heat pump systems is that the discharge pipe on the heating cycle becomes the suction pipe on the cooling cycle. As a result, proper sizing of the refrigerant lines is an important consideration, because a minimum flow velocity (FPM) must be maintained to insure a satisfactory movement of the oil and refrigerant vapor in vertical risers. If the sizing is incorrect, the oil will be trapped. The manufacturer should provide sizing data in the installation instructions. The *1965 ASHRAE Guide* also provides sizing data.

Always locate the compressor section as close to the air handler section as possible. By keeping the refrigerant tubing lengths to a minimum, the capacity loss usually caused by long lines is also minimized.

INSTALLATION RECOMMENDATIONS

The following recommendations are offered as a guide line and checklist for installing a heat pump.

1. Install unit level or slightly slanted toward drain for proper condensation drainage.

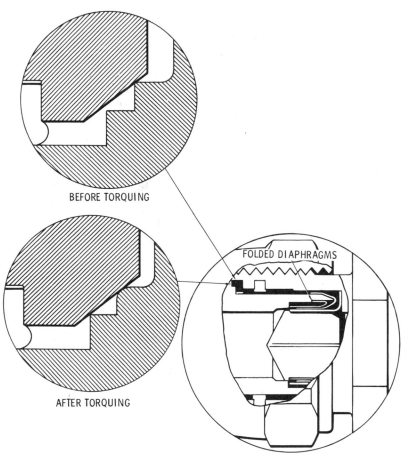

BEFORE TORQUING

AFTER TORQUING

FOLDED DIAPHRAGMS

Fig. 12-21. Diaphragm completely folded and entirely out of flow path. Metal seal engaged. Lines 100 percent connected.

2. Provide for free air travel to and from the condenser.
3. Check unit wiring for compliance with wiring diagram and local codes and regulations.
4. Make sure all wiring connections are tight.
5. Ground unit by grounded waterproof conduit or with separate ground wire.

473

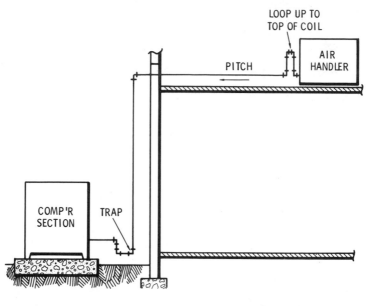

Fig. 12-22. Vertical loop on suction line.

6. Check condenser fan and evaporator blower for unobstructed and quiet movement.
7. Check condensation drain line for proper slope and drainage.
8. Fasten and seal all ducts and fittings with a suitable duct tape.

OPERATING INSTRUCTIONS

The heat pump manufacturer should provide operating instructions with the installation literature. If no copy is available, contact the manufacturer's local field representative for one.

If no operating instructions are available for the unit, follow those contained in this section. These instructions are suitable for most electric heat pumps used in residential installations.

The first step, of course, is to turn on the power supply at the disconnect switch. This is a very simple but important step that is often overlooked. In remote heat pump installations, both the indoor and outdoor sections may have a disconnect switch, or a breaker or fuse in the house box. Both switches must be in the *on* position to start the system.

Move the thermostat setting as high as it will go, and switch the fan selection switch to the *on* setting. This should start the blower. The remainder of the operating instructions will depend on whether heating or cooling is desired.

HEATING INSTRUCTIONS

For the heating cycle, move the fan selector switch to the *auto* position and slowly *raise* the heating temperature setting on the thermostat. Stop moving the lever as soon as the first (upper) stage mercury bulb makes contact. The blower, compressor, and condenser fan should start at this point.

Under normal operating conditions, the unit may automatically trip on its high-pressure cutout and stop the compressor and outdoor fan if the outdoor ambient temperature exceeds approximately 80°F. If the outdoor ambient temperature is too low for automatic cutout, block the return air until the unit trips. In a cold climate, it may take 5 minutes or more to trip.

Check the thermostat heat anticipator setting to make sure it matches the current draw of the heating relays (see Chapter 4 of Volume 2, Thermostats and Humidistats), and make certain the contactors and heaters are operating correctly.

The defrost timer can be checked in the winter when the outdoor coil is cold enough to activate it. Observe at least one defrost cycle to make sure the unit defrosts properly.

Turn the thermostat off. If the unit is set at the *on* position, only the blower should operate. Turn the thermostat on again and proceed as follows:

1. Adjust discharge air grilles for suitable air flow.
2. Check the system for proper balance and correct if necessary.
3. Check for air leaks in the ductwork and correct if necessary.

COOLING INSTRUCTIONS

For the cooling cycle, move the fan selector switch to the *auto* position and *lower* the heating temperature setting on the thermostat to below room temperature. This should start the blower, compressor, and condenser fan. If these components are operating satisfactorily, turn the thermostat off. The unit should stop with the exception of the blower if the thermostat is set at the *on* position.

If a combination heating and cooling thermostat is used, switch it to the *heat* position and check for correct heating operation. Make certain the thermostat heat anticipator is set to match the current draw of the heating relays (see Chapter 4 of Volume 2, Thermostats and Humidistats).

After waiting approximately 5 minutes, turn the air conditioner on and proceed as follows:

1. Adjust discharge air grilles.
2. Check the system for proper balancing.
3. Check for air leaks in the ductwork.

HEAT PUMP MAINTENANCE

The heat pump manufacturer will provide the necessary operation and maintenance literature with the unit. This literature contains nontechnical instructions that the average homeowner can understand and follow with little or no difficulty. By following these instructions, the operational life span of the heat pump will be prolonged and it will operate at maximum efficiency. These instructions can be summarized as follows:

1. Periodically inspect and clean (or replace) the air filters.
2. Lubricate blower motor annually according to the manufacturer's instructions.
3. Contact your local serviceman if there is an excessive frost buildup on the coils.
4. Periodically clean the coils in the outdoor unit (remote system) by washing with water hose.

5. Check condensation drain line during cooling season for free-flow condition. Water should flow freely.

TROUBLESHOOTING HEAT PUMPS

The following troubleshooting charts list the most common problems associated with the operation of a heat pump. For each symptom of a problem, a possible cause and remedy are suggested.

Symptom and Possible Cause   *Possible Remedy*

Noisy Operation

1. Loose parts.

1. Check all setscrews on blower and fan blade. Adjust and tighten all thrust collars.

2. Loose belts.

2. Adjust all belts and check drives.

Heat Pump Will Not Operate

1. Disconnect switch off.

1. Turn switch on. Both disconnect switches must be on in remote system.

2. Incorrect thermostat setting.

2. Change to proper setting.

3. Tripped circuit breaker or unit fuse.

3. Turn unit disconnect switch (es) off. Check unit fuses and replace with same size and type of fuse.

Unit Runs But Little Heat or No Cooling

1. Obstructed outdoor coil.
2. Dirty or plugged air filter.

1. Remove obstruction.
2. Clean or change if necessary.

Symptom and Possible Cause   *Possible Remedy*

Reversing Valve Will Not Shift from Heat to Cool

1. No voltage to coil.	1. Repair electrical current.
2. Defective coil.	2. Replace coil.
3. Low refrigerant charge.	3. Repair leak and recharge system.
4. Pressure differential too high.	4. Reset differential.
5. Pilot valve operating correctly. Dirt in one bleeder hole.	5. Deenergize solenoid, raise head pressure, reenergize solenoid to break dirt loose. If unsuccessful, remove valve and wash out. Check on air before installing. If no movement, replace reversing valve, add strainer to discharge tube, mount valve horizontally.
6. Piston cup leak.	6. Stop unit. After pressure equalizes, restart with valve solenoid energized. If valve shifts, reattempt with compressor running. ⋅ If still no shift, replace reversing valve.
7. Clogged pilot tubes.	7. Raise head pressure and operate solenoid to free tube of obstruction. If still no shift, replace reversing valve.
8. Both ports of pilot open. Back seat port did not close.	8. Raise head pressure and operate solenoid to free partially clogged port. If still no shift, replace reversing valve.

478

Symptom and Possible Cause   *Possible Remedy*

Reversing Valve Starts to Shift but Does Not Complete Reversal

1. Not enough pressure differential at start of stroke, or not enough flow to maintain pressure differential.
2. Body damage.
3. Both ports of pilot open.

4. Valve hung up at midstroke. Pumping volume of compressor not sufficient to maintain reversal.

1. Check unit for correct operating pressures and charge. Raise head pressure. If no shift, use valve with smaller ports.
2. Replace reversing valve.
3. Raise head pressure and operating solenoid. If no shift, replace reversing valve.
4. Raise head pressure and operate solenoid. If no shift, use a reversing valve with smaller ports.

Reversing Valve Has Apparent Leak in Heating Position

1. Piston needle on end of slide leaking.

2. Pilot needle and piston needle leaking.

1. Operate reversing valve several times, then recheck. If excessive leak, replace valve.
2. Operate reversing valve several times, then recheck. If excessive leak, replace valve.

Reversing Valve Will Not Shift from Heat to Cool

1. Pressure differential too high.

1. Stop unit. Valve will reverse during equalization period. Recheck system.

479

Symptom and Possible Cause	*Possible Remedy*
2. Clogged pilot tube.	2. Raise head pressure. Operate solenoid to free dirt. If still no shift, replace reversing valve.
3. Dirt in bleeder.	3. Raise head pressure and operate solenoid. Remove reversing valve and wash it out. Check on air before reinstalling. If no movement, replace valve. Add strainer to discharge tube. Mount valve horizontally.
4. Piston cup leak.	4. Stop unit. After pressure equalizes, restart with solenoid deenergized. If valve shifts, reattempt with compressor running. If it still will not reverse while running, replace reversing valve.
5. Defective pilot.	5. Replace valve.

CHAPTER 13

Humidifiers and Dehumidifiers

The air around us has a capacity for holding a specific amount of water at given temperatures. For example, 10,000 cu. ft. of air at 70°F will hold 10.95 pt. of water, no more. This means the air in a home 25 ft. by 50 ft. with 8-ft. ceilings (10,000 cu. ft.) could hold nearly 11 pt. of water when temperature inside is 70°F. This would represent 100 percent relative humidity conditions. If there were only 2 pt. of water in this same home at 70°F, the relative humidity would be $\frac{2}{11}$ or 18 percent.

The amount of water air can hold is dependent upon the temperature of the air. For example, air at 70°F can hold 16 times as much water as air at 0°F. This means that 10,000 cu. ft. of air at 0° will hold $\frac{2}{3}$ pt. of water. When that same air is heated to 70°F, it could hold nearly 11 pt. of water, or 16 times as much water. Thus, if 0°F air that had a relative humidity of 96 percent was brought into a house through a door, window, or any other crack and warmed up to the inside temperature of 70°F, this same air would now have a relative humidity of 96/16, or 6 percent. This relationship between temperature and relative humidity can be illustrated by the data in Table. 13-1.

As you can see, warming the fresh air will reduce the relative humidity in the house until the air becomes extremely dry. In some situations, the air in a house may be even drier than the Sahara Desert. The Sahara has a relative humidity average of 20 percent. The average home in the winter months (i.e., the heating season) without humidity control or other means of adding moisture to the air maintains a relative humidity of approximately 12 to 15 percent.

Air that has a low relative humidity will absorb water vapor from any available source. Most important from a personal standpoint is the evaporation of moisture from the membranes of the nose, mouth, and throat. These are our protective zones, and excessive dryness of these membranes will cause discomfort.

Because dry air will pick up moisture from any available source, the furnishings and other contents of the house are also affected. Furniture wood shrinks, and the joints may crack. If plaster is used, it may also develop cracks. The wall paint will usually peel, and carpet materials may become dry and brittle.

Table 13-1. Relationship Between Temperature and Relative Humidity

Outside Temperature	Outside Relative Humidity (%)	Inside Temperature	Inside Relative Humidity (%)
20°F	0 20 40 60 80	70°F	0 3 6 8 11

If the air in the house is excessively dry, there is more evaporation from the skin as moisture from the body is absorbed by the drier air. This evaporation generally makes the individual feel so cool that temperature settings of 70°F (or even higher) are not warm enough for comfort. Moving the temperature setting to a higher point may obtain the desired comfort level, but the higher temperature setting also increases the fuel cost. Each degree that the thermostat setting is raised increases the fuel cost approximately 2 percent for the same period of time. Obviously it is impractical and more costly to obtain comfort in this manner. It is far more economical to devise a means of reducing the rate of evaporation of moisture from the body and thereby eliminate the chilling effect. This can be done by increasing the relative humidity in the home (see below, "Humidifiers").

Excessive moisture in a home can also be a problem. When there is too much moisture in the air, mold and mildew form on surfaces, and there is generally an unpleasant feeling of dampness. If the condition persists, wood will warp and metal surfaces may begin to rust. The solution here is to reduce the relative humidity by removing moisture from the air (see "Dehumidifiers" in this chapter).

HUMIDIFIERS

A *humidifier* is a device used to add moisture to the air (Fig. 13-1). This function is accomplished either by evaporation or by breaking water down into fine particles and spraying them into the air. The latter method of adding moisture is used by spray humidifiers. The three principal types of humidifiers that use the evaporation method are:

1. Pan humidifiers.
2. Stationary-pad humidifiers.
3. Rotary humidifiers.

It is difficult to pinpoint the most desirable level of relative humidity, but it is generally agreed that the range between 30 and 50 percent is the best from both health and comfort standpoints.

The upper part of this range (40 to 50 percent) is impractical during many very cold days of the winter because of condensation on the windows. Therefore, it is recommended that a relative humidity of between 30 and 40 percent be maintained during the heating season.

SPRAY HUMIDIFIERS

A *spray humidifier* (or *air washer*) can be used for either humidification or dehumidification. Essentially this type of humidifier consists of a chamber containing a spray nozzle system, a recirculating water pump, and a collection tank (Fig. 13-2). As the air passes through the chamber, it comes into contact with the water spray from the nozzles, resulting in heat transfer between the air and water. This, in turn, results in *either* humidification (adding moisture) or dehumidification (removing moisture) depending upon the relative temperatures of the sprayed water and the air passing through the chamber. Dehumidification occurs when the temperature of the water is lower than the dew point of the air; humidification when it is higher (see "Spray Dehumidifiers" in this chapter).

Fig. 13-1. Typical power humidifier.

In some spray humidifiers, the water is sprayed on a heating coil through which steam or hot water is passed (Fig. 13-3).The heat causes the water to evaporate and thereby increase the moisture content of the air.

An *atomizing humidifier* is a form of spray humidifier that uses

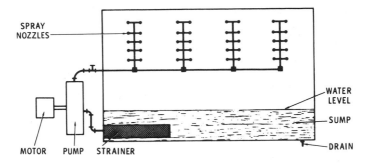

Fig. 13-2. Components of a spray humidifier.

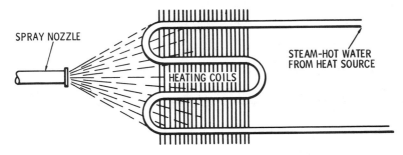

Fig. 13-3. Water sprayed on heating coil.

compressed air to reduce the water particles to a fine mist (Fig. 13-4). The mist is converted to a vapor, which is absorbed by the drier room air.

PAN HUMIDIFIERS

A *pan humidifier* consists of a water tank (pan), heating coils, a fan, and a fan motor (Fig. 13-5). The heating coils are installed in the water tank. Heat is supplied to the pan heating coils either by low-pressure steam or forced hot water where a water temperature of 200°F or higher is maintained.

The pan humidifier is completely automatic; the water level in the tank is controlled by means of a float control. When the relative humidity drops below the setting on the humidistat, the fan blows air over the surface of the heated water in the tank.

485

Fig. 13-4. Atomizing humidifier.

Courtesy American Moistening Co.

The air picks up moisture as it travels over the surface of the water and is blown into the space to be humidified. When the humidistat is satisfied, the fan is shut off.

STATIONARY-PAD HUMIDIFIERS

A *stationary-pad humidifier* contains a stationary evaporator pad over which warm air from a supply duct is drawn by a motor-operated blower or fan. The air picks up moisture as it passes over the pad. The humidified air is then returned to the room or space.

ROTARY HUMIDIFIERS

A *rotary humidifier* consists basically of an evaporator pad attached to a rotating device such as a disc, wheel, drum, or belt. The evaporator pad is first rotated through a pan of water where the pad absorbs moisture and then into the airstream where the moisture is released. Because rotary humidifers are relatively inexpensive to install and operate, they are the most common type of humidifier used in residences and small structures.

The principal components of a rotary humidifier are:

1. Motor.
2. Water tank or reservoir.

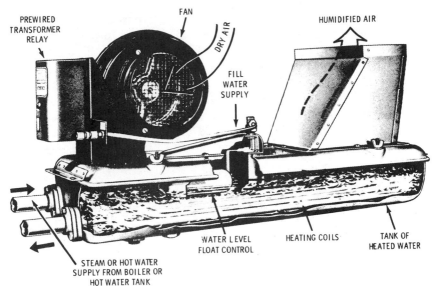

Fig. 13-5. Pan humidifier.

3. Rotating disc, wheel, drum, or belt.
4. Automatic drain valve.
5. Float-operated fill valve.
6. Overflow line.
7. Fill-water connection.
8. Drain connection.

Some of these components are illustrated by the humidifier in Figs. 13-6 and 13-7. This humidifier uses bronze wire-mesh discs that rotate at an angle in the reservoir water. The water-level float is isolated in a separate compartment.

AUTOMATIC CONTROLS

The operation of a humidifier is controlled by a room or furnace-mounted humidistat (Fig. 13-8). The sensing element of the furnace-mounted humidistat should be installed in the return air duct with the open side down (Fig. 13-9). If the humidifier is installed with a plenum adapter, the sensing element of the

487

humidistat should be mounted above the bypass duct. Never use a furnace-mounted humidistat with a horizontal furnace.

A room humidistat should be mounted on a wall 4 to 5 ft. above the floor in a location that has free air circulation of average temperature and humidity for the entire space to be controlled. Avoid spots near air ducts or supply air grilles.

Depending upon the type of installation, either a low-voltage or line voltage humidistat may be used. Wiring diagrams for both types of control circuits are illustrated in Figs. 13-10 and 13-11.

INSTALLATION INSTRUCTIONS

A common location for a rotating-type humidifier is on the underside of a horizontal warm-air supply duct as close to the furnace as convenient working conditions will permit. The duct should be at least 12 inches wide (Fig. 13-12). As shown in Fig.

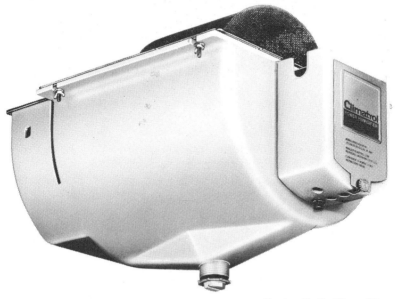

Courtesy Mueller Climatrol Corp.

Fig. 13-6. Rotary power humidifier.

Fig. 13-7. Components of a rotary humidifier.

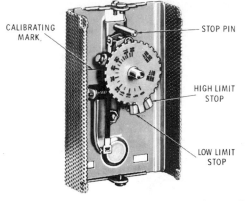

Fig. 13-8. Typical humidistat used to control both humidifying and dehumidifying equipment.

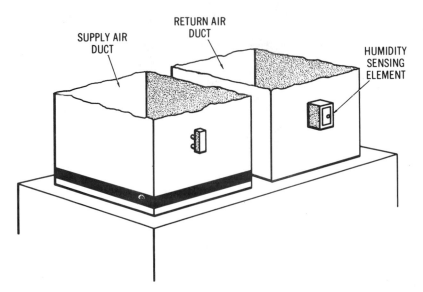

Fig. 13-9. Location of furnace-mounted humidistat sensing element.

13-12, a deflector should be installed in the duct to guide the airflow into the humidifier. The deflector should be about 4 to 5 in. wide. It can be cut from sheet metal and screwed to the duct wall with sheet-metal screws.

490

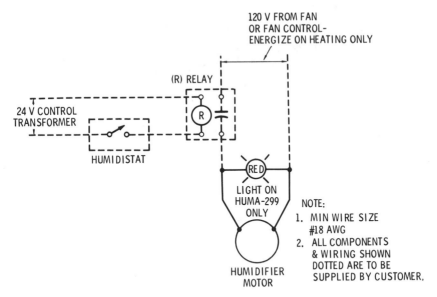

Fig. 13-10. Wiring diagram of a low-voltage humidistat.

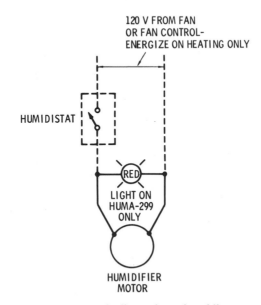

Fig. 13-11. Wiring diagram of a line voltage humidistat.

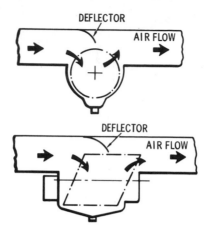

Fig. 13-12. Deflector installed to guide airflow.

If the humidifier is to be installed in a furnace plenum that also contains a fan and limit control, care must be taken to keep the sensing element of the fan and limit control a suitable distance from the inlet of the humidifier (Fig. 13-13); otherwise, the return air may be drawn through the humidifier and over the element and cause improper operation of the fan control.

Humidifiers are also installed on the sides of furnace plenums. The manufacturer usually supplies a plenum adapter kit to mount the humidifier. As shown in Fig. 13-14, a typical kit consists of a

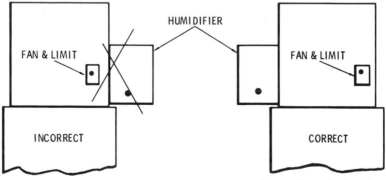

Courtesy Thermo-Products, Inc.

Fig. 13-13. Furnace plenum location of humidifier, fan, and limit controls.

492

plenum adapter hood, closing panel adapter collar, and round cold-air collar. A length of flexible or galvanized ductwork is installed between the adapter hood and the cold-air collar (Fig. 13-15).

The most desirable location for a humidifier is the warm air'-plenum. Some special installations are illustrated in Figs. 13-16, 13-17, and 13-18.

If the adapter is installed on the cold-air return plenum, the distance that the hot air is ducted must be as short as possible. In an air conditioning system, a damper must be installed to close off the cold-air return opening during the cooling season. If the opening is not closed, sufficient air can circulate through the shut-down humidifier to frost the evaporator coil.

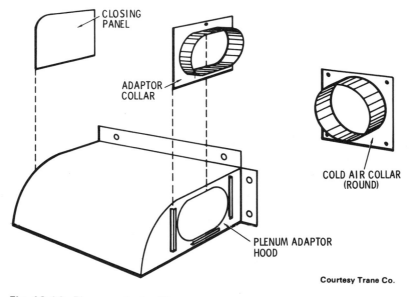

Courtesy Trane Co.

Fig. 13-14. Plenum adapter kit.

The following recommendations should be followed when installing a rotating-type humidifier:

1. Carefully read the manufacturer's installation instructions and any local codes and regulations that would apply to the installation of a humidifier.

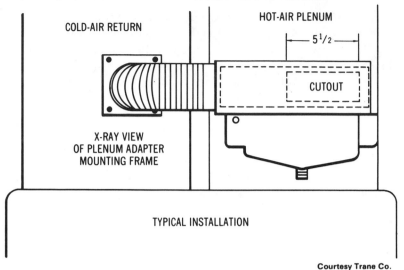

COLD-AIR RETURN

HOT-AIR PLENUM

5¹/₂

CUTOUT

X-RAY VIEW
OF PLENUM ADAPTER
MOUNTING FRAME

TYPICAL INSTALLATION

Courtesy Trane Co.

Fig. 13-15. Use of flexible ductwork.

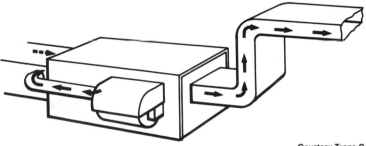

Courtesy Trane Co.

Fig. 13-16. Horizontal furnace installation.

2. Unpack the humidifier, and examine it for any shipping damage. Check the equipment and parts off against the inventory list.
3. Place the humidifier base or template against the duct or plenum wall, and mark the position where it will be mounted.
4. Cut out the opening for the humidifier, and mount it according to the manufacturer's instructions.

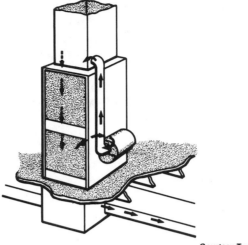

Fig. 13-17. Counterflow installation slab mounted.

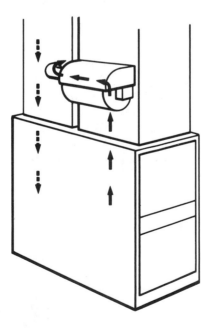

Fig. 13-18. Plenum installation.

5. Connect the saddle valve to a convenient water line. *Never* connect to a water line supplying chemically softened water.
6. Install the required drain fittings and connections in the reservoir.
7. Turn on the water supply, and fill the reservoir. The float valve arm should be adjusted to maintain a 2-in. water level in the reservoir.
8. Place the drum or wheel in the reservoir so that the axles are properly seated on the sloped supports. The gears will mesh automatically. Do *not* attempt to rotate the drum by hand because you may damage the gears inside the motor.

OPERATING AND MAINTENANCE SUGGESTIONS

A humidifier should be regularly drained to remove lime and other residue that has collected in the reservoir. The first draining of the reservoir is recommended for two weeks after the unit has been installed or two weeks after the seasonal startup. Subsequent drainings should follow the schedule suggested by the manufacturer.

A major problem with any humidifier is the buildup of lime or other mineral deposits in the water reservoir and on the evaporator pad. Manufacturers have attempted to minimize this problem by constructing the reservoir out of an extremely smooth material, such as an acrylic, so that these mineral deposits can be easily flushed loose during maintenance.

Some humidifiers are equipped with *automatic* flush system to clean the mineral deposits from the unit. The components of a Thermo-Pride automatic flush system are illustrated by the schematic of a humidifier in Fig. 13-19. The drum *(A)* is the rotating pad with the flush wheel attached. As the drum rotates, water in the reservoir pan *(C)* is scooped up into the flush wheel. The float assembly *(B)* then permits fresh water to enter the pan to replace that which was scooped out (and which evaporated from the pad). As the flush wheel continues to rotate, the mineral-laden water scooped out is conveyed into the drain cavity *(D)* of the pan and out the drain hose *(E)*. The mineral content of the water in the reservoir pan will be maintained at the lowest possible level with the operation of the automatic flush system.

A commonly used method for removing lime and other mineral deposits from the inside surface of the reservoir and the rotating evaporator pad is by cleaning these parts with muriatic acid (*never* with a detergent). Muriatic acid is an extremely efficient cleaning agent for this purpose, but it generates toxic fumes as it works on the deposits. For this reason, it is *absolutely* mandatory that the cleaning take place outdoors.

TROUBLESHOOTING HUMIDIFIERS

The troubleshooting chart that follows lists the most common problems associated with the operation of a humidifier. For each operating problem, a possible cause is suggested and a remedy proposed.

Symptom and Possible Cause   *Possible Remedy*

Humidifier Fails to Maintain Proper Humidification

1. Humidistat set too low.	1. Raise setting.
2. Humidistat broken.	2. Repair or replace.
3. Water valve off.	3. Open valve.
4. Water valve clogged.	4. Clean valve.
5. Limed unit.	5. Clean lime from discs or replace.

Humidifier Fails to Operate

1. Drum motor not receiving current.	1. Check wiring to motor.
2. Inoperative motor due to lack of lubrication.	2. Lubricate or replace motor.

Excess Condensation on Windows

1. Humidistat setting too high.	1. Lower the humidistat setting.

DEHUMIDIFIERS

Dehumidification is the name given to the process of removing moisture from the air. The device used for this purpose is called a *dehumidifier* (Fig. 13-20). Dehumidifiers can be classified on the basis of *how* they remove moisture from the air into the following three categories:

1. Adsorption dehumidifiers.
2. Spray dehumidifiers.
3. Refrigeration dehumidifiers.

Both spray and refrigeration dehumidifiers remove moisture from the air by cooling it. Dehumidifiers operating on the refrigeration principle are the most common type used in residential heating and cooling systems (see below, "Refrigeration Dehumidifiers").

ADSORPTION DEHUMIDIFIERS

An *adsorption dehumidifier* extracts moisture from the air by means of a sorbent material. This type of dehumidifier is very

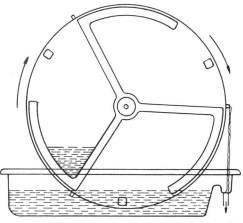

Courtesy Thermo-Products, Inc.

Fig. 13-19. Operating principles of an automatic flush system.

common in commercial and industrial installations but is rarely found in residences.

A sorbent material is one that contains a vast number of microscopic pores. These pores afford great internal surface to which water adheres or is absorbed. Moisture is removed from the air as a result of the low vapor pressure of the sorbent material.

Courtesy Westinghouse Electric Corp.

Fig. 13-20. Cabinet-model electric dehumidifier.

Fig. 13-21 illustrates the operating principle of a rotating bed dehumidifier. The unit consists of a cylinder or drum filled with a dehumidifying or drying agent. In operation, airflow through the drum is directed by baffles, which form three independent air streams to flow through the adsorbing material. One airstream consists of the wet air to be dehumidified. The second airstream is heated drying air used to dry that part of the dehumidifying material that has become saturated. The third airstream precools the bed to permit an immediate pickup of moisture when that part of the bed returns to the dehydration cycle. In the rotating bed dehumidifier, the baffle sheets are stationary and the screened bed rotates at a definite speed to permit the proper time of contact in the drying, cooling, and dehumidifying cycles.

The operating principles of a stationary bed solid adsorbent

499

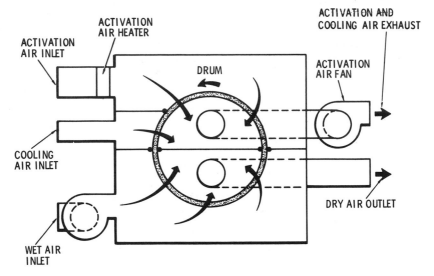

Fig. 13-21. Rotating-bed solid adsorbent dehumidifier.

dehumidifier are illustrated in Fig. 13-22. It has two sets of stationary adsorbing beds arranged so that one set is dehumidifying the air while the other set is drying. With dampers in position as shown, air to be dried flows through one set of beds and is dehumidified while the drying air is heated and circulated through the other set. After completion of drying, the beds are cooled by shutting off the drying air heaters and allowing unheated air to circulate through them. An automatic timer controller is provided to cause the dampers to rotate to the opposite side when the beds have adsorbed moisture to a degree that begins to impair performance.

The *liquid* adsorbents most frequently used in dehumidifiers are chloride brines or bromides of various inorganic elements, such as lithium chloride and calcium chloride.

A typical liquid adsorbent dehumidifier is shown in Fig. 13-23. It includes an external interchamber having essential parts consisting of a liquid contactor, a solution heater, and a cooling coil, as shown. In operation, the air to be conditioned is brought into contact with an aqueous brine solution having a vapor pressure below that of the entering air. This results in a conversion of latent heat to sensible heat, which raises the solution temperature

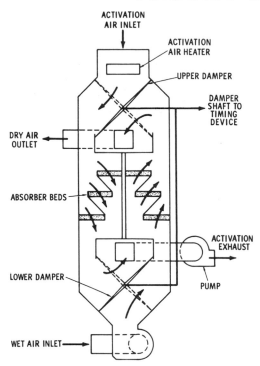

Fig. 13-22. Stationary-bed solid-adsorbent dehumidifier.

and consequently the air temperature. The temperature change of the air being processed is determined by the cooling water temperature and the amount of moisture removed in the equipment.

SPRAY DEHUMIDIFIERS

Dehumidification can be accomplished by means of an air washer as long as the temperature of the spray is *lower* than the dew point of the air passing through the unit. This is an important fact to remember because condensation will not take place if the temperature of the spray is higher than the dew point. Sensible heat is removed from the air during the time it is in contact with the water spray. Latent heat removal occurs during condensation.

Spray dehumidifiers, or air washers, usually have their own recirculating pumps. These pumps deliver a mixture of water from the washer sump (which has not been cooled) and refrigerated water. The mixture of sump water and refrigerated water is proportioned by a three-way or mixing valve actuated by a dew point thermostat located in the washer air outlet or by humidity controllers located in the conditioned space.

A spray dehumidifier results in greater odor absorption and cleaner air than is possible with those using a cooling coil. A principal disadvantage of this type of dehumidifier is that it sometimes experiences problems with the water-level control and may flood.

REFRIGERATION DEHUMIDIFIERS

A *refrigeration dehumidifier* removes moisture from the air by passing it over a cooling coil. The cool surfaces of the coil cause

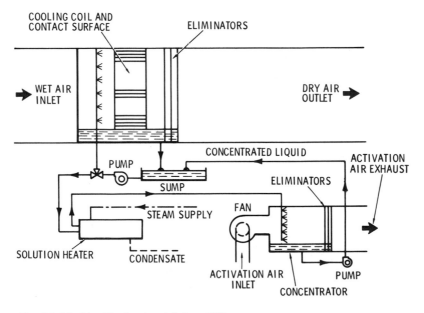

Fig. 13-23. Liquid adsorbent dehumidifier.

the moisture in the air to condense. This moisture then collects on the coils and eventually runs into a collection tray or pan located below the unit, or through a hose into a nearby drain.

The amount of moisture removed from the air by a refrigeration dehumidifier will depend on the volume of air and its relative humidity. The initial amount of moisture removed will be relatively large in comparison to the amount removed at later stages in the operation of the dehumidifier. This reduction in the amount of moisture removal is *not* an indication that the dehumidifier is not operating properly. This is a normal operating characteristic. As the relative humidity approaches the desired level, the amount of moisture being removed from the air will be considerably less.

Dehumidifying coils depend upon the dew point of the air entering and leaving the coil for removal of moisture from the air. To accomplish moisture removal, the dew point of the air entering the coil must be *higher* than the dew point of the air leaving the coil.

AUTOMATIC CONTROLS

A refrigeration dehumidifier consists of a motor-driven compressor, a condenser or cooling coil, and a receiver. A refrigerant, such as Freon, circulates through the cooling coil of the unit, the refrigerant flow being controlled by a capillary tube circuit.

The room air is drawn over the cooling coil by means of a motor-operated fan or blower. When the moisture-laden air comes in contact with the cool surfaces of the cooling coil, it condenses and runs off the coil into a collection tray or pan, or through a hose into a drain.

Dehumidifier operation is controlled by a humidistat, which starts and stops the unit to maintain a selected humidity level. The humidistat accomplishes this function by switching the compressor and unit fan on and off in response to changes in the moisture content of the air.

Humidistat control settings will generally range from *dry* to *extra dry* to *continuous* to *off*. During the initial period of operation (usually three to four weeks), the humidistat should be set at *extra dry.* If the moisture content of the air has been noticeably reduced, the humidistat setting can be moved to *dry.* Minor adjustments may be required from time to time.

Never purchase a dehumidifier that does not have a humidistat. Without a humidistat, a dehumidifier will run long after the humidity of the air has dropped to a satisfactory level. Operating a dehumidifier can prove to be very costly and wasteful of energy.

Dehumidifiers that empty the condensate into a container are equipped with an integral switch to turn the unit off when the condensate container is full. This action avoids overflow conditions. Most of these dehumidifiers will also have a signal light that indicates when the unit container needs emptying.

INSTALLATION SUGGESTIONS

A dehumidifier is most effective in an enclosed area where good air circulation is found. For maximum effectiveness, the unit should be located as close to the center of room, space, or structure as possible.

OPERATING AND MAINTENANCE SUGGESTIONS

A dehumidifier generally will not operate very satisfactorily at temperatures below 65°F. The reason for this is obvious. When the ambient temperature is below 65°F, the cooling coil must operate at below-freezing temperatures in order to cause the moisture in the air to condense. Unfortunately, operating at these temperatures also causes ice to form on the coils, and this ice formation may eventually damage the dehumidifier. The ice is removed by running the dehumidifier through a defrost cycle at least once every hour. The dehumidifier cannot remove moisture from the air while it is in the defrost cycle.

Fungus will sometimes collect on the cooling coil. This accumulation of fungus can be removed by loosening it with a soft brush and washing away the residue with water.

The principal components of a refrigeraton dehumidifier are hermetically sealed at the factory. These components are permanently lubricated and should not require any further servicing.

TROUBLESHOOTING REFRIGERATION DEHUMIDIFIERS

The troubleshooting chart that follows lists the most common problems associated with the operation of a refrigeration dehumidifier.

Symptom and Possible Cause　　*Possible Remedy*

Condensation Container Fills to Brim or Flows Over

1. Overflow prevention control malfunctioning.	1. Replace control.
2. Shutoff level too high.	2. Lower shutoff level.

Dehumidifier Runs Continuously

1. Humidity too high for unit.	1. Use dehumidifier with higher capacity.
2. Defective humidistat.	2. Replace humidistat.

CHAPTER 14

Air Cleaners
and Filters

All indoor air contains a certain amount of microscopic air-borne dust and dirt particles. Cooking and tobacco smoke also contribute to the pollution of the indoor air, and pollen is a factor during some months of the year.

Excessive air pollution stains furniture and fabrics and causes a dust film to form on glass surfaces such as windows or mirrors. It can also be a health problem, especially to those having dust or pollen allergies.

A number of devices are used to remove dust, dirt, smoke, pollen, and other contaminants from the indoor air. They are not all equally effective. For example, an ordinary air filter in a furnace only removes approximately 10 percent of all airborne contaminants. An electronic air cleaner, on the other hand, can remove as much as nine times that amount.

The air cleaning devices described in this chapter are: (1) electronic air cleaners, (2) air washers, and (3) conventional air filters.

ELECTRONIC AIR CLEANERS

Electronic air cleaners are devices designed to electrically remove airborne particles from the air. The best electronic air cleaners remove 70 to 90 percent of all air contaminants. These standards are met in testing methods devised by the National Bureau of Standards and the Air-Conditioning and Refrigeration Institute. Claims by manufacturers for higher rates of airborne particle removal should be attributed to enthusiasm for their product. In any event, an electronic air cleaner is vastly more effective than the conventional filter used in a warm-air furnace. They are available as permanently mounted units for use in central heating and/or cooling systems or as independent cabinet units.

Permanently mounted electronic air cleaners are installed at the furnace, air handler, or air conditioning unit, or in wall or ceiling return air grilles. Some typical installations are shown in Fig. 14-1.

An electronic air cleaner installed at the furnace, air handler, or air conditioning unit is either mounted against the surface of the unit or a short distance from it on the return air duct. These electronic air cleaners are sometimes referred to as *multiposition* models because they can be installed in a number of locations with equal effectiveness. Some typical examples of multiposition electronic air cleaners are shown in Figs. 14-2, 14-3, and 14-4.

A design feature of some *return grille* electronic air cleaners is a hinged cell carrier, which swings out to allow the cells to removed (Fig. 14-5). Return grille electronic air cleaners can be installed in the wall or ceiling openings of return air ducts, but not in a floor return. These units are also referred to as *wall (or ceiling) electronic air cleaners* or as *through-the-wall electronic air cleaners*. Typical installations are shown in Figs. 14-6, 14-7, and 14-8.

The independent cabinet units (Figs. 14-9 and 14-10) are designed for use in installations where a permanently mounted

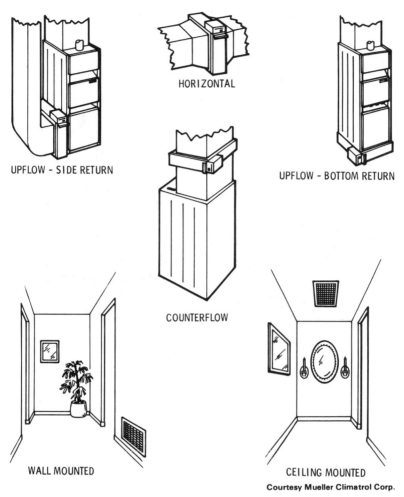

UPFLOW - SIDE RETURN

HORIZONTAL

UPFLOW - BOTTOM RETURN

COUNTERFLOW

WALL MOUNTED

CEILING MOUNTED

Courtesy Mueller Climatrol Corp.

Fig. 14-1. Some typical electronic air cleaner installations.

unit is impractical or where *selective* air cleaning is desired. These units can be installed anywhere in the structure.

Based on their operating principle, electronic air cleaners can be divided into the following two principal types:

1. Charged-media air cleaners.
2. Two-stage air cleaners.

509

CONTROL PANEL
PERFORMANCE INDICATOR
POWER SWITCH

IONIZING
SECTION

PRE-FILTER FOR
LARGE PARTICLES

COLLECTING CELLS

Courtesy Lennox Air Conditioning and Heating

Fig. 14-2. Multiposition electronic air cleaner.

Charged Media Air Cleaners

The basic working components of a *charged-media electronic air cleaner* are an electrically charged grid operating in conjunction with a media pad or mat. Media pads are commonly made of fiberglass, cellulose, or a similar material.

The charged-media air cleaner operates on the electrostatic principle. When voltage is applied to the grid, an intense electrostatic field is created. Dust particles passing through this field are polarized and caught by the media pads in much the same way that metal filings adhere to a magnet. When these media pads are filled, they must be removed and replaced with clean ones.

Two-Stage Air Cleaners

The *two-stage* (or *ionizing*) *electronic air cleaner* also operates on the electrostatic principle, but the airborne particles pass through *two* electrical fields rather than the single field used in

Fig. 14-3. Climatrol multipurpose electronic air cleaner.

charged-media air cleaners. The effectiveness of this type of air cleaner is indicated in test results from the National Bureau of Standards (Fig. 14-11).

Air entering a two-stage air cleaner must first pass through a permanent screen or prefilter, which catches the larger airborne

511

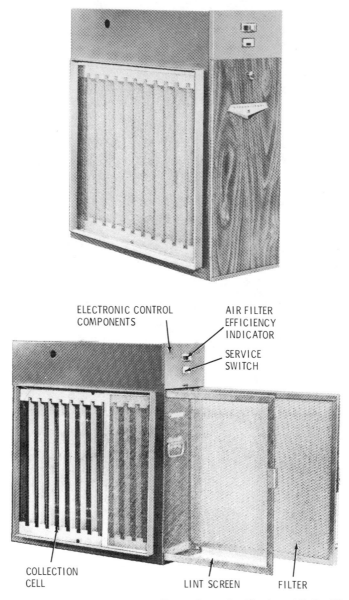

ELECTRONIC CONTROL
COMPONENTS

AIR FILTER
EFFICIENCY
INDICATOR

SERVICE
SWITCH

COLLECTION
CELL

LINT SCREEN

FILTER

Fig. 14-4. Utica International multiposition electronic air cleaner.

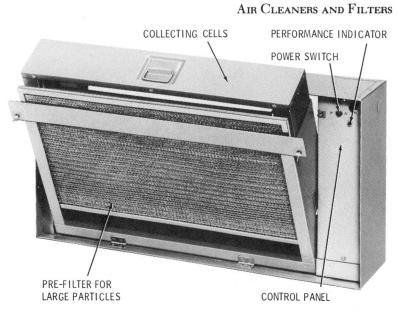

COLLECTING CELLS   PERFORMANCE INDICATOR

POWER SWITCH

PRE-FILTER FOR
LARGE PARTICLES

CONTROL PANEL

Fig. 14-5. Wall-mounted electronic air cleaner.

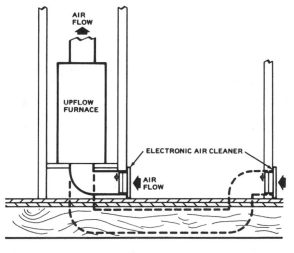

AIR
FLOW

UPFLOW
FURNACE

ELECTRONIC AIR CLEANER

AIR
FLOW

Fig. 14-6. Typical application on a platform-mounted upflow furnace.

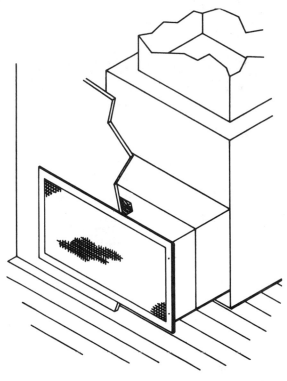

Fig. 14-7. Typical installation on an upflow (highboy) furnace.

particles. After passing through the prefilter, the air enters the so-called ionizer, or first stage, where the airborne particles receive an intense positive electrical charge. The positively charged airborne particles subsequently enter the collection, or second stage, which consists of a series of collector plates. These collector plates are metal plates or screens alternately charged with positive and negative high voltages. Because the airborne dust and dirt particles received a positive charge when they passed through the first stage of the electronic air cleaner, they are repelled by the *positively* charged plates in the second stage and propelled against the negatively charged collector plates where they adhere until washed away. The airborne particles are removed from the negative collector plates by periodic vacuum-

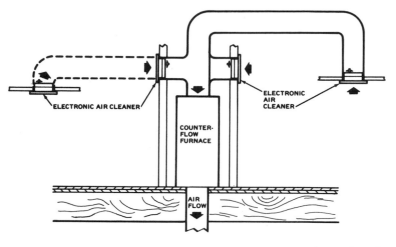

Fig. 14-8. Typical application on a high-capacity counterflow (downflow) furnace.

ing or washing. Some electronic air cleaners are equipped with washing systems that flush the particles off the plates.

The first stage (ionizing section) and second stage (collector section) are referred to collectively as the *electronic cell,* or the *electronic air cleaning cell.*

AUTOMATIC CONTROLS

A built-in electronic air cleaner can be connected electrically to the system blower motor or directly through a disconnect switch to the 120-volt line voltage power source. If the unit is connected to the system blower motor, the electronic cell will energize each time the blower motor operates.

Because the electronic air cleaner can be wired to operate either automatically or continuously in conjunction with fan operation, there is no need for a special wall-mounted air cleaner control. Thus, the fan control on a room thermostat, or combination thermostat and humidistat, is used to control both the system fan and the electronic air cleaner. A typical unit combining all

515

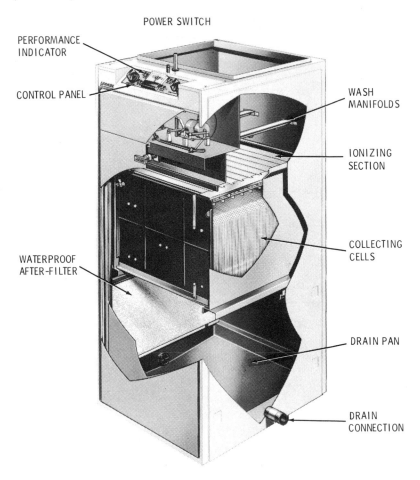

POWER SWITCH

PERFORMANCE
INDICATOR

CONTROL PANEL

WASH
MANIFOLDS

IONIZING
SECTION

COLLECTING
CELLS

WATERPROOF
AFTER-FILTER

DRAIN PAN

DRAIN
CONNECTION

Fig. 14-9. Cabinet-model electronic air cleaner.

heating and/or cooling system controls under a single cover is shown in Fig. 14-12. When the fan control switch is set on *auto*, the air is cleaned automatically whenever the heating and/or cooling system is operating. Continuous air cleaning (when extra air cleaning is required) is obtained by setting the fan control switch at *on*.

Fig. 14-10. Cabinet-model electronic air cleaner with built-in automatic water-wash system.

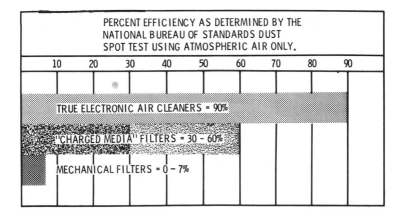

Fig. 14-11. National Bureau of Standards test results.

517

CLOGGED FILTER INDICATOR

The *clogged filter indicator* shown in Figs. 14-13 and 14-14 is used with Trane electronic air cleaners to sense pressure conditions in the blower chamber between the unit and the blower. An increase in pressure indicates a clogged filter, and this condition will be indicated by the light on the room thermostat control (see

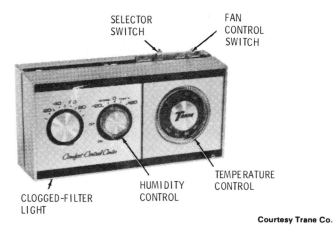

SELECTOR SWITCH

FAN CONTROL SWITCH

TEMPERATURE CONTROL

HUMIDITY CONTROL

CLOGGED-FILTER LIGHT

Courtesy Trane Co.

Fig. 14-12. Combination thermostat and humidistat used to control electronic air cleaner.

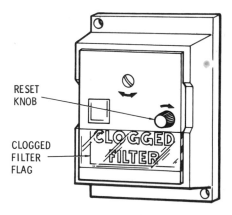

RESET KNOB

CLOGGED FILTER FLAG

Courtesy Trane Co.

Fig. 14-13. Furnace-mounted clogged filter indicator.

518

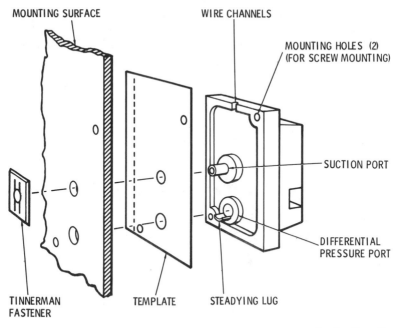

MOUNTING SURFACE

WIRE CHANNELS

MOUNTING HOLES (2)
(FOR SCREW MOUNTING)

SUCTION PORT

DIFFERENTIAL
PRESSURE PORT

TINNERMAN
FASTENER

TEMPLATE

STEADYING LUG

Courtesy Trane Co.

Fig. 14-14. Clogged filter indicator details.

Fig. 14-12) or when the clogged filter indicator on the furnace shows red.

The clogged filter indicator should be located on a rigid sheet metal mounting surface to prevent bumps or other vibrations from tripping the indicator. It should also be located where it can properly sense the pressure conditions.

PERFORMANCE LIGHTS

Most electronic air cleaners are equipped with performance lights to indicate how the unit is operating. How these lights will be used will depend upon the individual manufacturer. Read the manufacturer's operating instructions concerning the use of these lights. As the following paragraphs make clear, they are not always used in the same way.

The built-in performance indicator light on the Trane electronic air cleaner shown in Fig. 14-15 operates in conjunction with the on-off switch. If the electronic air cleaner is operating correctly, the performance indicator light will be on whenever the system fan is running and the on-off switch is in the *on* position.

The performance indicator light on the Lennox electronic air cleaner shown in Fig. 14-16 glows red when the unit is operating

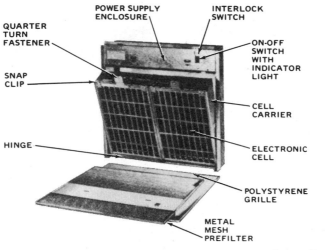

Courtesy Trane Co.

Fig. 14-15. Trane multiposition electronic air cleaner.

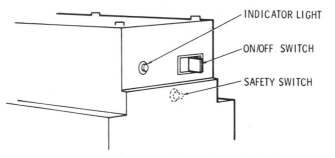

Courtesy Lennox Air Conditioning and Heating

Fig. 14-16. Lennox performance indicator light.

520

correctly. This light also operates in conjunction with the on-off switch which must be *on*. An optional performance light is available with a Lennox electronic air cleaner for installation in the living spaces (Fig. 14-17). It remains off when the unit is operating correctly.

Thermo-Pride electronic air cleaners use both an amber-colored operating light and a red performance light with their units. The operating light indicates line voltage is on. The red performance light indicates that the electronic cell is operating properly. Both lights must be on during normal operation.

SAIL SWITCH

A *sail switch* (Fig. 14-18) is designed to complete circuit power to auxiliary equipment in a forced warm-air system when the duct air velocity is increased. Consequently, it provides on-off control of electronic air cleaners, humidifiers, odor-control systems, and other equipment that is energized when the fan is operating.

In operation, the air movement pushing against the sail actuates the switching device, which then energizes the power supply. Using a sail switch allows the auxiliary equipment to be wired independently of the system blower motor.

The manufacturer's installation instructions should be carefully read and followed because the sail is installed at the site. The switch mechanism or sail switch body is mounted on the back of the electronic air cleaner usually before it is installed in the return air duct opening (Fig. 14-18). The sail is mounted on the switch body after installation of the air cleaner to prevent damage of the sail. A typical wiring diagram is shown in Fig. 14-19.

IN-PLACE WATER-WASH CONTROLS

Some electronic air cleaners are equipped with a water-wash system for in-place cleaning of the electronic cells. Operation of the water-wash system is governed by a control unit that includes a sequencing timer, water valve, detergent aspirator and valve,

Fig. 14-17. Optional performance light.

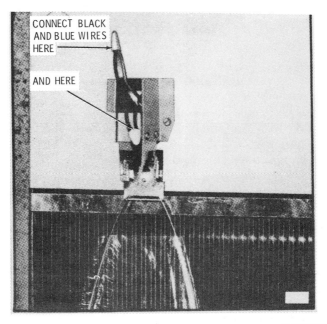

CONNECT BLACK
AND BLUE WIRES
HERE

AND HERE

Fig. 14-18. Sail switch mounted on back of electronic air cleaner.

and a fan interlock switch to control the system fan during the wash cycle (Fig. 14-20).The timer controls the internal water and detergent valves to provide automatic wash and rinse of the electronic cells. Typical hookups for a Honeywell W922 Wash Control are shown in Figs. 14-21 and 14-22.

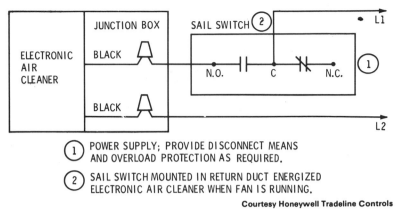

① POWER SUPPLY; PROVIDE DISCONNECT MEANS
AND OVERLOAD PROTECTION AS REQUIRED.

② SAIL SWITCH MOUNTED IN RETURN DUCT ENERGIZED
ELECTRONIC AIR CLEANER WHEN FAN IS RUNNING.

Courtesy Honeywell Tradeline Controls

Fig. 14-19. Wiring diagram of an electronic air cleaner controlled by a sail switch.

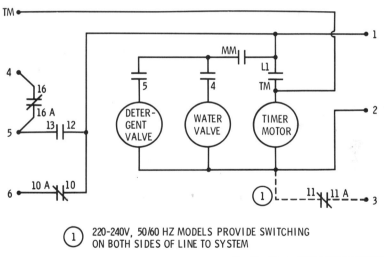

① 220-240V, 50/60 HZ MODELS PROVIDE SWITCHING
ON BOTH SIDES OF LINE TO SYSTEM

Courtesy Honeywell Tradeline Controls

Fig. 14-20. Water-wash control wiring diagram.

CABINET MODEL CONTROL PANELS

The cabinet model electronic air cleaner shown in Fig. 14-9 is equipped with a control panel that contains a meter, operating controls, and water-wash controls.

523

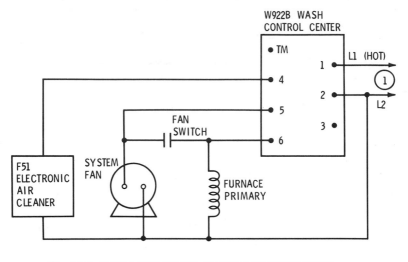

120V, 60 HZ POWER SUPPLY. PROVIDE DISCONNECT MEANS AND OVERLOAD PROTECTION AS REQUIRED.

Courtesy Honeywell Tradeline Controls

Fig. 14-21. Typical hookup of wash control and electronic air cleaner for in-place washing of the electronic cell.

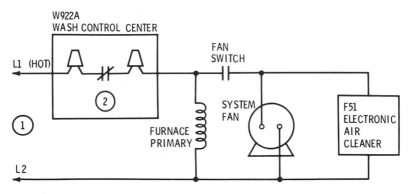

① 120V, 60Hz POWER SUPPLY. PROVIDE DISCONNECT MEANS AND OVERLOAD PROTECTION AS REQUIRED.

② THE NORMALLY CLOSED CONTACTS OPEN WHEN COVER IS OPENED.

Courtesy Honeywell Tradeline Controls

Fig. 14-22. Typical hookup of wash control and electronic air cleaner for automatic wash and rinse of electronic cells.

524

The meter is used to check the performance of the electronic air cleaner (Fig. 14-23). The meter is divided into three sections. If the unit is operating properly, the indicator needle will be steady and remain in the center section (the "normal operating range" on the meter). Fluctuations of the needle outside the normal operating range indicate that the unit is not operating properly. The specific problem is determined by the position of the needle.

The system/blower controls are located to the right of the meter; the washer/cleaner controls to the left. The voltage-control knob operates in conjunction with the meter and is used to adjust the unit for line voltage variations. The basic procedure for operating the electronic air cleaner shown in Fig. 14-9 is as follows:

1. Turn cleaner time switch to *on* position.
2. Push the *off* wash button.
3. Push the *on* cleaner button.
4. Push the *cont* blower button for continuous operation or the *auto* blower button for intermittent operation.
5. Push the *on* system button.
6. Adjust the voltage-control knob so that the needle remains in the normal operating range section on the meter.

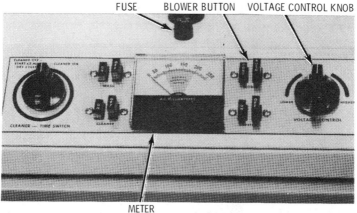

FUSE   BLOWER BUTTON   VOLTAGE CONTROL KNOB

METER

Courtesy Lennox Air Conditioning and Heating

Fig. 14-23. Cabinet-model control panel.

INSTALLATION INSTRUCTIONS

When installing a built-in electronic air cleaner, make certain there is enough clearance to allow easy access to the filters and collector cells. These components must be cleaned periodically and should not be obstructed. Most manufacturers recommend at least 30 in. clearance for servicing.

Pay particular attention to the airflow arrow printed on the side of the electronic cell. This arrow is used to indicate the correct direction of the airflow through the unit. If the electronic cell is installed so that airflow arrow is facing the wrong direction, excessive arcing will occur and the unit will not operate efficiently.

The conventional air filter should be removed after the electronic air cleaner has been installed in order to help reduce pressure drop through the furnace.

Air volume adjustment is another factor that must be taken into consideration when installing one of these units. These adjustments should be made in accordance with the manufacturer's instructions. They will involve determining the required temperature rise through the heat exchanger, making the necessary fan speed adjustments, and calibrating the filter flag setting (when used).

Some existing installations may require transitions or duct turns. If the ductwork makes an abrupt turn at the air cleaner, turning vanes should be installed in the duct to help provide an equal distribution of air across the entire surface of the filters. Do not install an atomizing humidifier upstream from an electronic air cleaner.

ELECTRICAL WIRING

Read and carefully follow the manufacturer's installation instructions before attempting to make any wiring connections. All wiring should be done in accordance with local codes and regulations. *Always* disconnect the power source before beginning any work in order to prevent electric shock or damage to the equipment.

Electronic air cleaners operate on regular 120-volt current and use less electricity than a 60-watt light bulb. Typical wiring connections for a Trane Model EAP-12A Electrostatic Air Cleaner are shown in Fig. 14-24. A ground wire for this unit is required. No ground wire is required for the return grille electronic air cleaner shown in Fig. 14-25. Other than installing the ground wire (where required), wiring an electronic air cleaner generally consists of simply hooking the unit up to the power source.

When making external circuit connections to the line voltage lead wires of an electronic air cleaner, only connectors listed by Underwriters' Laboratories should be used.

An electronic air cleaner can be connected electrically to the system blower motor or directly through a disconnect switch to the 120-volt power source. If the unit is connected to the system blower motor, the electronic cell will energize each time the blower operates.

As shown in Fig. 14-25, resistors are installed in the circuit to bleed off the electrical charges that the collector plates, acting as capacitors, are capable of storing. As a result, the serviceman and homeowner are protected against shock. On a unit that does *not*

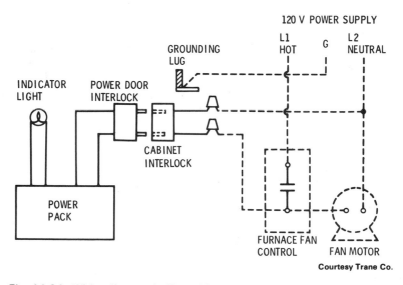

Courtesy Trane Co.

Fig. 14-24. Wiring diagram for Trane Model EAP-12A electronic air cleaner.

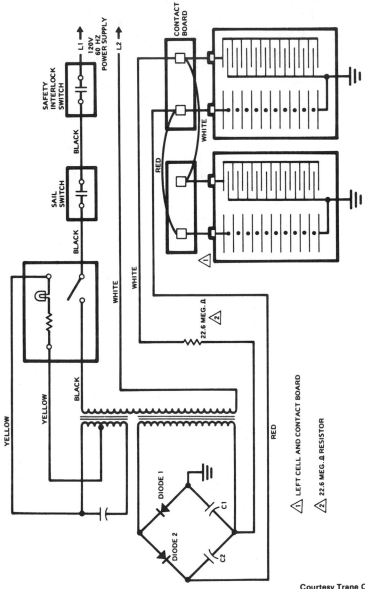

Courtesy Trane Co.

Fig. 14-25. Wiring diagram for Trane return grille electronic air cleaner.

include bleed-off resistors in its circuit, the cell must be grounded out before it is touched.

MAINTENANCE INSTRUCTIONS

The function of the filter in an electronic air cleaner is to eliminate the need for frequent cleaning of the electronic cell. Particles build up on the collecting plates until they reach a size large enough to be affected by the air velocity. When this point is reached, the particles are blown off the plates onto the after filter.

The prefilter (lint filter) and the after filter (so named because it follows the collecting plates in the unit) should be cleaned every 4 to 6 weeks. The required frequency of cell washing varies from one installation to the next and will depend largely on the level of air pollution each unit is expected to handle. In any event, the normal period will range from 1 to 6 months. Homes with large families, heavy and frequent cooking and laundering, and several tobacco smokers usually require monthly cleaning of the cell.

An automatic dishwasher can be used to wash an electronic cell without any fear of damaging the cell itself. Place the cells on the lower rack of the dishwasher with the airflow arrows facing up. If the cells are too large for the dishwasher, they will have to be washed by hand.

The procedure for manually cleaning the filters is as follows:

1. Turn off the current to the unit.
2. Note direction in which airflow arrows are pointing on the cell.
3. Remove the cell and both filters.
4. Soak cell and filters in a tub of water and electric dishwater detergent for about 30 minutes.
5. Rinse both sides of cell and filters with clear, clean water until all traces of dirt and detergent have been removed.
6. Shake excess water from cell and filters.
7. Replace cell and filters in the same order as before.

This last step is very important. If the cell and filters are not replaced so that the air flows in the direction of the arrow, the unit will not function properly.

Excess arcing or flickering of the red performance light immediately after washing the cell and filters is normal and is caused by water droplets on the surface. If these conditions continue, check to make certain the cell and filters were replaced in the correct order. If there is no problem with the direction of airflow, the unit is shorting (see "Electrical Wiring" and "Troubleshooting Electronic Air Cleaners" in this chapter).

REPLACING TUNGSTEN IONIZING WIRES

From time to time, the tungsten ionizing wires in the charging section of the electronic cell may break or become damaged. A broken or damaged wire generally causes a short to ground, which may or may not be accompanied by visible arcing or sparking. All parts of these broken or damaged wires should be removed from the unit immediately to prevent further shorting of the circuit. The unit will be able to operate on a *temporary basis* with one wire missing, but a new wire should be installed as soon as possible.

Some electronic air cleaners require the disassembly of the electronic cell in order to replace a broken or damaged ionizing wire. The procedure is as follows:

1. Place the cell on a flat surface and remove the screws from the terminal end of the cell (Fig. 14-26). Pull off the terminal end.
2. Remove the screws from the rear end and then pull off the end and the two sides.
3. Remove the top screen and carefully lift off the ionizer section. Turn the ionizer section upside down as shown in Fig. 14-27.
4. Remove all parts of the broken or damaged wire.
5. Compress both tension members and string the new ionizing wire.
6. *Slowly and carefully* release the compression force first on one tension member and then on the other (sudden release may snap the wire).

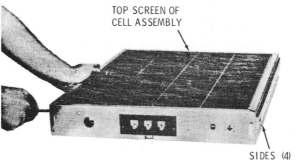

TOP SCREEN OF
CELL ASSEMBLY

SIDES (4)

Courtesy Trane Co.

Fig. 14-26. Cell placed on a flat surface.

IONIZER SECTION

COLLECTOR SECTION

Courtesy Trane Co.

Fig. 14-27. Ionizer section turned upside down.

7. Reassemble the cell.

The complete disassembly of the cell in order to replace an ionizing wire is not necessary on all electronic air cleaners. On the return grille electronic air cleaner, shown in Fig. 14-15, the replacement wires are cut to length and installed as follows:

1. Remove all parts of the broken or damaged ionizing wire.

531

2. Hook one eyelet of the ionizing wire over the spring connector at one end of the cell (Fig. 14-28).
3. Hold the opposite eyelet of the ionizing wire with a needle-nose pliers and stretch the wire the length of the cell.
4. Depress the opposite spring connector and hook the eyelet over it.

TROUBLESHOOTING ELECTRONIC AIR CLEANERS

Most electronic air cleaners are equipped with a performance indicator light, which is usually mounted in the control panel with the other controls. If the *unit* performance light is off, check the first section of the troubleshooting list for possible causes (see "Air Cleaner Will Not Operate").

A flickering performance indicator light is not necessarily an indication that the air cleaner is malfunctioning. It is not unusual for the light to flicker when the unit needs cleaning.

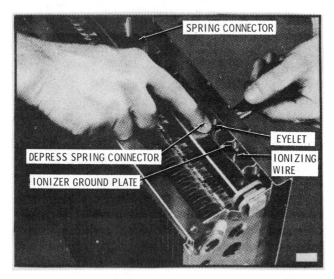

Courtesy Trane Co.

Fig. 14-28. Installation of new ionizing wires.

The troubleshooting chart that follows lists the most common problems associated with operating an electronic air cleaner. For each operating problem, a possible cause and its remedy has been suggested.

Symptom and Possible Cause	Possible Remedy

Air Cleaner Will Not Operate

1. Unit power switch off.	1. Turn switch on.
2. System fuse blown.	2. Check and replace with same-size fuse.
3. Blower unit or furnace power (disconnect) switch off.	3. Turn switch on.
4. Unit rectifier malfunctioning.	4. Replace rectifier.
5. Unit transformer malfunctioning.	5. Replace transformer.

Unit Performance Indicator Light Flickers

1. Dirty electronic cells.	1. Remove and clean cells.
2. Air flowing through cell in wrong direction.	2. Rearrange components in proper order (airflow in direction of arrow on cell).

Excess Arcing

1. Broken ionizing wires.	1. Replace wires.
2. Bent collecting plates.	2. Repair or replace plates.
3. Foreign object wedged between plates or ionizers.	3. Inspect and remove.

AIR WASHERS

Air washers operate by first passing the air through fine sprays of water and then past baffle plates upon the wetted surface on which is deposited whatever dust and dirt were not caught by the sprays.

An air washer functions as a filter, humidifier, and dehumidifier. Using it to regulate the moisture in the air decreases its efficiency as a filter for removing airborne dust and dirt particles.

Air washers are classified as either two-unit or three-unit types. The two-unit air washer consists simply of sprays and eliminators as shown in Fig. 14-29. A three-unit air washer includes a filter located between the sprays and eliminators (Fig. 14-30). Excluding the air filter, the basic components of a typical air washer are:

1. Cabinet.
2. Spray nozzles.
3. Eliminators or baffles.
4. Sump.
5. Pump.
6. Blower.

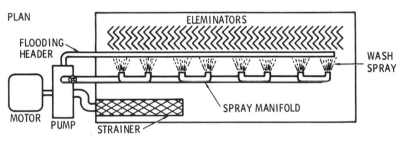

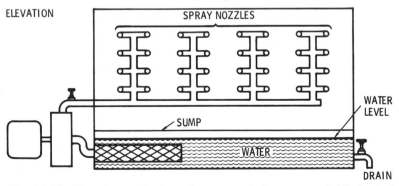

Fig. 14-29. Elevation and plan of a two-unit (sprayer and eliminator) washer.

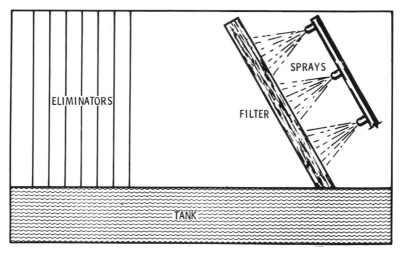

Fig. 14-30. Three-unit air washer consisting of sprays, filter, and eliminator plates.

The major advantage of using an air washer instead of a conventional air filter or an electronic air cleaner is that it does not become clogged with dust and dirt and is therefore not subject to loss of efficiency; however, air washers are no longer used in air cleaning installations because of the following disadvantages:

1. Equipment bulk.
2. Higher operating expense.
3. Inefficiency in removing fine particles.
4. Tendency to add moisture to the air when the water is not cool.

AIR FILTERS

Conventional air filters are commonly used in forced warm-air furnaces to trap and remove airborne particles and other contaminants from the air (Fig. 14-31). These filters are generally placed in the return air duct at a point just before the air supply enters the furnace or in the outdoor air intake ducts. In air conditioners, filters are properly placed ahead of heating or cooling coils and

535

Fig. 14-31. Conventional permanent washable air filter.

other air conditioning equipment in the system to protect them from dust and dirt. Air filters are *not* used in gravity warm-air furnaces because they obstruct the flow of air. Conventional air filters will effectively clean the air that is being circulated through the structure, but they will not prevent dirt from leaking into the house. On the average, a conventional air filter installed in the furnace can be expected to remove approximately 10 percent of all airborne particles from the air.

Filters should be inspected *at least* twice a year. Throwaway filters must be replaced with new ones, and washable filters should be cleaned when they become loaded with foreign matter. If this action is not taken, the efficiency of the furnace or air conditioner will be greatly reduced. As the dirt builds up on the filter, it increases the resistance of the passage of the air.

In a new house, the first set of filters may become clogged after a short time due to the presence of dust and dirt in the air created by the building operation. Check the filters after the first month of operation.

In an older house in which a winter air conditioning system has been installed, the dust and dirt that accompanies the dismantling of the old heating system may also clog the first set of filters in a short time. The new filters should also be checked after about a month of furnace operation.

If new rugs or carpets have been installed in the house, considerable lint will be given off at first. Under such conditions, replacement or cleaning of the filters will be necessary.

DRY AIR FILTERS

A *dry air filter* consists of a dry filtering medium such as cloth, porous paper, pads of loosely held cellulose fiber, wool felt, or some similar material held together in a lightweight metal or wire frame.

Both washable and disposable (throwaway) dry air filters are used in forced warm-air furnaces. Disposable filters are constructed of inexpensive materials and are designed to be discarded after one use. The frame is frequently a combination of cardboard and wire. Washable filters usually have metal frames. Various cleaning methods have been recommended, such as air jet, water jet, steam jet, or washing in kerosene and dipping in oil. The latter method may serve both to clean the filter and add the necessary adhesive.

VISCOUS AIR FILTERS

A *viscous air filter* (or *viscous-impingement air filter*) contains a filtering material consisting of coarse fibers coated with a sticky substance. This sticky substance catches the dust and dirt as the air passes through the mat. Viscous air filters can be reconditioned by washing them and recoating their surface with fresh liquid.

An oil or grease, sometimes referred to as the *adhesive* or *saturant*, is used as the viscous substance in these filters. The arrangement of the filter mat is such that the airstream is broken up into many small airstreams, and these are caused to abruptly change direction a number of times in order to throw the dust and dirt particles by centrifugal force against the adhesive.

The method used for cleaning a viscous filter will depend on the filter and the dust and dirt particles trapped by it. Most dry dust and dirt particles, as well as lint, can often be removed by rapping the filter frame.

FILTER INSTALLATION AND MAINTENANCE

Access to filters must be provided through a service panel in the furnace. Inspection and replacement of the filter by the user must be made possible without the use of special tools. When a new furnace is installed, care must be taken to provide sufficient clearance to the filter service panel. The furnace manufacturer will usually specify the minimum clearance in the installation instructions. An additional set of filter instructions should be attached to the filter service panel.

Always replace a disposable filter with one having the *same* dimensional size. The filter dimensions are printed on the filter frame.

APPENDIX A

Professional and Trade Associations

Many professional and trade associations have been organized to develop and provide materials and/or services in connection with heating, ventilating, and air conditioning. These materials and services include:

1. Formulating and establishing specifications and professional standards.
2. Certifying equipment and materials.
3. Conducting product research.
4. Promoting interest in the product.

Much useful information can be obtained by contacting these associations. With that in mind, the names and addresses of the principal organizations have been included in this appendix. They are listed in alphabetical order.

AIR-CONDITIONERS AND REFRIGERATION WHOLESALERS (ARW)
P.O. Box 640
1351 South Federal Highway
Deerfield Beach, FL 33441
Air conditioning and refrigeration equipment wholesalers.

AIR CONDITIONING CONTRACTORS OF AMERICA (ACCA)
1228 17th Street, N.W.
Washington, DC 20036
Formerly (until 1978) National Environmental Systems Contractors Association. Heating, air conditioning, and refrigeration systems contractors.

AIR-CONDITIONING AND REFRIGERATION INSTITUTE (ARI)
1815 North Fort Meyer Drive
Arlington, VA 22209
Manufacturers of air conditioning, warm-air heating, and commercial and industrial refrigeration equipment.

AIR DIFFUSION COUNCIL (ADC)
435 North Michigan Avenue
Chicago, IL 60611
Manufacturers of registers, grilles, diffusers, and related equipment.

AIR FILTER INSTITUTE. *See* Air-Conditioning and Refrigeration Institute

AMERICAN GAS ASSOCIATION (AGA)
1515 Wilson Boulevard
Arlington, VA 22209
Residential gas operating and performance standards from distributors and transporters of natural, manufactured, and mixed gas.

AIR MOVEMENT AND CONTROL ASSOCIATION
(AMCA)
30 West University Drive
Arlington Heights, IL 60004
Formerly (until 1977) Air Moving And Conditioning Association. Manufacturers of air moving and conditioning equipment.

AMERICAN NATIONAL STANDARDS INSTITUTE
(ANSI)
1430 Broadway
New York, NY 10018
Clearinghouse for nationally coordinated safety, engineering, and industrial standards.

AMERICAN SOCIETY FOR TESTING AND MATERIALS
(ASTM)
1916 Race Street
Philadelphia, PA 19103
Engineering standards for materials.

AMERICAN SOCIETY OF HEATING, REFRIGERATION
AND AIR-CONDITIONING ENGINEERS (ASHRAE)
1791 Tullie Circle, N.E.
Atlanta, GA 30329

AMERICAN SOCIETY OF MECHANICAL ENGINEERS
(ASME)
345 East 47th Street
New York, NY 10017
Develops safety codes and standards for various types of equipment (boiler and pressure vessel codes, etc.)

AMERICAN VENTILATION ASSOCIATION (AVA)
Box 7464
Houston, TX 77008
Association of residential ventilating equipment manufacturers and dealers.

BETTER HEATING-COOLING COUNCIL. *See* Hydronics
Institute

EDISON ELECTRIC INSTITUTE (EEI)
1111 19th Street, N.W.
Washington, DC 20036
Association of investor-owned electric utility companies.

ELECTRICAL ENERGY ASSOCIATION. *See* Edison Electric Institute

FIREPLACE INSTITUTE. Merged with Wood Energy Institute in 1980 to form Wood Heating Alliance.

HEATING AND PIPING CONTRACTORS NATIONAL
ASSOCIATION. *See* Mechanical Contractors Association of
America

HOME VENTILATING INSTITUTE (HVI)
4300-L Lincoln Avenue
Rolling Meadows, IL 60008
Develops performance standards for residential ventilating
equipment.

HYDRONICS INSTITUTE
35 Russo Place
Berkley Heights, NJ 07922
Manufacturers, installers, etc., of hot-water and steam heating
and cooling equipment. Formed by a merger of the Better
Heating-Cooling Council (1956–1970) and the Institute of
Boiler and Radiator Manufacturers (1915–1970).

INSTITUTE OF BOILER AND RADIATOR MANUFAC-
TURERS. *See* Hydronics Institute

INTERNATIONAL SOLAR ENERGY SOCIETY
National Science Center
P.O. Box 52
Parkville, Victoria 3052
Australia.
Educational and research organization.

MECHANICAL CONTRACTORS ASSOCIATION OF
AMERICA (MCAA)
5530 Wisconsin Avenue, N.W.
Suite 750
Washington, DC 20015
Contractors of piping and related equipment used in heating,
cooling, refrigeration, ventilating, and air conditioning.

NATIONAL ASSOCIATION OF PLUMBING-HEATING-
COOLING CONTRACTORS
1016 20th Street, N.W.
Washington, DC 20005
Local plumbing, heating, and cooling contractors association.

NATIONAL BOARD OF FIRE UNDERWRITERS (NBFU)
85 John Street
New York, NY 10036
Safety standards.

NATIONAL BUREAU OF STANDARDS
U.S. Department of Commerce
Washington, DC 20025
Government regulatory agency.

NATIONAL ENVIRONMENTAL SYSTEMS
CONTRACTORS ASSOCIATION. *See* Air Conditioning
Contractors of America

NATIONAL FIRE PROTECTION ASSOCIATION (NFPA)
Batterymarch Park
Quincy, MA 02269
Fire safety standards.

NATIONAL LP-GAS ASSOCIATION (NLPGA)
1301 West 22nd Street
Oak Brook, IL 60521

NATIONAL OIL FUEL INSTITUTE. *See* National Oil Jobbers Council

NATIONAL OIL JOBBERS COUNCIL
1707 H Street, N.W.
Washington, DC 20006
Absorbed National Oil Fuel Institute in 1974. Independent wholesale petroleum marketers and retail fuel oil dealers.

NATIONAL WARM AIR HEATING AND AIR CONDITIONING ASSOCIATION. *See* Air Conditioning Contractors of America

PLUMBING-HEATING-COOLING INFORMATION BUREAU (PHCIB)
35 East Wacker Drive
Chicago, IL 60601
Association of plumbing, heating, and cooling manufacturers, contractors, wholesalers, etc.

REFRIGERATION AND AIR CONDITIONING CONTRACTORS ASSOCIATION. *See* Air Conditioning Contractors of America

REFRIGERATION SERVICE ENGINEERS SOCIETY (RSES)
960 Rand Road
Des Plaines, IL 60018
Association of refrigeration, air conditioning, and heating equipment installers, servicemen, and salesmen.

SHEET METAL AND AIR CONDITIONING CONTRAC-
TORS NATIONAL ASSOCIATION (SMACNA)
8224 Old Courthouse Road
Vienna, VA 22180
Sheet-metal contractors who install ventilating, warm-air heat-
ing, and air handling equipment and systems.

STEAM HEATING EQUIPMENT MANUFACTURERS
ASSOCIATION. Defunct.

STEEL BOILER INSTITUTE. Defunct.

UNDERWRITERS' LABORATORIES
333 Pfingsten Road
Northbrook, IL 60062
Testing laboratory. Promotes safety standards for equipment.

WOOD ENERGY INSTITUTE. Merged with Fireplace Institute
in 1980 to form Wood Heating Alliance.

WOOD HEATING ALLIANCE (WHA)
1101 Connecticut Avenue, S.W.
Suite 700
Washington, DC 20036
Manufacturers, dealers, and suppliers of fireplaces, fireplace
components, and related equipment.

APPENDIX B

Table B-1. Equivalent Length of New Straight Pipe for Valves and Fittings for Turbulent Flow

Fittings			1/4	3/8	1/2	3/4	1	1 1/4	1 1/2	2	2 1/2	3	4	5	6	8	10	12	14	16	18	20	24
Regular 90° Ell	Screwed	Steel	2.3	3.1	3.6	4.4	5.2	6.6	7.4	8.5	9.3	11	13	—	—	—	—	—	—	—	—	—	—
		C.I.	—	—	—	—	—	—	—	—	—	9.0	11	—	—	—	—	—	—	—	—	—	—
	Flanged	Steel	—	—	—	—	—	—	—	—	—	—	5.9	7.3	8.9	12	14	17	18	21	23	25	30
		C.I.	—	—	—	—	—	—	—	—	—	—	4.8	—	7.2	9.8	12	15	17	19	22	24	28
Long Radius 90° Ell	Screwed	Steel	1.5	2.0	2.2	2.3	2.7	3.2	3.4	3.6	3.6	4.0	4.6	—	—	—	—	—	—	—	—	—	—
		C.I.	—	—	1.1	1.3	1.6	2.0	2.3	2.7	2.9	3.4	—	—	—	—	—	—	—	—	—	—	—
	Flanged	Steel	—	—	—	—	—	—	—	—	—	—	3.7	5.0	5.7	7.0	8.0	9.0	9.4	10	11	12	14
		C.I.	—	—	—	—	—	—	—	—	—	3.3	3.4	—	4.7	5.7	6.8	7.8	8.6	9.6	11	11	13
Regular 45° Ell	Screwed	Steel	.34	.52	.71	.92	1.3	1.7	2.1	2.7	3.2	4.0	5.5	—	—	—	—	—	—	—	—	—	—
		C.I.	—	—	—	—	—	—	—	—	—	3.3	4.5	—	—	—	—	—	—	—	—	—	—
	Flanged	Steel	—	—	—	—	—	—	—	—	—	—	3.5	4.5	5.6	7.7	9.0	11	13	15	16	18	22
		C.I.	—	—	—	—	—	—	—	—	—	—	2.9	—	4.5	6.3	8.1	9.7	12	13	15	17	20
Tee-Line Flow	Screwed	Steel	.79	1.2	1.7	2.4	3.2	4.6	5.6	7.7	9.3	12	17	—	—	—	—	—	—	—	—	—	—
		C.I.	—	—	.69	.82	1.0	1.3	1.5	1.8	1.9	9.9	14	—	—	—	—	—	—	—	—	—	—
	Flanged	Steel	—	—	—	—	—	—	—	—	—	—	4.7	5.2	5.6	7.7	9.0	11	13	15	16	18	22
		C.I.	—	—	—	—	—	—	—	—	—	—	3.9	4.6	4.5	6.3	8.1	9.7	12	13	15	17	20
Tee-Branch Flow	Screwed	Steel	2.4	3.5	4.2	5.3	6.6	8.7	9.9	12	13	17	21	—	—	—	—	—	—	—	—	—	—
		C.I.	—	—	2.0	2.6	3.3	4.4	5.2	6.6	7.5	14	17	—	—	—	—	—	—	—	—	—	—
	Flanged	Steel	—	—	—	—	—	—	—	—	—	9.4	12	15	18	24	30	34	37	43	47	52	62
		C.I.	—	—	—	—	—	—	—	—	—	7.7	10	—	15	20	25	30	35	39	44	49	57

Table B-1. Equivalent Length of New Straight Pipe for Valves and Fittings for Turbulent Flow (Cont'd)

Fitting	Connection	Material	¼	⅜	½	¾	1	1¼	1½	2	2½	3	4	5	6	8	10	12	14	16	18	20	24
180° Return Bend	Screwed	Steel	2.3	3.1	3.6	4.4	5.2	6.6	7.4	8.5	9.3	11	13	—	—	—	—	—	—	—	—	—	—
		C.I.	—	—	—	—	—	—	—	—	—	9.0	11	—	—	—	—	—	—	—	—	—	—
	Reg. Flanged	Steel	—	—	.92	1.2	1.6	2.1	2.4	3.1	3.6	4.4	5.9	7.3	8.9	12	14	17	18	21	23	25	30
		C.I.	—	—	—	—	—	—	—	—	—	3.6	4.8	—	7.2	9.8	12	15	17	19	22	24	28
	Long Rad. Flanged	Steel	—	—	1.1	1.3	1.6	2.0	2.3	2.7	2.9	3.4	4.2	5.0	5.7	7.0	8.0	9.0	9.4	10	11	12	14
		C.I.	—	—	—	—	—	—	—	—	—	2.8	3.4	—	4.7	5.7	6.8	7.8	8.6	9.6	11	11	13
Globe Valve	Screwed	Steel	21	22	22	24	29	37	42	54	62	79	110	—	—	—	—	—	—	—	—	—	—
		C.I.	—	—	—	—	—	—	—	—	—	65	86	—	—	—	—	—	—	—	—	—	—
	Flanged	Steel	—	—	—	—	—	—	—	—	—	94	120	150	190	260	310	390	—	—	—	—	—
		C.I.	—	—	—	—	—	—	—	—	—	77	99	—	150	210	270	330	—	—	—	—	—
Gate Valve	Screwed	Steel	.32	.45	.56	.67	.84	1.1	1.2	1.5	1.7	1.9	2.5	—	—	—	—	—	—	—	—	—	—
		C.I.	—	—	—	—	—	—	—	—	—	1.6	2.0	—	—	—	—	—	—	—	—	—	—
	Flanged	Steel	—	—	—	—	—	—	—	2.6	2.7	2.8	2.9	3.1	3.2	3.2	3.2	3.2	3.2	3.2	3.2	3.2	3.2
		C.I.	—	—	—	—	—	—	—	—	—	2.3	2.4	—	2.6	2.7	2.8	2.9	3.0	3.0	3.0	3.0	3.0
Angle Valve	Screwed	Steel	12.8	15	15	15	17	18	18	18	18	18	18	—	—	—	—	—	—	—	—	—	—
		C.I.	—	—	—	—	—	—	—	—	—	15	15	—	—	—	—	—	—	—	—	—	—
	Flanged	Steel	—	—	—	—	—	—	—	—	—	28	38	50	63	90	120	140	160	190	210	240	300
		C.I.	—	—	—	—	—	—	—	—	—	23	31	—	52	74	98	120	150	170	200	230	280
	Screwed	Steel	7.2	7.3	8.0	8.8	11	13	15	19	22	27	38	—	—	—	—	—	—	—	—	—	—
		C.I.	—	—	—	—	—	—	—	—	—	22	31	—	—	—	—	—	—	—	—	—	—

Table B-1. Equivalent Length of New Straight Pipe for Valves and Fittings for Turbulent Flow (Cont'd)

Fittings	Type	Material	1/4	3/8	1/2	3/4	1	1 1/4	1 1/2	2	2 1/2	3	4	5	6	8	10	12	14	16	18	20	24
Swing Check Valve	Flanged	Steel	—	—	3.8	5.3	7.2	10	12	17	21	27	38	50	63	90	120	140	—	—	—	—	—
		C.I.	—	—	—	—	—	—	—	—	—	22	31	—	52	74	98	120	—	—	—	—	—
Coupling or Union	Screwed	Steel	.14	.18	.21	.24	.29	.36	.39	.45	.47	.53	.65	.95	—	—	—	—	—	—	—	—	—
		C.I.	—	—	—	—	—	—	—	—	—	.44	.62	—	—	—	—	—	—	—	—	—	—
Bell Mouth Inlet		Steel	.04	.07	.10	.13	.18	.26	.31	.43	.52	.67	.95	1.3	1.6	2.3	2.9	3.5	4.0	4.7	5.3	6.1	7.6
		C.I.	—	—	—	—	—	—	—	—	—	.55	.77	—	1.3	1.9	2.4	3.0	3.6	4.3	5.0	5.7	7.0
Square Mouth Inlet		Steel	.44	.68	.96	1.3	1.8	2.6	3.1	4.3	5.2	6.7	9.5	13	16	23	29	35	40	47	53	61	76
		C.I.	—	—	—	—	—	—	—	—	—	5.5	7.7	—	13	19	24	30	36	43	50	57	70
Reentrant Pipe		Steel	.88	1.4	1.9	2.6	3.6	5.1	6.2	8.5	10	13	19	25	32	45	58	70	80	95	110	120	150
		C.I.	—	—	—	—	—	—	—	—	—	11	15	—	26	37	49	61	73	86	100	110	140
Y-Strainer			—	4.6	5.0	6.6	7.7	18	20	27	29	34	42	53	61	—	—	—	—	—	—	—	—

Sudden Enlargement

$$h = \frac{(V_1 - V_2)^2}{2g} \text{ Feet of Liquid; IF } V_2 = 0 \quad h = \frac{V^2}{2g} \text{ Feet of Liquid}$$

Courtesy The Hydraulic Institute (reprinted from the *Standards of the Hydraulic Institute*, Eleventh Edition. Copyright 1965)

Table B-2. Schedule 80 Pipe Dimensions

Size Inches	Diameters		Nominal Thickness Inches	Transverse Areas			Length of Pipe Per Square Foot of		Cubic Feet per Foot of Pipe	Weight per Foot Pounds	Number Threads per Inch of Screw
	External Inches	Internal Inches		External Sq. In.	Internal Sq. In.	Metal Sq. In.	External Surface Feet	Internal Surface Feet			
1/8	0.405	0.215	0.095	0.129	0.036	0.093	9.431	17.750	0.00025	0.314	27
1/4	0.540	0.302	0.119	0.229	0.072	0.157	7.073	12.650	0.00050	0.535	18
3/8	0.675	0.423	0.126	0.358	0.141	0.217	5.658	9.030	0.00098	0.738	18
1/2	0.840	0.546	0.147	0.554	0.234	0.320	4.547	7.000	0.00163	1.00	14
3/4	1.050	0.742	0.154	0.866	0.433	0.433	3.637	5.15	0.00300	1.47	14
1	1.315	0.957	0.179	1.358	0.719	0.639	2.904	3.995	0.00500	2.17	11 1/2
1 1/4	1.660	1.278	0.191	2.164	1.283	0.881	2.301	2.990	0.00891	3.00	11 1/2
1 1/2	1.900	1.500	0.200	2.835	1.767	1.068	2.010	2.542	0.01227	3.65	11 1/2
2	2.375	1.939	0.218	4.430	2.953	1.477	1.608	1.970	0.02051	5.02	11 1/2
2 1/2	2.875	2.323	0.276	6.492	4.238	2.254	1.328	1.645	0.02943	7.66	8
3	3.500	2.900	0.300	9.621	6.605	3.016	1.091	1.317	0.04587	10.3	8
3 1/2	4.000	3.364	0.318	12.56	8.888	3.678	0.954	1.135	0.06172	12.5	8
4	4.500	3.826	0.337	15.90	11.497	4.407	0.848	0.995	0.0798	14.9	8
5	5.563	4.813	0.375	24.30	18.194	6.112	0.686	0.792	0.1263	20.8	8
6	6.625	5.761	0.432	34.47	26.067	8.300	0.576	0.673	0.1810	28.6	8
8	8.625	7.625	0.500	58.42	46.663	12.76	0.442	0.501	0.3171	43.4	8
10	10.750	9.564	0.593	90.76	71.84	18.92	0.355	0.400	0.4989	64.4	
12	12.750	11.376	0.687	127.64	101.64	26.00	0.299	0.336	0.7058	88.6	
14	14.000	12.500	0.750	153.94	122.72	31.22	0.272	0.306	0.8522	107.0	
16	16.000	14.314	0.843	201.05	160.92	40.13	0.238	0.263	1.112	137.0	
18	18.000	16.126	0.937	254.85	204.24	50.61	0.212	0.237	1.418	171.0	
20	20.000	17.938	1.031	314.15	252.72	61.43	0.191	0.208	1.755	209.0	
24	24.000	21.564	1.218	452.40	365.22	87.18	0.159	0.177	2.536	297.0	

Courtesy Sarco Company, Inc.

Table B-3. Schedule 40 Pipe Dimensions

Size Inches	Diameters External Inches	Diameters Internal Inches	Nominal Thickness Inches	Transverse Areas External Sq. In.	Transverse Areas Internal Sq. In.	Transverse Areas Metal Sq. In.	Length of Pipe per Square Foot of External Surface Feet	Length of Pipe per Square Foot of Internal Surface Feet	Cubic Feet per Foot of Pipe	Weight per Foot Pounds	Number Threads per Inch of Screw
1/8	0.405	0.269	0.068	0.129	0.057	0.072	9.431	14.199	0.00039	0.244	27
1/4	0.540	0.364	0.088	0.229	0.104	0.125	7.073	10.493	0.00072	0.424	18
3/8	0.675	0.493	0.091	0.358	0.191	0.167	5.658	7.747	0.00133	0.567	18
1/2	0.840	0.622	0.109	0.554	0.304	0.250	4.547	6.141	0.00211	0.850	14
3/4	1.050	0.824	0.113	0.866	0.533	0.333	3.637	4.635	0.00370	1.130	14
1	1.315	1.049	0.133	1.358	0.864	0.494	2.904	3.641	0.00600	1.678	11½
1¼	1.660	1.380	0.140	2.164	1.495	0.669	2.301	2.767	0.01039	2.272	11½
1½	1.900	1.610	0.145	2.835	2.036	0.799	2.010	2.372	0.01414	2.717	11½
2	2.375	2.067	0.154	4.430	3.355	1.075	1.608	1.847	0.02330	3.652	11½
2½	2.875	2.469	0.203	6.492	4.788	1.704	1.328	1.547	0.03325	5.793	8
3	3.500	3.068	0.216	9.621	7.393	2.228	1.091	1.245	0.05134	7.575	8
3½	4.000	3.548	0.226	12.56	9.886	2.680	0.954	1.076	0.06866	9.109	8
4	4.500	4.026	0.237	15.90	12.73	3.174	0.848	0.948	0.08840	10.790	8
5	5.563	5.047	0.258	24.30	20.00	4.300	0.686	0.756	0.1389	14.61	8
6	6.625	6.065	0.280	34.47	28.90	5.581	0.576	0.629	0.2006	18.97	8
8	8.625	7.981	0.322	58.42	50.02	8.399	0.442	0.478	0.3552	28.55	8
10	10.750	10.020	0.365	90.76	78.85	11.90	0.355	0.381	0.5476	40.48	
12	12.750	11.938	0.406	127.64	111.9	15.74	0.299	0.318	0.7763	53.6	
14	14.000	13.125	0.437	153.94	135.3	18.64	0.272	0.280	0.9354	63.0	
16	16.000	15.000	0.500	201.05	176.7	24.35	0.238	0.254	1.223	78.0	
18	18.000	16.874	0.563	254.85	224.0	30.85	0.212	0.226	1.555	105.0	
20	20.000	18.814	0.593	314.15	278.0	36.15	0.191	0.203	1.926	123.0	
24	24.000	22.626	0.687	452.40	402.1	50.30	0.159	0.169	2.793	171.0	

Courtesy Sarco Company, Inc.

Table B-4. Properties of Saturated Steam

Gauge Pressure psig	Temperature °F	Heat in Btu/lb. Sensible	Latent	Total	Specific Volume Cu. ft. per lb.	Gauge Pressure psig	Temperature °F	Heat in Btu/lb. Sensible	Latent	Total	Specific Volume Cu. ft. per lb.
25 (In Vac.)	134	102	1017	1119	142.0	150	366	339	857	1196	2.74
20	162	129	1001	1130	73.9	155	368	341	885	1196	2.68
15	179	147	990	1137	51.3	160	371	344	853	1197	2.60
10	192	160	982	1142	39.4	165	373	346	851	1197	2.54
5	203	171	976	1147	31.8	170	375	348	849	1197	2.47
0	212	180	970	1150	26.8	175	377	351	847	1198	2.41
1	215	183	968	1151	25.2	180	380	353	845	1198	2.34
2	219	187	966	1153	23.5	185	382	355	843	1198	2.29
3	222	190	964	1154	22.3	190	384	358	841	1199	2.24
4	224	192	962	1154	21.4	195	386	360	839	1199	2.19
5	227	195	960	1155	20.1	200	388	362	837	1199	2.14
6	230	198	959	1157	19.4	205	390	364	836	1200	2.09
7	232	200	957	1157	18.7	210	392	366	834	1200	2.05
8	233	201	956	1157	18.4	215	394	368	832	1200	2.00
9	237	205	954	1159	17.1	220	396	370	830	1200	1.96
10	239	207	953	1160	16.5	225	397	372	828	1200	1.92
12	244	212	949	1161	15.3	230	399	374	827	1201	1.89
14	248	216	947	1163	14.3	235	401	376	825	1201	1.85
16	252	220	944	1164	13.4	240	403	378	823	1201	1.81
18	256	224	941	1165	12.6	245	404	380	822	1202	1.78

Table B-4. Properties of Saturated Steam (Cont'd)

Gauge Pressure psig	Temperature °F	Heat in Btu/lb.			Specific Volume Cu. ft. per lb.
		Sensible	Latent	Total	
20	259	227	939	1166	11.9
22	262	230	937	1167	11.3
24	265	233	934	1167	10.8
26	268	236	933	1169	10.3
28	271	239	930	1169	9.85
30	274	243	929	1172	9.46
32	277	246	927	1173	9.10
34	279	248	925	1173	8.75
36	282	251	923	1174	8.42
38	284	253	922	1175	8.08
40	286	256	920	1176	7.82
42	289	258	918	1176	7.57
44	291	260	917	1177	7.31
46	293	262	915	1177	7.14
48	295	264	914	1178	6.94
50	298	267	912	1179	6.68
55	300	271	909	1180	6.27
60	307	277	906	1183	5.84
65	312	282	901	1183	5.49
70	316	286	898	1184	5.18

Gauge Pressure psig	Temperature °F	Heat in Btu/lb.			Specific Volume Cu. ft. per lb.
		Sensible	Latent	Total	
250	406	382	820	1202	1.75
255	408	383	819	1202	1.72
260	409	385	817	1202	1.69
265	411	387	815	1202	1.66
270	413	389	814	1203	1.63
275	414	391	812	1203	1.60
280	416	392	811	1203	1.57
285	417	394	809	1203	1.55
290	418	395	808	1203	1.53
295	420	397	806	1203	1.49
300	421	398	805	1203	1.47
305	423	400	803	1203	1.45
310	425	402	802	1204	1.43
315	426	404	800	1204	1.41
320	427	405	799	1204	1.38
325	429	407	797	1204	1.36
330	430	408	796	1204	1.34
335	432	410	794	1204	1.33
340	433	411	793	1204	1.31
345	434	413	791	1204	1.29

Table B-4. Properties of Saturated Steam (Cont'd)

Gauge Pressure psig	Temperature °F	Heat in Btu/lb. Sensible	Heat in Btu/lb. Latent	Heat in Btu/lb. Total	Specific Volume Cu. ft. per lb.
75	320	290	895	1185	4.91
80	324	294	891	1185	4.67
85	328	298	889	1187	4.44
90	331	302	886	1188	4.24
95	335	305	883	1188	4.05
100	338	309	880	1189	3.89
105	341	312	878	1190	3.74
110	344	316	875	1191	3.59
115	347	319	873	1192	3.46
120	350	322	871	1193	3.34
125	353	325	868	1193	3.23
130	356	328	866	1194	3.12
140	361	333	861	1194	2.92
145	363	336	859	1195	2.84

Gauge Pressure psig	Temperature °F	Heat in Btu/lb. Sensible	Heat in Btu/lb. Latent	Heat in Btu/lb. Total	Specific Volume Cu. ft. per lb.
350	435	414	790	1204	1.28
355	437	416	789	1205	1.26
360	438	417	788	1205	1.24
365	440	419	786	1205	1.22
370	441	420	785	1205	1.20
375	442	421	784	1205	1.19
380	443	422	783	1205	1.18
385	445	424	781	1205	1.16
390	446	425	780	1205	1.14
395	447	427	778	1205	1.13
400	448	428	777	1205	1.12
450	460	439	766	1205	1.00
500	470	453	751	1204	0.89
550	479	464	740	1204	0.82
600	489	475	728	1203	0.74

Table B-5. Friction Loss for Water in Feet per 100 Feet of Schedule 40 Steel Pipe

U.S. Gal./Min.	Vel. Ft./Sec.	hf Friction	U.S. Gal./Min.	Vel. Ft./Sec.	hf Friction
³⁄₈" PIPE			¹⁄₂" PIPE		
1.4	2.25	9.03	2	2.11	5.50
1.6	2.68	11.6	2.5	2.64	8.24
1.8	3.02	14.3	3	3.17	11.5
2.0	3.36	17.3	3.5	3.70	15.3
2.5	4.20	26.0	4.0	4.22	19.7
3.0	5.04	36.6	5	5.28	29.7
3.5	5.88	49.0	6	6.34	42.0
4.0	6.72	63.2	7	7.39	56.0
5.0	8.40	96.1	8	8.45	72.1
6	10.08	136	9	9.50	90.1
7	11.8	182	10	10.56	110.6
8	13.4	236	12	12.7	156
9	15.1	297	14	14.8	211
10	16.8	364	16	16.9	270
³⁄₄" PIPE			1" PIPE		
4	2.41	4.85	6	2.23	3.16
5	3.01	7.27	8	2.97	5.20
6	3.61	10.2	10	3.71	7.90
7	4.21	13.6	12	4.45	11.1
8	4.81	17.3	14	5.20	14.7
9	5.42	21.6	16	5.94	19.0
10	6.02	26.5	18	6.68	23.7
12	7.22	37.5	20	7.42	28.9
14	8.42	50.0	22	8.17	34.8
16	9.63	64.8	24	8.91	41.0
18	10.8	80.9	26	9.65	47.8
20	12.0	99.0	28	10.39	55.1
22	13.2	120	30	11.1	62.9
24	14.4	141	35	13.0	84.4
26	15.6	165	40	14.8	109
28	16.8	189	45	16.7	137
			50	18.6	168

Table B-5. Friction Loss for Water in Feet per 100 Feet of Schedule 40 Steel Pipe (Cont'd)

U.S. Gal./Min.	Vel. Ft./Sec.	hf Friction	U.S. Gal./Min.	Vel. Ft./Sec.	hf Friction
1¼" PIPE			1½" PIPE		
12	2.57	2.85	16	2.52	2.26
14	3.00	3.77	18	2.84	2.79
16	3.43	4.83	20	3.15	3.38
18	3.86	6.00	22	3.47	4.05
20	4.29	7.30	24	3.78	4.76
22	4.72	8.72	26	4.10	5.54
24	5.15	10.27	28	4.41	6.34
26	5.58	11.94	30	4.73	7.20
28	6.01	13.7	35	5.51	9.63
30	6.44	15.6	40	6.30	12.41
35	7.51	21.9	45	7.04	15.49
40	8.58	27.1	50	7.88	18.9
45	9.65	33.8	55	8.67	22.7
50	10.7	41.4	60	9.46	26.7
55	11.8	49.7	65	10.24	31.2
60	12.9	58.6	70	11.03	36.0
65	13.9	68.6	75	11.8	41.2
70	15.0	79.2	80	12.6	46.6
75	16.1	90.6	85	13.4	52.4
			90	14.2	58.7
			95	15.0	65.0
			100	15.8	71.6
2" PIPE			2½" PIPE		
25	2.39	1.48	35	2.35	1.15
30	2.87	2.10	40	2.68	1.47
35	3.35	2.79	45	3.02	1.84
40	3.82	3.57	50	3.35	2.23
45	4.30	4.40	60	4.02	3.13
50	4.78	5.37	70	4.69	4.18
60	5.74	7.58	80	5.36	5.36
70	6.69	10.2	90	6.03	6.69
80	7.65	13.1	100	6.70	8.18
90	8.60	16.3	120	8.04	11.50
100	4.34	2.72	200	5.04	12.61
120	11.5	28.5	160	10.7	20.0
140	13.4	38.2	180	12.1	25.2
160	15.3	49.5	200	13.4	30.7
			220	14.7	37.1
			240	16.1	43.8

Table B-5. Friction Loss for Water in Feet per 100 Feet of Schedule 40 Steel Pipe (Cont'd)

U.S. Gal./Min.	Vel. Ft./Sec.	hf Friction	U.S. Gal./Min.	Vel. Ft./Sec.	hf Friction
3" PIPE			4" PIPE		
50	2.17	0.762	100	2.52	0.718
60	2.60	1.06	120	3.02	1.01
70	3.04	1.40	140	3.53	1.35
80	3.47	1.81	160	4.03	1.71
90	3.91	2.26	180	4.54	2.14
100	3.34	2.75	200	5.04	2.61
120	5.21	3.88	220	5.54	3.13
140	6.08	5.19	240	6.05	3.70
160	6.94	6.68	260	6.55	4.30
180	7.81	8.38	280	7.06	4.95
200	8.68	10.2	300	7.56	5.63
220	9.55	12.3	350	8.82	7.54
240	10.4	14.5	400	10.10	9.75
260	11.3	16.9	450	11.4	12.3
280	12.2	19.5	500	12.6	14.4
300	13.0	22.1	550	13.9	18.1
350	15.2	30	600	15.1	21.4
5" PIPE			6" PIPE		
160	2.57	0.557	220	2.44	0.411
180	2.89	0.698	240	2.66	0.482
200	3.21	0.847	260	2.89	0.560
220	3.53	1.01	300	3.33	0.733
240	3.85	1.19	350	3.89	0.980
260	4.17	1.38	400	4.44	1.25
300	4.81	1.82	450	5.00	1.56
350	5.61	2.43	500	5.55	1.91
400	6.41	3.13	600	6.66	2.69
450	7.22	3.92	700	7.77	3.60
500	8.02	4.79	800	8.88	4.64
600	9.62	6.77	900	9.99	5.81
700	11.2	9.13	1000	11.1	7.10
800	12.8	11.8	1100	12.2	8.52
900	14.4	14.8	1200	13.3	10.1
1000	16.0	18.2	1300	14.4	11.7
			1400	15.5	13.6

Courtesy Sarco Company, Inc.

Table B-6. Flow of Water through Schedule 40 Steel Pipe

Pressure Drop per 1000 Feet of Schedule 40 Steel Pipe, in Pounds per Square Inch

Vel. = Velocity, Ft. per Sec.; Pres. = Pressure Drop. Values listed under each nominal pipe size.

Discharge Gal. per Min.	1" Vel.	1" Pres.	1¼" Vel.	1¼" Pres.	1½" Vel.	1½" Pres.	2" Vel.	2" Pres.	2½" Vel.	2½" Pres.	3" Vel.	3" Pres.	3½" Vel.	3½" Pres.	4" Vel.	4" Pres.	5" Vel.	5" Pres.	6" Vel.	6" Pres.	8" Vel.	8" Pres.
1	0.37	0.49																				
2	0.74	1.70	0.43	0.45																		
3	1.12	3.53	0.64	0.94	0.47	0.44																
4	1.49	5.94	0.86	1.55	0.63	0.74																
5	1.86	9.02	1.07	2.36	0.79	1.12																
6	2.24	12.25	1.28	3.30	0.95	1.53	0.57	0.46														
8	2.98	21.1	1.72	5.52	1.26	2.63	0.76	0.75														
10	3.72	30.8	2.14	8.34	1.57	3.86	0.96	1.14	0.67	0.48												
15	5.60	64.6	3.21	17.6	2.36	8.13	1.43	2.33	1.00	0.99												
20	7.44	110.5	4.29	29.1	3.15	13.5	1.91	3.86	1.34	1.64	0.87	0.59										
25			5.36	43.7	3.94	20.2	2.39	5.81	1.68	2.48	1.08	0.67	0.81	0.42								
30			6.43	62.9	4.72	29.1	2.87	8.04	2.01	3.43	1.30	1.21	0.97	0.60								
35			7.51	82.5	5.51	38.2	3.35	10.95	2.35	4.49	1.52	1.58	1.14	0.79	0.88	0.42						
40					6.30	47.8	3.82	13.7	2.68	5.88	1.74	2.06	1.30	1.00	1.01	0.53						
45					7.08	60.6	4.30	17.4	3.00	7.14	1.95	2.51	1.46	1.21	1.13	0.67						
50					7.87	74.7	4.78	20.6	3.35	8.82	2.17	3.10	1.62	1.44	1.26	0.80						
60							5.74	29.6	4.02	12.2	2.60	4.29	1.95	2.07	1.51	1.10						
70							6.69	38.6	4.69	15.3	3.04	5.84	2.27	2.71	1.76	1.50	1.12	0.48				
80							7.65	50.3	5.37	21.7	3.48	7.62	2.59	3.53	2.01	1.87	1.28	0.63				
90							8.60	63.6	6.04	26.1	3.91	9.22	2.92	4.46	2.26	2.37	1.44	0.80	1.11	0.39		
100							9.56	75.1	6.71	32.3	4.34	11.4	3.24	5.27	2.52	2.81	1.60	0.95	1.39	0.56		
125									8.38	48.2	5.45	17.1	4.05	7.86	3.15	4.38	2.00	1.48	1.67	0.78		
150									10.06	60.4	6.51	23.3	4.86	11.3	3.78	6.02	2.41	2.04	1.94	1.06		
175									11.73	90.0	7.59	32.0	5.67	14.7	4.41	8.20	2.81	2.78	2.22	1.32		
200											8.68	39.7	6.48	19.2	5.04	10.2	3.21	3.46	2.22	1.32		
225											9.77	50.2	7.29	23.1	5.67	12.9	3.61	4.37	2.50	1.66	1.44	0.44
250											10.85	61.9	8.10	28.5	6.30	15.9	4.01	5.14	2.78	2.05	1.60	0.55
275											11.94	75.0	8.91	34.4	6.93	18.3	4.41	6.22	3.06	2.36	1.76	0.63
300											13.02	84.7	9.72	40.9	7.56	21.8	4.81	7.41	3.33	2.80	1.92	0.75
325													10.53	45.5	8.18	25.5	5.21	8.25	3.61	3.29	2.08	0.88
350													11.35	52.7	8.82	29.7	5.61	9.57	3.89	3.62	2.24	0.97
375													12.17	60.7	9.45	32.3	6.01	11.0	4.16	4.16	2.40	1.11
400													12.97	68.9	10.08	39.7	6.41	12.9	4.44	4.72	2.56	1.27
425													13.78	77.8	10.70	41.5	6.82	14.1	4.72	5.34	2.73	1.43

Table B-6. Flow of Water through Schedule 40 Steel Pipe (Cont'd)

Pressure Drop per 1000 Feet of Schedule 40 Steel Pipe, in Pounds per Square Inch

Discharge Gal. per Min.	6"		8"		10"		12"		14"		16"		18"		20"		24"	
	Vel. Ft. per Sec.	Pres. Drop	Vel. Ft. per Sec.	Pres. Drop	Vel. Ft. per Sec.	Pres. Drop	Vel. Ft. per Sec.	Pres. Drop	Vel. Ft. per Sec.	Pres. Drop	Vel. Ft. per Sec.	Pres. Drop	Vel. Ft. per Sec.	Pres. Drop	Vel. Ft. per Sec.	Pres. Drop	Vel. Ft. per Sec.	Pres. Drop
450	5.00	5.96	2.88	1.60														
475	5.27	6.66	3.04	1.69	1.93	0.53												
500	5.55	7.39	3.20	1.87	2.04	0.63												
550	6.11	8.94	3.53	2.26	2.24	0.70												
600	6.66	10.6	3.85	2.70	2.44	0.86												
650	7.21	11.8	4.17	3.16	2.65	1.01												
700	7.77	13.7	4.49	3.69	2.85	1.18	2.01	0.48										
750	8.32	15.7	4.81	4.21	3.05	1.35	2.15	0.55										
800	8.88	17.8	5.13	4.79	3.26	1.54	2.29	0.62										
850	9.44	20.2	5.45	5.11	3.46	1.74	2.44	0.70	2.02	0.43								
900	10.00	22.6	5.77	5.73	3.66	1.94	2.58	0.79	2.14	0.48								
950	10.55	23.7	6.09	6.38	3.87	2.23	2.72	0.88	2.25	0.53								
1,000	11.10	26.3	6.41	7.08	4.07	2.40	2.87	0.98	2.38	0.59								
1,100	12.22	31.8	7.05	8.56	4.48	2.74	3.16	1.18	2.61	0.68								
1,200	13.32	37.8	7.69	10.2	4.88	3.27	3.45	1.40	2.85	0.81	2.18	0.40						
1,300	14.43	44.4	8.33	11.3	5.29	3.86	3.73	1.56	3.09	0.95	2.36	0.47						
1,400	15.54	51.5	8.97	13.0	5.70	4.44	4.02	1.80	3.32	1.10	2.54	0.54						
1,500	16.65	55.5	9.62	15.0	6.10	5.11	4.30	2.07	3.55	1.19	2.73	0.62						
1,600	17.76	63.1	10.26	17.0	6.51	5.46	4.59	2.36	3.80	1.35	2.91	0.71						
1,800	19.98	79.8	11.54	21.6	7.32	6.91	5.16	2.98	4.27	1.71	3.27	0.85	2.58	0.48				
2,000	22.20	98.5	12.83	25.0	8.13	8.54	5.73	3.47	4.74	2.11	3.63	1.05	2.88	0.56				
2,500			16.03	39.0	10.18	12.5	7.17	5.41	5.92	3.09	4.54	1.63	3.59	0.88				
3,000			19.24	52.4	12.21	18.0	8.60	7.31	7.12	4.45	5.45	2.21	4.31	1.27	3.45	0.73		
3,500			22.43	71.4	14.25	22.9	10.03	9.95	8.32	6.18	6.35	3.00	5.03	1.52	4.03	0.94		
4,000			25.65	93.3	16.28	29.9	11.48	13.0	9.49	7.92	7.25	3.92	5.74	2.12	4.61	1.22	3.19	0.51
4,500					18.31	37.8	12.90	15.4	10.67	9.36	8.17	4.97	6.47	2.50	5.19	1.55	3.59	0.60
5,000					20.35	46.7	14.34	18.9	11.84	11.6	9.08	5.72	7.17	3.08	5.76	1.78	3.99	0.74
6,000					24.42	67.2	17.21	27.3	14.32	15.4	10.88	8.24	8.62	4.45	6.92	2.57	4.80	1.00
7,000					28.50	85.1	20.08	37.2	16.60	21.0	12.69	12.2	10.04	6.06	8.06	3.50	5.68	1.36
8,000							22.95	45.1	18.98	27.4	14.52	13.6	11.48	7.34	9.23	4.57	6.38	1.78
9,000							25.80	57.0	21.35	34.7	16.32	17.2	12.92	9.20	10.37	5.36	7.19	2.25
10,000							28.63	70.4	23.75	42.9	18.16	21.2	14.37	11.5	11.53	6.63	7.96	2.78
12,000							34.38	93.6	28.50	61.8	21.80	30.9	17.23	16.5	13.83	9.54	9.57	3.71
14,000									33.20	84.0	25.42	41.6	20.10	20.7	16.14	12.0	11.18	5.05
16,000											29.05	54.4	22.96	27.1	18.43	15.7	12.77	6.60

Courtesy Sarco Company, Inc.

Table B-7. Warmup Load in Pounds of Steam per 100 Feet of Steam Main (Ambient Temperature 70°F)*

Steam Pressure (psig)	MAIN SIZE														0°F Correction Factor†
	2"	2½"	3"	4"	5"	6"	8"	10"	12"	14"	16"	18"	20"	24"	
0	6.2	9.7	12.8	18.2	24.6	31.9	48	68	90	107	140	176	207	208	1.50
5	6.9	11.0	14.4	20.4	27.7	35.9	48	77	101	120	157	198	233	324	1.44
10	7.5	11.8	15.5	22.0	29.9	38.8	58	83	109	130	169	213	251	350	1.41
20	8.4	13.4	17.5	24.9	33.8	43.9	66	93	124	146	191	241	284	396	1.37
40	3.9	15.8	20.6	29.3	39.7	51.6	78	110	145	172	225	284	334	465	1.32
60	11.0	17.5	22.9	32.6	44.2	57.3	86	122	162	192	250	316	372	518	1.29
80	12.0	19.0	24.9	35.3	47.9	62.1	93	132	175	208	271	342	403	561	1.27
100	12.8	20.3	26.6	37.8	51.2	66.5	100	142	188	222	290	366	431	600	1.26
125	13.7	21.7	28.4	40.4	54.8	71.1	107	152	200	238	310	391	461	642	1.25
150	14.5	23.0	30.0	42.8	58.0	75.2	113	160	212	251	328	414	487	679	1.24
175	15.3	24.2	31.7	45.1	61.2	79.4	119	169	224	265	347	437	514	716	1.23
200	16.0	25.3	33.1	47.1	63.8	82.8	125	177	234	277	362	456	537	748	1.22
250	17.2	27.3	35.8	50.8	68.9	89.4	134	191	252	299	390	492	579	807	1.21
300	25.0	38.3	51.3	74.8	104.0	142.7	217	322	443	531	682	854	1045	1182	1.20
400	27.8	42.6	57.1	83.2	115.7	158.7	241	358	493	590	759	971	1163	1650	1.18
500	30.2	46.3	62.1	90.5	125.7	172.6	262	389	535	642	825	1033	1263	1793	1.17
600	32.7	50.1	67.1	97.9	136.0	186.6	284	421	579	694	893	1118	1367	1939	1.16

Courtesy Sarco Company, Inc.

*Loads based on Schedule 40 pipe for pressures up to and including 250 psig and on Schedule 80 pipe for pressures above 250 psig.

†For outdoor temperature of 0°F, multiply load value in table for each main size by correction factor corresponding to steam pressure.

Table B-8. Condensation Load in Pounds per Hour per 100 Feet of Insulated Steam Main (Ambient Temperature 70°F; Insulation 80% Efficient)*

Steam Pressure (psig)	MAIN SIZE														0°F Correction Factor†
	2"	2½"	3"	4"	5"	6"	8"	10"	12"	14"	16"	18"	20"	24"	
10	6	7	9	11	13	16	20	24	29	32	36	39	44	53	1.58
30	8	9	11	14	17	20	26	32	38	42	48	51	57	68	1.50
60	10	12	14	18	24	27	33	41	49	54	62	67	74	89	1.45
100	12	15	18	22	28	33	41	51	61	67	77	83	93	111	1.41
125	13	16	20	24	30	36	45	56	66	73	84	90	101	121	1.39
175	16	19	23	26	33	38	53	66	78	86	98	107	119	142	1.38
250	18	22	27	34	42	50	62	77	92	101	116	126	140	168	1.36
300	20	25	30	37	46	54	68	85	101	111	126	138	154	184	1.35
400	23	28	34	43	53	63	80	99	118	130	148	162	180	216	1.33
500	27	33	39	49	61	73	91	114	135	148	170	185	206	246	1.32
600	30	37	44	55	68	82	103	128	152	167	191	208	232	277	1.31

*Chart loads represent losses due to radiation and convection for saturated steam.
†For outdoor temperature of 0°F, multiply load value in table for each main size by correction factor corresponding to steam pressure.

Courtesy Sarco Company, Inc.

Table B-9. Flange Standards
(All dimensions are in inches)

125-lb. CAST IRON

ASA B16.1

Pipe Size	½	¾	1	1¼	1½	2	2½	3	3½	4	5	6	8	10	12
Diameter of Flange			4¼	4⅝	5	6	7	7½	8½	9	10	11	13½	16	19
Thickness of Flange (min)[1]			⁷/₁₆	½	⁹/₁₆	⅝	¹¹/₁₆	¾	¹³/₁₆	¹⁵/₁₆	¹⁵/₁₆	1	1⅛	1³/₁₆	1¼
Diameter of Bolt Circle			3⅛	3½	3⅞	4¾	5½	6	7	7½	8½	9½	11¾	14¼	17
Number of Bolts			4	4	4	4	4	4	8	8	8	8	8	12	12
Diameter of Bolts			½	½	½	⅝	⅝	⅝	⅝	⅝	¾	¾	¾	⅞	⅞

[1] 125-lb. flanges have plain faces.

250-lb. CAST IRON

ASA B16.2

Pipe Size	½	¾	1	1¼	1½	2	2½	3	3½	4	5	6	8	10	12
Diameter of Flange			4⅞	5¼	6⅛	6½	7½	8¼	9	10	11	12½	15	17½	20½
Thickness of Flange (min)[2]			¹¹/₁₆	¾	¹³/₁₆	⅞	1	1⅛	1³/₁₆	1¼	1⅜	1⁷/₁₆	1⅝	1⅞	2
Diameter of Raised Face			2¹¹/₁₆	3¹/₁₆	3⁹/₁₆	4³/₁₆	4¹⁵/₁₆	6⁵/₁₆	6⁵/₁₆	6¹⁵/₁₆	8⁵/₁₆	9¹¹/₁₆	11¹⁵/₁₆	14¹/₁₆	16⁷/₁₆
Diameter of Bolt Circle			3½	3⅞	4½	5	5⅞	6⅝	7¼	7⅞	9¼	10⅝	13	15¼	17¾
Number of Bolts			4	4	4	8	8	8	8	8	8	12	12	16	16
Diameter of Bolts			⅝	⅝	¾	⅝	¾	¾	¾	¾	¾	¾	⅞	1	1⅛

[2] 250-lb. flanges have a ¹/₁₆″ raised face, which is included in the flange thickness dimensions.

Table B-9. Flange Standards (Cont'd)

(All dimensions are in inches)

150-lb. BRONZE

								ASA B16.24							
Pipe Size	½	¾	1	1¼	1½	2	2½	3	3½	4	5	6	8	10	12
Diameter of Flange	3½	3⅞	4¼	4⅝	5	6	7	7½	8½	9	10	11	13½	16	19
Thickness of Flange (min)[3]	5/16	11/32	3/8	13/32	7/16	½	9/16	5/8	11/16	11/16	¾	13/16	15/16	1	1 1/16
Diameter of Bolt Circle	2⅜	2¾	3⅛	3½	3⅞	4¾	5½	6	7	7½	8½	9½	11¾	14¼	17
Number of Bolts	4	4	4	4	4	4	4	4	8	8	8	8	8	12	12
Diameter of Bolts	½	½	½	½	½	⅝	⅝	⅝	⅝	⅝	¾	¾	¾	⅞	⅞

[3]150-lb. bronze flanges have plain faces with two concentric gasket-retaining grooves between the port and the bolt holes.

300-lb. BRONZE

								ASA B16.24							
Pipe Size	½	¾	1	1¼	1½	2	2½	3	3½	4	5	6	8	10	12
Diameter of Flange	3¾	4⅝	4⅞	5¼	6⅛	6½	7½	8¼	9	10	11	12½	15		
Thickness of Flange (min)[4]	½	17/32	19/32	5/8	11/16	¾	13/16	29/32	31/32	1 1/16	1⅛	1 3/16	1 3/8		
Diameter of Bolt Circle	2⅝	3¼	3½	3⅞	4½	5	5⅞	6⅝	7¼	7⅞	9¼	10⅝	13		
Number of Bolts	4	4	4	4	4	8	8	8	8	8	8	12	12		
Diameter of Bolts	½	⅝	⅝	⅝	¾	⅝	¾	¾	¾	¾	¾	¾	⅞		

[4]300-lb. bronze flanges have plain faces with two concentric gasket-retaining grooves between the port and the bolt holes.

Table B-9. Flange Standards (Cont'd)
(All dimensions are in inches)

150-lb. STEEL

ASA B16.5

Pipe Size	1/2	3/4	1	1 1/4	1 1/2	2	2 1/2	3	3 1/2	4	5	6	8	10	12
Diameter of Flange			4 1/4	4 5/8	5	6	7	7 1/2	8 1/2	9	10	11	13 1/2	16	19
Thickness of Flange (min)[5]			7/16	1/2	9/16	5/8	11/16	3/4	13/16	15/16	15/16	1	1 1/8	1 3/16	1 1/4
Diameter of Raised Face			2	2 1/2	2 7/8	3 5/8	4 1/8	5	5 1/2	6 3/16	7 5/16	8 1/2	10 5/8	12 3/4	15
Diameter of Bolt Circle			3 1/8	3 1/2	3 7/8	4 3/4	5 1/2	6	7	7 1/2	8 1/2	9 1/2	11 3/4	14 1/4	17
Number of Bolts			4	4	4	4	4	4	8	8	8	8	8	12	12
Diameter of Bolts			1/2	1/2	1/2	5/8	5/8	5/8	5/8	5/8	3/4	3/4	3/4	7/8	7/8

[5]150-lb. steel flanges have a 1/16" raised face, which is included in the flange thickness dimensions.

300-lb. STEEL

ASA B16.5

Pipe Size	1/2	3/4	1	1 1/4	1 1/2	2	2 1/2	3	3 1/2	4	5	6	8	10	12
Diameter of Flange			4 7/8	5 1/4	6 1/8	6 1/2	7 1/2	8 1/4	9	10	11	12 1/2	15	17 1/2	20 1/2
Thickness of Flange (min)[6]			11/16	3/4	13/16	7/8	1	1 1/8	1 3/16	1 1/4	1 3/8	1 7/16	1 5/8	1 7/8	2
Diameter of Raised Face			2	2 1/2	2 7/8	3 5/8	4 1/8	5	5 1/2	6 3/16	7 5/16	8 1/2	10 5/8	12 3/4	15
Diameter of Bolt Circle			3 1/2	3 7/8	4 1/2	5	5 7/8	6 5/8	7 1/4	7 7/8	9 1/4	10 5/8	13	15 1/4	17 3/4
Number of Bolts			4	4	4	8	8	8	8	8	8	12	12	16	16
Diameter of Bolts			5/8	5/8	3/4	5/8	3/4	3/4	3/4	3/4	3/4	3/4	7/8	1	1 1/8

[6]300-lb. steel flanges have a 1/16" raised face, which is included in the flange thickness dimensions.

Table B-9. Flange Standards (Cont'd)

(All dimensions are in inches)

400-lb. STEEL

Pipe Size	½	¾	1	1¼	1½	2	2½	3	ASA B16.5 3½	4	5	6	8	10	12
Diameter of Flange	3¾	4⅝	4⅞	5¼	6⅛	6½	7½	8¼	9	10	11	12½	15	17½	20½
Thickness of Flange (min)[7]	9/16	⅝	11/16	13/16	⅞	1	1⅛	1¼	1⅜	1⅜	1½	1⅝	1⅞	2⅛	2¼
Diameter of Raised Face	1⅜	1¹¹/₁₆	2	2½	2⅞	3⅝	4⅛	5	5½	6³/₁₆	7⁵/₁₆	8½	10⅝	12¾	15
Diameter of Bolt Circle	2⅝	3¼	3½	3⅞	4½	5	5⅞	6⅝	7¼	7⅞	9¼	10⅝	13	15¼	17¾
Number of Bolts	4	4	4	4	4	8	8	8	8	8	8	12	12	16	16
Diameter of Bolts	½	⅝	⅝	⅝	¾	⅝	¾	¾	⅞	⅞	⅞	⅞	1	1⅛	1¼

[7]400-lb. steel flanges have a ¼" raised face, which is NOT included in the flange dimensions.

600-lb. STEEL

Pipe Size	½	¾	1	1¼	1½	2	2½	3	ASA B16.5 3½	4	5	6	8	10	12
Diameter of Flange	3¾	4⅝	4⅞	5¼	6⅛	6½	7½	8¼	9	10¾	13	14	16½	20	22
Thickness of Flange (min)[8]	9/16	⅝	11/16	13/16	⅞	1	1⅛	1¼	1⅜	1½	1¾	1⅞	2³/₁₆	2½	2⅝
Diameter of Raised Face	1⅜	1¹¹/₁₆	2	2½	2⅞	3⅝	4⅛	5	5½	6³/₁₆	7⁵/₁₆	8½	10⅝	12¾	15
Diameter of Bolt Circle	2⅝	3¼	3½	3⅞	4½	5	5⅞	6⅝	7¼	8½	10½	11½	13¼	17	19½
Number of Bolts	4	4	4	4	4	8	8	8	8	8	8	12	12	16	20
Diameter of Bolts	½	⅝	⅝	⅝	¾	⅝	¾	¾	⅞	⅞	1	1	1⅛	1¼	1¼

[8]600-lb. steel flanges have a ¼" raised face, which is NOT included in the flange dimensions.

Table B-10. Pressure Drop in Schedule 40 Pipe

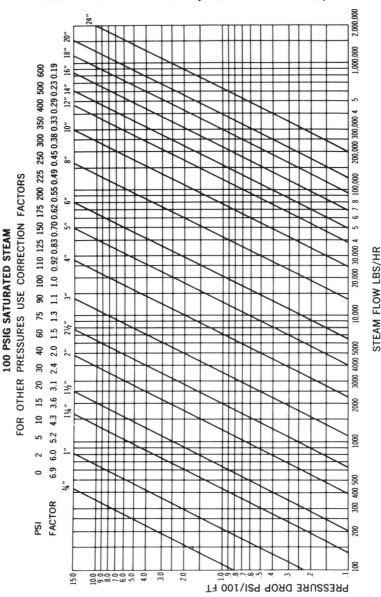

Table B-11. Steam Velocity Chart

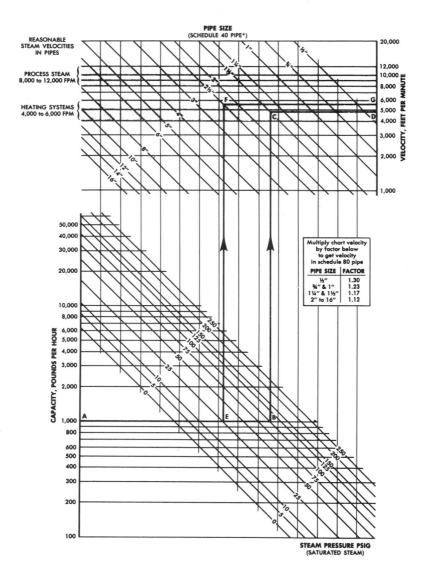

Table B-12. Pressure Drop in Schedule 80 Pipe

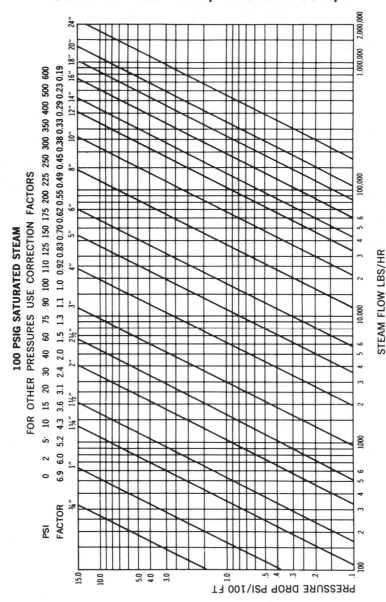

Courtesy Sarco Company, Inc.

Table B-13. Chimney Connector and Vent Connector Clearance from Combustible Materials

Description of Appliance	Minimum Clearance, Inches*
RESIDENTIAL APPLIANCES	
Single-Wall, Metal Pipe Connectors	
Electric, Gas, and Oil Incinerators	18
Oil and Solid-Fuel Appliances	18
Oil Appliances Listed as Suitable for Use with Type L	
Venting Systems, but only when connected to chimneys	9
Type L Venting-System Piping Connectors	
Electric, Gas, and Oil Incinerators	9
Oil and Solid-Fuel Appliances	9
Oil Appliances Listed as Suitable for Use with	
Type L Venting Systems	†
COMMERCIAL AND INDUSTRIAL APPLIANCES	
Low-Heat Appliances	
Single-Wall, Metal Pipe Connectors	
Gas, Oil, and Solid-Fuel Boilers, Furnaces,	
and Water Heaters	18
Ranges, Restaurant Type	18
Oil Unit Heaters	18
Other Low-Heat Industrial Appliances	18
Medium-Heat Appliances	
Single-Wall, Metal Pipe Connectors	
All Gas, Oil, and Solid-Fuel Appliances	36

*These clearances apply except if the listing of an appliance specifies different clearance, in which case the listed clearance takes precedence.

†If listed Type L venting-system piping is used, the clearance may be in accordance with the venting-system listing.

If listed Type B or Type L venting-system piping is used, the clearance may be in accordance with the venting-system listing.

The clearances from connectors to combustible materials may be reduced if the combustible material is protected in accordance with Table 1C.

Courtesy National Oil Fuel Institute

Table B-14. Standard Clearances for Heat-Producing Appliances in Residential Installations

Residential Type Appliances for Installation in Rooms that are Large*		APPLIANCE				
		Above Top of Casing or Appliance	From Top and Sides of Warm-Air Bonnet or Plenum	From Front†	From Back	From Sides
Boilers and Water Heaters Steam Boilers — 15 psi; Water Boilers — 250°F; Water Heaters — 200°F; All Water Walled or Jacketed	Automatic Oil or Comb. Gas-Oil	6	—	24	6	6
Furnaces — Central Gravity, Upflow, Downflow, Horizontal and Duct. Warm-Air — 250°F Max.	Automatic Oil or Comb. Gas-Oil	6‡	6‡	24	6	6
Furnaces — Floor For Mounting in Combustible Floors	Automatic Oil or Comb. Gas-Oil	36	—	12	12	12

* Rooms that are large in comparison to the size of the appliance are those having a volume equal to at least 12 times the total volume of a furnace and at least 16 times the total volume of a boiler. If the actual ceiling height of a room is greater than 8 ft., the volume of a room shall be figured on the basis of a ceiling height of 8 ft.

† The minimum dimension should be that necessary for servicing the appliance including access for cleaning and normal care, tube removal, etc.

‡ For a listed oil, combination gas-oil, gas, or electric furnace, this dimension may be 2 in. if the furnace limit control cannot be set higher than 250°F or 1 in. if the limit control cannot be set higher than 200°F

Courtesy National Oil Fuel Institute

570

Table B-15. Standard Clearances for Heat-Producing Appliances in Commercial and Industrial Installations

Commercial-Industrial Type Low Heat Appliances (Any and All Physical Sizes Except As Noted)		APPLIANCE				
		Above Top of Casing or Appliance*	From Top and Sides of Warm-Air Bonnet or Plenum	From Front	From Back*	From Sides*
Boiler and Water Heaters						
100 cu. ft. or less, Any psi, Steam	All Fuels	18	—	48	18	18
50 psi or less, Any Size	All Fuels	18	—	48	18	18
Unit Heaters						
Floor Mounted or Suspended — Any Size	Steam or Hot Water	1	—	—	1	1
Suspended — 100 cu. ft. or less	Oil or Comb. Gas-Oil	6	—	24	18	18
Suspended — Over 100 cu. ft.	All Fuels	18	—	48	18	18
Floor Mounted Any Size	All Fuels	18	—	48	18	18

*If the appliance is encased in brick, the 18 in. clearance above and at sides and rear may be reduced to not less than 12 in.

Courtesy National Oil Fuel Institute

Table B-16. Clearance (in Inches) with Specified Forms of Protection

Input Heat Units	Eff. %	Usable Btuh	gph 100°F Rise	Tank Size, Gal.	AVAILABLE HOT-WATER STORAGE PLUS RECOVERY 100°F RISE				Continuous Draw, gph
					15 Min.	30 Min.	45 Min.	60 Min.	
ELECTRICITY, kWh									
1.5	92.5	4,750	5.7*	20	21.4	22.8	24.3	25.7	5.7
2.5	92.5	7,900	9.5*	20	32.4	34.8	37.1	39.5	9.5
4.5	92.5	14,200	17.1*	30	44.3	48.6	52.9	57.1	17.1
4.5	92.5	14,200	17.1*	50	54.3	58.6	62.9	67.1	17.1
6.0	92.5	19,000	22.8*	66	71.6	77.2	82.8	88.8	22.8
7.0	92.5	22,100	26.5	80	86.6	93.2	99.8	106.5	26.5
GAS, btuh									
34,000	75	25,500	30.6	30	37.7	45.3	53.0	55.6†	25.6†
42,000	75	31,600	38.0	30	39.5	49.0	58.8	61.7†	31.7†
50,000	75	37,400	45.0	40	51.3	62.6	73.9	77.6†	37.6†
60,000	75	45,000	54.0	50	63.5	77.0	90.5	95.0†	45.0†
OIL, gph									
0.50	75	52,500	63.0	30	45.8	61.6	77.4	82.5†	52.5†
0.75	75	78,700	94.6	30	53.6	77.2	100.8	109.0†	79.0†
0.85	75	89,100	107.0	30	57.7	83.4	110.1	119.1†	89.0†
1.00	75	105,000	126.0	50	81.5	113.0	144.5	155.0†	105.0†
1.20	75	126,000	151.5	50	87.9	125.8	163.7	176.0†	126.0†
1.35	75	145,000	174.0	50	93.5	137.0	180.5	195.0†	145.0†
1.50	75	157,000	188.5	85	132.1	179.2	226.3	242.0†	157.0†
1.65	75	174,000	204.5	85	136.1	187.2	238.4	259.0†	174.0†

*Assumes simultaneous operation of upper and lower elements.
†Based on 50 minute-per-hour operation.

Courtesy National Oil Fuel Institute

Table B-17. Performance of Storage Water Heaters

TYPE OF PROTECTION — Applied to the combustible material unless otherwise specified and covering all surfaces within the distance specified as the required clearance with no protection (see Fig. 1A). Thicknesses are minimum.	Where the required clearance with no protection is*														
	36 Inches			18 Inches			12 Inches			9 Inches			6 Inches		
	Above	Sides & Rear	Chimney or Vent Connector	Above	Sides & Rear	Chimney or Vent Connector	Above	Sides & Rear	Chimney or Vent Connector	Above	Sides & Rear	Chimney or Vent Connector	Above	Sides & Rear	Chimney or Vent Connector
(a) 1/4 in. asbestos millboard spaced out 1 in.†	30	18	30	15	9	12	9	6	6	6	6	6	3	2	3
(b) 28-gauge sheet metal on 1/4 in. asbestos millboard	24	18	24	12	9	12	9	6	4	6	6	4	3	2	2
(c) 28-gauge sheet metal spaced out 1 in.†	18	12	18	9	6	9	6	4	4	6	4	4	2	2	2
(d) 28-gauge sheet metal on 1/8 in. asbestos millboard spaced out 1 in.†	18	12	18	9	6	9	6	4	4	4	4	4	2	2	2
(e) 1 1/2 in. asbestos cement covering on heating appliance	18	12	36	9	6	18	6	4	9	4	4	9	2	1	6
(f) 1/4 in. asbestos millboard on 1 in. mineral fiber bats reinforced with wire mesh or equivalent	18	12	18	6	6	6	4	4	4	4	4	4	2	2	2
(g) 22-gauge sheet metal on 1 in. mineral fiber bats reinforced with wire or equivalent	18	12	12	4	3	3	2	2	2	2	2	2	2	2	2
(h) 1/4 in. asbestos millboard	36	36	36	18	18	18	12	12	9	9	9	9	4	4	4
(i) 1/4 in. cellular asbestos	36	36	36	18	18	18	12	12	9	9	9	9	3	3	3

*Except for the protection described in (e), all clearances should be measured from the outer surface of the appliance to the combustible material, disregarding any intervening protection applied to the combustible material.

†Spacers shall be of noncombustible material.

NOTE: Asbestos millboard referred to above is a different material from asbestos cement board. It is not intended that asbestos cement board be used in complying with these requirements when asbestos millboard is specified.

Courtesy National Oil Fuel Institute

Index

**Over a Century of Excellence
for the Professional
and
Vocational Trades and the Crafts**

Order now from your local bookstore
or use the convenient order form
at the back of this book.

AUDEL

These fully illustrated, up-to-date guides and manuals mean a better job done for mechanics, engineers, electricians, plumbers, carpenters, and all skilled workers.

CONTENTS

ELECTRICAL

HOUSE WIRING (Sixth Edition)
ROLAND E. PALMQUIST
5 1/2 x 8 1/4   Hardcover   256 pp.   150 Illus.
ISBN: 0-672-23404-1 $14.95

The rules and regulations of the National Electrical Code as they apply to residential wiring fully detailed with examples and illustrations.

PRACTICAL ELECTRICITY
(Fifth Edition)
ROBERT G. MIDDLETON;
revised by L. DONALD MEYERS
5 1/2 x 8 1/4   Hardcover   512 pp.   335 Illus.
ISBN: 0-02-584561-6 $19.95

The fundamentals of electricity for electrical workers, apprentices, and others requiring concise information about electric principles and their practical applications.

GUIDE TO THE 1987 NATIONAL ELECTRICAL CODE
ROLAND E. PALMQUIST
5 1/2 x 8 1/4   Hardcover   664 pp.   225 Illus.
ISBN: 0-02-594560-2 $22.50

The most authoritative guide available to interpreting the National Electrical Code for electricians, contractors, electrical inspectors, and homeowners. Examples and illustrations.

MATHEMATICS FOR ELECTRICIANS AND ELECTRONICS TECHNICIANS
REX MILLER
5 1/2 x 8 1/4   Hardcover   312 pp.   115 Illus.
ISBN: 0-8161-1700-4 $14.95

Mathematical concepts, formulas, and problem-solving techniques utilized on-the-job by electricians and those in electronics and related fields.

FRACTIONAL-HORSEPOWER ELECTRIC MOTORS
REX MILLER and
MARK RICHARD MILLER
5 1/2 x 8 1/4   Hardcover   436 pp.   285 Illus.
ISBN: 0-672-23410-6 $15.95

The installation, operation, maintenance, repair, and replacement of the small-to-moderate-size electric motors that power home appliances and industrial equipment.

ELECTRIC MOTORS (Fourth Edition)
EDWIN P. ANDERSON;
revised by REX MILLER
5 1/2 x 8 1/4   Hardcover   656 pp.   405 Illus.
ISBN: 0-672-23376-2 $14.95

Installation, maintenance, and repair of all types of electric motors.

HOME APPLIANCE SERVICING (Fourth Edition)
EDWIN P. ANDERSON;
revised by REX MILLER
5 1/2 x 8 1/4   Hardcover   640 pp.   345 Illus.
ISBN: 0-672-23379-7 $22.50

The essentials of testing, maintaining, and repairing all types of home appliances.

TELEVISION SERVICE MANUAL (Fifth Edition)

ROBERT G. MIDDLETON;
revised by JOSEPH G. BARRILE

5 1/2 x 8 1/4   Hardcover   512 pp.   395 Illus.
ISBN: 0-672-23395-9 $16.95

A guide to all aspects of television transmission and reception, including the operating principles of black and white and color receivers. Step-by-step maintenance and repair procedures.

ELECTRICAL COURSE FOR APPRENTICES AND JOURNEYMEN (Third Edition)

ROLAND E. PALMQUIST

5 1/2 x 8 1/4   Hardcover   478 pp.   290 Illus.
ISBN: 0-02-594550-5 $19.95

This practical course in electricity for those in formal training programs or learning on their own provides a thorough understanding of operational theory and its applications on the job.

QUESTIONS AND ANSWERS FOR ELECTRICIANS EXAMINATIONS (Ninth Edition)

ROLAND E. PALMQUIST

5 1/2 x 8 1/4   Hardcover   316 pp.   110 Illus.
ISBN: 0-02-594691-9 $18.95

Based on the 1987 National Electrical Code, this book reviews the subjects included in the various electricians examinations—apprentice, journeyman, and master. Question and Answer format.

MACHINE SHOP AND MECHANICAL TRADES

MACHINISTS LIBRARY
(Fourth Edition, 3 Vols.)

REX MILLER

5 1/2 x 8 1/4   Hardcover   1352 pp.   1120 Illus.
ISBN: 0-672-23380-0 $52.95

An indispensable three-volume reference set for machinists, tool and die makers, machine operators, metal workers, and those with home workshops. The principles and methods of the entire field are covered in an up-to-date text, photographs, diagrams, and tables.

Volume I: Basic Machine Shop
REX MILLER

5 1/2 x 8 1/4   Hardcover   392 pp.   375 Illus.
ISBN: 0-672-23381-9 $17.95

Volume II: Machine Shop
REX MILLER

5 1/2 x 8 1/4   Hardcover   528 pp.   445 Illus.
ISBN: 0-672-23382-7 $19.95

Volume III: Toolmakers Handy Book
REX MILLER

5 1/2 x 8 1/4   Hardcover   432 pp.   300 Illus.
ISBN: 0-672-23383-5 $14.95

MATHEMATICS FOR MECHANICAL TECHNICIANS AND TECHNOLOGISTS

JOHN D. BIES

5 1/2 x 8 1/4   Hardcover   342 pp.   190 Illus.
ISBN: 0-02-510620-1 $17.95

The mathematical concepts, formulas, and problem-solving techniques utilized on the job by engineers, technicians, and other workers in industrial and mechanical technology and related fields.

MILLWRIGHTS AND MECHANICS GUIDE
(Fourth Edition)

CARL A. NELSON

5 1/2 x 8 1/4   Hardcover   1,040 pp.   880 Illus.
ISBN: 0-02-588591-x $29.95

The most comprehensive and authoritative guide available for millwrights, mechanics, maintenance workers, riggers, shop workers, foremen, inspectors, and superintendents on plant installation, operation, and maintenance.

WELDERS GUIDE (Third Edition)

JAMES E. BRUMBAUGH

5 1/2 x 8 1/4   Hardcover   960 pp.   615 Illus.
ISBN: 0-672-23374-6 $23.95

The theory, operation, and maintenance of all welding machines. Covers gas welding equipment, supplies, and process; arc welding equipment, supplies, and process; TIG and MIG welding; and much more.

WELDERS/FITTERS GUIDE

JOHN P. STEWART

8 1/2 x 11   Paperback   160 pp.   195 Illus.
ISBN: 0-672-23325-8 $7.95

Step-by-step instruction for those training to become welders/fitters who have some knowledge of welding and the ability to read blueprints.

SHEET METAL WORK

JOHN D. BIES

5 1/2 x 8 1/4   Hardcover   456 pp.   215 Illus.
ISBN: 0-8161-1706-3 $19.95

An on-the-job guide for workers in the manufacturing and construction industries and for those with home workshops. All facets of sheet metal work detailed and illustrated by drawings, photographs, and tables.

POWER PLANT ENGINEERS GUIDE (Third Edition)

FRANK D. GRAHAM;
revised by CHARLIE BUFFINGTON

5 1/2 x 8 1/4   Hardcover   960 pp.   530 Illus.
ISBN: 0-672-23329-0 $27.50

This all-inclusive, one-volume guide is perfect for engineers, firemen, water tenders, oilers, operators of steam and diesel-power engines, and those applying for engineer's and firemen's licenses.

MECHANICAL TRADES POCKET MANUAL
(Second Edition)

CARL A. NELSON

4 x 6   Paperback   364 pp.   255 Illus.
ISBN: 0-672-23378-9   10.95

A handbook for workers in the industrial and mechanical trades on methods, tools, equipment, and procedures. Pocket-sized for easy reference and fully illustrated.

PLUMBING

PLUMBERS AND PIPE FITTERS LIBRARY
(Fourth Edition, 3 Vols.)

CHARLES N. McCONNELL

5 1/2 x 8 1/4   Hardcover   952 pp.   560 Illus.
ISBN: 0-02-582914-9 $68.45

This comprehensive three-volume set contains the most up-to-date information available for master plumbers, journeymen, apprentices, engineers, and those in the building trades. A detailed text and clear diagrams, photographs, and charts and tables treat all aspects of the plumbing, heating, and air conditioning trades.

Volume I: Materials, Tools, Roughing-In

CHARLES N. McCONNELL;
revised by TOM PHILBIN

5 1/2 x 8 1/4   Hardcover   304 pp.   240 Illus.
ISBN: 0-02-582911-4 $20.95

Volume II: Welding, Heating, Air Conditioning

CHARLES N. McCONNELL;
revised by TOM PHILBIN

5 1/2 x 8 1/4   Hardcover   384 pp.   220 Illus.
ISBN: 0-02-582912-2 $22.95

Volume III: Water Supply, Drainage, Calculations

CHARLES N. McCONNELL;
revised by TOM PHILBIN

5 1/2 x 8 1/4   Hardcover   264 pp.   100 Illus.
ISBN: 0-02-582913-0 $20.95

HOME PLUMBING HANDBOOK
(Third Edition)

CHARLES N. McCONNELL

8 1/2 x 11   Paperback   200 pp.   100 Illus.
ISBN: 0-672-23413-0 $13.95

An up-to-date guide to home plumbing installation and repair.

THE PLUMBERS HANDBOOK
(Seventh Edition)

JOSEPH P. ALMOND, SR.

4 x 6   Paperback   352 pp.   170 Illus.
ISBN: 0-672-23419-x $11.95

A handy sourcebook for plumbers, pipe fitters, and apprentices in both trades. It has a rugged binding suited for use on the job, and fits in the tool box or conveniently in the pocket.

QUESTIONS AND ANSWERS FOR PLUMBERS EXAMINATIONS (Second Edition)

JULES ORAVITZ

5 1/2 x 8 1/4   Paperback   256 pp.   145 Illus.
ISBN: 0-8161-1703-9 $9.95

A study guide for those preparing to take a licensing examination for apprentice, journeyman, or master plumber. Question and answer format.

HVAC

AIR CONDITIONING: HOME AND COMMERCIAL
(Second Edition)

EDWIN P. ANDERSON;
revised by REX MILLER

5 1/2 x 8 1/4   Hardcover   528 pp.   180 Illus.
ISBN: 0-672-23397-5 $15.95

A guide to the construction, installation, operation, maintenance, and repair of home, commercial, and industrial air conditioning systems.

HEATING, VENTILATING, AND AIR CONDITIONING LIBRARY
(Second Edition, 3 Vols.)
JAMES E. BRUMBAUGH
5 1/2 x 8 1/4 Hardcover 1,840 pp. 1,275 Illus.
ISBN: 0-672-23388-6 $53.95
An authoritative three-volume reference library for those who install, operate, maintain, and repair HVAC equipment commercially, industrially, or at home.

Volume I: Heating Fundamentals, Furnaces, Boilers, Boiler Conversions
JAMES E. BRUMBAUGH
5 1/2 x 8 1/4 Hardcover 656 pp. 405 Illus.
ISBN: 0-672-23389-4 $17.95

Volume II: Oil, Gas and Coal Burners, Controls, Ducts, Piping, Valves
JAMES E. BRUMBAUGH
5 1/2 x 8 1/4 Hardcover 592 pp. 455 Illus.
ISBN: 0-672-23390-8 $17.95

Volume III: Radiant Heating, Water Heaters, Ventilation, Air Conditioning, Heat Pumps, Air Cleaners
JAMES E. BRUMBAUGH
5 1/2 x 8 1/4 Hardcover 592 pp. 415 Illus.
ISBN: 0-672-23391-6 $17.95

OIL BURNERS (Fourth Edition)
EDWIN M. FIELD
5 1/2 x 8 1/4 Hardcover 360 pp. 170 Illus.
ISBN: 0-672-23394-0 $16.95
An up-to-date sourcebook on the construction, installation, operation, testing, servicing, and repair of all types of oil burners, both industrial and domestic.

REFRIGERATION: HOME AND COMMERCIAL (Second Edition)
EDWIN P. ANDERSON;
revised by REX MILLER
5 1/2 x 8 1/4 Hardcover 768 pp. 285 Illus.
ISBN: 0-672-23396-7 $19.95
A reference for technicians, plant engineers, and the home owner on the installation, operation, servicing, and repair of everything from single refrigeration units to commercial and industrial systems.

PNEUMATICS AND HYDRAULICS

HYDRAULICS FOR OFF-THE-ROAD EQUIPMENT (Second Edition)
HARRY L. STEWART;
revised by TOM PHILBIN
5 1/2 x 8 1/4 Hardcover 256 pp. 175 Illus.
ISBN: 0-8161-1701-2 $13.95

This complete reference manual on heavy equipment covers hydraulic pumps, accumulators, and motors; force components; hydraulic control components; filters and filtration, lines and fittings, and fluids; hydrostatic transmissions; maintenance; and troubleshooting.

PNEUMATICS AND HYDRAULICS (Fourth Edition)
HARRY L. STEWART;
revised by TOM STEWART
5 1/2 x 8 1/4 Hardcover 512 pp. 315 Illus.
ISBN: 0-672-23412-2 $19.95
The principles and applications of fluid power. Covers pressure, work, and power; general features of machines; hydraulic and pneumatic symbols; pressure boosters; air compressors and accessories; and much more.

PUMPS (Fourth Edition)
HARRY STEWART;
revised by TOM PHILBIN
5 1/2 x 8 1/4 Hardcover 508 pp. 360 Illus.
ISBN: 0-672-23400-9 $17.95
The principles and day-to-day operation of pumps, pump controls, and hydraulics are thoroughly detailed and illustrated.

CARPENTRY AND CONSTRUCTION

CARPENTERS AND BUILDERS LIBRARY (Fifth Edition, 4 Vols.)
JOHN E. BALL; revised by TOM PHILBIN
5 1/2 x 8 1/4 Hardcover 1,224 pp. 1,010 Illus.
ISBN: 0-672-23369-x $43.95

Also available as a boxed set at no extra cost:
ISBN: 0-02-506450-9 $43.95

This comprehensive four-volume library has set the professional standard for decades for carpenters, joiners, and woodworkers.

Volume I: Tools, Steel Square, Joinery
JOHN E. BALL; revised by TOM PHILBIN
5 1/2 x 8 1/4 Hardcover 384 pp. 345 Illus.
ISBN: 0-672-23365-7 $10.95

Volume II: Builders Math, Plans, Specifications
JOHN E. BALL; revised by TOM PHILBIN
5 1/2 x 8 1/4 Hardcover 304 pp. 205 Illus.
ISBN: 0-672-23366-5 $10.95

Volume III: Layouts, Foundations, Framing
JOHN E. BALL; revised by TOM PHILBIN
5 1/2 x 8 1/4 Hardcover 272 pp. 215 Illus.
ISBN: 0-672-23367-3 $10.95

Volume IV: Millwork, Power Tools, Painting

JOHN E. BALL; revised by TOM PHILBIN

5 1/2 x 8 1/4   Hardcover   344 pp.   245 Illus.
ISBN: 0-672-23368-1 $10.95

COMPLETE BUILDING CONSTRUCTION (Second Edition)

JOHN PHELPS; revised by TOM PHILBIN

5 1/2 x 8 1/4   Hardcover   744 pp.   645 Illus.
ISBN: 0-672-23377-0 $22.50

Constructing a frame or brick building from the footings to the ridge. Whether the building project is a tool shed, garage, or a complete home, this single fully illustrated volume provides all the necessary information.

COMPLETE ROOFING HANDBOOK

JAMES E. BRUMBAUGH

5 1/2 x 8 1/4   Hardcover   536 pp.   510 Illus.
ISBN: 0-02-517850-4 $29.95

Covers types of roofs; roofing and reroofing; roof and attic insulation and ventilation; skylights and roof openings; dormer construction; roof flashing details; and much more.

COMPLETE SIDING HANDBOOK

JAMES E. BRUMBAUGH

5 1/2 x 8 1/4   Hardcover   512 pp.   450 Illus.
ISBN: 0-02-517880-6 $24.95

This companion volume to the *Complete Roofing Handbook* includes comprehensive step-by-step instructions and accompanying line drawings on every aspect of siding a building.

MASONS AND BUILDERS LIBRARY (Second Edition, 2 Vols.)

LOUIS M. DEZETTEL;
revised by TOM PHILBIN

5 1/2 x 8 1/4   Hardcover   688 pp.   500 Illus.
ISBN: 0-672-23401-7 $27.95

This two-volume set provides practical instruction in bricklaying and masonry. Covers brick; mortar; tools; bonding; corners, openings, and arches; chimneys and fireplaces; structural clay tile and glass block; brick walls; and much more.

Volume I: Concrete, Block, Tile, Terrazzo

LOUIS M. DEZETTEL;
revised by TOM PHILBIN

5 1/2 x 8 1/4   Hardcover   304 pp.   190 Illus.
ISBN: 0-672-23402-5 $13.95

Volume 2: Bricklaying, Plastering, Rock Masonry, Clay Tile

LOUIS M. DEZETTEL;
revised by TOM PHILBIN

5 1/2 x 8 1/4   Hardcover   384 pp.   310 Illus.
ISBN: 0-672-23403-3 $13.95

WOODWORKING

WOOD FURNITURE: FINISHING, REFINISHING, REPAIRING (Second Edition)

JAMES E. BRUMBAUGH

5 1/2 x 8 1/4   Hardcover   352 pp.   185 Illus.
ISBN: 0-672-23409-2 $12.95

A fully illustrated guide to repairing furniture and finishing and refinishing wood surfaces. Covers tools and supplies; types of wood; veneering; inlaying; repairing, restoring, and stripping; wood preparation; and much more.

WOODWORKING AND CABINETMAKING

F. RICHARD BOLLER

5 1/2 x 8 1/4   Hardcover   360 pp.   455 Illus.
ISBN: 0-02-512800-0 $18.95

Essential information on all aspects of working with wood. Step-by-step procedures for woodworking projects are accompanied by detailed drawings and photographs.

MAINTENANCE AND REPAIR

BUILDING MAINTENANCE (Second Edition)

JULES ORAVETZ

5 1/2 x 8 1/4   Hardcover   384 pp.   210 Illus.
ISBN: 0-672-23278-2 $11.95

Professional maintenance procedures used in office, educational, and commercial buildings. Covers painting and decorating; plumbing and pipe fitting; concrete and masonry; and much more.

GARDENING, LANDSCAPING AND GROUNDS MAINTENANCE (Third Edition)

JULES ORAVETZ

5 1/2 x 8 1/4   Hardcover   424 pp.   340 Illus.
ISBN: 0-672-23417-3 $15.95

Maintaining lawns and gardens as well as industrial, municipal, and estate grounds.

HOME MAINTENANCE AND REPAIR: WALLS, CEILINGS AND FLOORS

GARY D. BRANSON

8 1/2 x 11   Paperback   80 pp.   80 Illus.
ISBN: 0-672-23281-2 $6.95

The do-it-yourselfer's guide to interior remodeling with professional results.

PAINTING AND DECORATING

REX MILLER and GLEN E. BAKER

5 1/2 x 8 1/4   Hardcover   464 pp.   325 Illus.
ISBN: 0-672-23405-x $18.95

A practical guide for painters, decorators, and homeowners to the most up-to-date materials and techniques in the field.

TREE CARE (Second Edition)

JOHN M. HALLER

8 1/2 x 11   Paperback   224 pp.   305 Illus.
ISBN: 0-02-062870-6 $9.95

The standard in the field. A comprehensive guide for growers, nursery owners, foresters, landscapers, and homeowners to planting, nurturing and protecting trees.

UPHOLSTERING (Updated)

JAMES E. BRUMBAUGH

5 1/2 x 8 1/4   Hardcover   400 pp.   380 Illus.
ISBN: 0-672-23372-x $15.95

The essentials of upholstering fully explained and illustrated for the professional, the apprentice, and the hobbyist.

AUTOMOTIVE AND ENGINES

DIESEL ENGINE MANUAL
(Fourth Edition)

PERRY O. BLACK;
revised by WILLIAM E. SCAHILL

5 1/2 x 8 1/4   Hardcover   512 pp.   255 Illus.
ISBN: 0-672-23371-1 $15.95

The principles, design, operation, and maintenance of today's diesel engines. All aspects of typical two- and four-cycle engines are thoroughly explained and illustrated by photographs, line drawings, and charts and tables.

GAS ENGINE MANUAL
(Third Edition)

EDWIN P. ANDERSON;
revised by CHARLES G. FACKLAM

5 1/2 x 8 1/4   Hardcover   424 pp.   225 Illus.
ISBN: 0-8161-1707-1 $12.95

How to operate, maintain, and repair gas engines of all types and sizes. All engine parts and step-by-step procedures are illustrated by photographs, diagrams, and troubleshooting charts.

SMALL GASOLINE ENGINES

REX MILLER and MARK RICHARD MILLER

5 1/2 x 8 1/4   Hardcover   640 pp.   525 Illus.
ISBN: 0-672-23414-9 $16.95

Practical information for those who repair, maintain, and overhaul two- and four-cycle engines—including lawn mowers, edgers, grass sweepers, snowblowers, emergency electrical generators, outboard motors, and other equipment with engines of up to ten horsepower.

TRUCK GUIDE LIBRARY (3 Vols.)

JAMES E. BRUMBAUGH

5 1/2 x 8 1/4   2,144 pp.   1,715 Illus.
ISBN: 0-672-23392-4 $45.95

This three-volume set provides the most comprehensive, profusely illustrated collection of information available on truck operation and maintenance.

Volume 1: Engines
JAMES E. BRUMBAUGH

5 1/2 x 8 1/4   Hardcover   416 pp.   290 Illus.
ISBN: 0-672-23356-8 $16.95

Volume 2: Engine Auxiliary Systems
JAMES E. BRUMBAUGH

5 1/2 x 8 1/4   Hardcover   704 pp.   520 Illus.
ISBN: 0-672-23357-6 $16.95

Volume 3: Transmissions, Steering, and Brakes
JAMES E. BRUMBAUGH

5 1/2 x 8 1/4   Hardcover   1,024 pp.   905 Illus.
ISBN: 0-672-23406-8 $16.95

DRAFTING

INDUSTRIAL DRAFTING

JOHN D. BIES

5 1/2 x 8 1/4   Hardcover   544 pp.   Illus.
ISBN: 0-02-510610-4 $24.95

Professional-level introductory guide for practicing drafters, engineers, managers, and technical workers in all industries who use or prepare working drawings.

ANSWERS ON BLUEPRINT
READING (Fourth Edition)
ROLAND PALMQUIST;
revised by THOMAS J. MORRISEY

5 1/2 x 8 1/4   Hardcover   320 pp.   275 Illus.
ISBN: 0-8161-1704-7 $12.95

Understanding blueprints of machines and tools, electrical systems, and architecture. Question and answer format.

HOBBIES

COMPLETE COURSE IN STAINED GLASS
PEPE MENDEZ

8 1/2 x 11   Paperback   80 pp.   50 Illus.
ISBN: 0-672-23287-1 $8.95

The tools, materials, and techniques of the art of working with stained glass.

Just select your books, fill out the card, and mail today.

Money-Back Guarantee